52/53 Advances in Polymer Science

Fortschritte der Hochpolymeren-Forschung

Crazing in Polymers

Editor: H. H. Kausch

With Contributions by
A. S. Argon, C. C. Chen, R. E. Cohen,
M. Dettenmaier, W. Döll, K. Friedrich,
O. S. Gebizlioglu, E. J. Kramer, J. A. Sauer,
C. E. Schwier

With 234 Figures and 12 Tables

Springer-Verlag
Berlin Heidelberg GmbH 1983

ISBN 978-3-662-15952-1   ISBN 978-3-540-38652-0 (eBook)
DOI 10.1007/978-3-540-38652-0

Library of Congress Catalog Card Number 61-642

© Springer-Verlag Berlin Heidelberg 1983
Originally published by Springer-Verlag Berlin Heidelberg New York in 1983.
Softcover reprint of the hardcover 1st edition 1983

2152/3020-543210

Editors

Preface

More than thirty years ago Hsiao and Sauer published the first important paper dealing with craze initiation in glassy polymers. In subsequent years several hundred publications have appeared, which were mostly concerned with the phenomenology of crazing (stress and environment effects, craze morphology and kinetics). The first two major reviews on crazing were written ten years ago by Kambour and by Rabinowitz and Beardmore, and they are still serving as a basis of reference.

Since then the research activity may even have increased. Central themes were the molecular mechanisms in craze initiation, the influence of molecular weight and presence of entanglements, the nature of plastic instabilities, the role of crazes as precursors to cracks, and last but certainly not least, the formation of crazes in semicrystalline and multiphase polymers. Although considerable progress has been made in the above mentioned fields, some important questions are still open today.

In such a situation, it seems adequate to try again to establish a frame of reference. In the editor's opinion this can best be done by presenting not only the crucial experimental results and the conclusions drawn from them, but by also displaying to a limited extent the new analytical and experimental tools used. This will help and encourage future researchers in applying these methods; it will also help clarifying apparently contradictory statements (for example why the Dugdale model describes very well a PMMA crack tip craze and less well a PS thin film craze). Evidently, such a wide-ranging task cannot be accomplished in a reasonable amount of time by one or two authors. The editor, however, has had the privilege to find the collaboration of a distinguished group of researchers who had a prominent role in furthering the above mentioned new techniques and fields of activity. Thus, it is hoped that a fairly comprehensive review can be presented, firstly on the formation and decay of fibrils in thin films, in homogeneous bulk specimens and at crack tips, and secondly on the role of crazes in deformation, fatigue and fracture of glassy homopolymers, block copolymers and blends and of semi-crystalline thermoplastics.

The electron beam imaging technique developed in Ithaca has proven to be an extremely powerful tool to determine the micro-mechanics, i.e. the shape and state of stress of a single craze in a thin film. E. J. Kramer reviews the considerable information gained with respect to craze nucleation, microstructure and growth. One of the most important craze parameters, the local fibril extension ratio, thus, has become amenable to study. The influence of molecular entanglements is being treated for a wide variety of glassy polymers with chains of different entanglement molecular weight. Kramer concludes that under special circumstances chain scission can have a significant role in the break down of craze fibrils.

In general, the individual surface craze owes its existence to the presence of some defect or impurity. That crazing can be an intrinsic phenomenon has been shown in Lausanne with poly-carbonate (PC). M. Dettenmaier gives a brief review of the structure and deformation of PC. He then concentrates on the newly discovered intrinsic craze phenomenon observed in PC stretched to high stresses and strains in a temperature region close to T_g. Studies on unoriented and pre-oriented material provided strong evidence that this phenomenon results from an instability of the entanglement network when stretched to its limiting extension. Small-angle X-ray scattering revealed the effect of the various experimental parameters on the craze microstructure. An outline of the scattering theory is included. The fundamental difference between conventional crazing (rotational instability of segments essentially conserving the existing network of entanglements) and intrinsic crazing (partial break down of the entanglement network) is discussed. Giving other examples, the review demonstrates that the intrinsic crazing of PC is related to the existence of a general mode of cavitational plasticity in glassy polymers.

A third method which recently provided considerable insight into the role of crazes in deformation and fracture of amorphous polymers is the optical interference measurement of crazes (preceding a crack). Since the pioneer work of Kambour, this method has been widely used to determine characteristic craze dimensions and critical displacements. W. Döll gives an overview on recent results and on their interpretation in terms of fracture mechanics parameters (stress intensity factor, plastic zone sizes, fracture surface morphology, fracture energy).

The second half of this volume is reserved to a discussion of specific craze problems encountered in practical application of polymer materials. J. A. Sauer and C. C. Chen analyze the fatigue behavior (mostly of rubber modified polymers). They show quantitatively the important effects of test variables and sample morphology on fatigue response. K. Friedrich gives an overview on the shear and craze phenomena in semicrystalline polymers, he points out the analogies existing with respect to crazing in amorphous glasses; distinctive differences are being introduced by the larger tendency to shear banding. A. S. Argon, R. Cohen, O. S. Gebizioglu and C. E. Schiwer review the principles of polymer toughening by crazing in block copolymers and blends. Here, evidently, the structure and geometry of the rubbery domains is a new and important parameter as is the thermal expansion mismatch. The authors described how these are instrumental in accelerating the molecular level precursor processes in craze initiation; they also discuss the mutual interference of adjacent growing crazes.

The editor of this volume wishes to thank wholeheartedly the authors who have so kindly accepted to collaborate on this book, for their hard and efficient work. He expresses his gratitude to the publisher for a pleasant cooperation and for the careful and fast production.

Lausanne, March 15, 1983 Hans-Henning Kausch

Table of Contents

Microscopic and Molecular Fundamentals of Crazing

Edward J. Kramer
Department of Materials Science & Engineering and the Materials Science Center,
Cornell University, Ithaca, NY, 14853 U.S.A.

Advances in Polymer Science 52/53
© Springer-Verlag Berlin Heidelberg 1983

List of Symbols

A', m_c	prefactor and exponent of K in crack growth equation
$\bar{D}$	average craze fibril diameter $\bar{D} = \langle D^2 \rangle / \langle D \rangle$
$\langle D \rangle$	mean fibril diameter
$\langle D^2 \rangle$	mean square fibril diameter
D_0	average fibril spacing as well as average fibril diameter before deformation (diameter of the "phantom" fibril cylinder)
D_0^*, $\nabla\sigma_0^*$	fibril spacing of fastest growing of craze interface and gradient of hydrostatic stress corresponding to this spacing
E^*	an effective Young's modulus of the polymer which equals Young's modulus for plane stress and $E/(1-v^2)$ for plane strain
G, G_N^0	shear modulus, shear modulus level of the rubbery plateau
G_{Ic}	critical strain energy release rate of crack (fracture toughness)
K	stress intensity factor for crack
M	axis of molecular orientation
M_0	molecular weight of stiff units along chain
M_c	critical molecular weight below which crazes do not form
M_e	entanglement molecular weight
M_n, M_n'	number average molecular weights of chains in the glass before crazing and in the craze fibrils respectively
M_w'	weight average molecular weight of chains in the craze fibrils
N_A	Avogadro's number
P	probability that there is at least one entangled chain in each entanglement length along a fibril
Q	SAXS invarient
R	radius of curvature at craze front
R	universal gas constant
S, S_t	tensile stress on craze surface and at the craze tip respectively averaged over several fibrils
$\Delta S(x)$	self stress of craze necessary to produce its measured displacement profile $w(x)$
T_A	absolute temperature
$T(x)$	craze thickness profile
T_g	glass transition temperature
U^*	activation energy for chain scission
U	energy needed to break a single primary polymer chain bond
V	crack growth velocity
χ	volume fraction of polymer in a blend
Y	tensile yield stress
Z	number of entanglement transfer lengths along a fibril
a	half length of craze (or of crack plus craze)
a_i	initial radius of a penny shaped crack
c	crack half length
d	root mean square end-to-end distance of a chain of molecular weight M_e, entanglement mesh size

f	force on a single entangled chain in a fibril
$i(s)$	scattered X-ray intensity
k'	Porod's law constant from high angle tail of SAXS intensity
k_0	rate constant prefactor
k_b	rate of bond scission
l	entanglement transfer length along fibril
l_0	average projected length of stiff units along chain
l_{act}	activation distance
l_e	chain contour length between entanglements
n	non Newtonian flow law exponent
n_c	refractive index of craze
p	probability that at least one entangled chain is left umbroken in a given entanglement transfer length
$\bar{p}(\zeta), \zeta$	Fourier transform of craze selfstress and the transform variable (spatial frequency) ζ
q	the ratio of entangled chain density before to that after crazing $(= v_E'/v_E)$
r	radius of curvature of steady state finger at craze tip
r_d	chain disentanglement rate
s	magnitude of the X-ray scattering vector
s_0	deviatoric stress
v	velocity of craze — bulk polymer interface in craze thickening
$w(x)$	displacement profile of craze interfaces
x, z	coordinates in craze plane normal to and parallel to the craze front
2θ	X-ray scattering angle
Λ	wave length of sinusoidal perturbation of craze front
Λ_c	fastest growing wave length of craze front perturbation
Λ_m	minimum wavelength of sinusoidal craze front perturbation that can grow
Λ_s	fastest growing spacing of steady state finger structure at craze tip
$\Phi_{craze},$ Φ_{film}, Φ_{hole}	optical density on TEM micrograph of craze, solid film and hole through film
Γ	energy to create new surface at craze tip or craze-bulk interface including an energy of primary chain rupture
α	proportionality constant between finger spacing and finger radius of curvature ($r = \alpha\Lambda_s$)
$\alpha(x)$	linear dislocation density of continuously distributed dislocations of Burger's vector b necessary to give the same displacement profile as the craze
γ	van der Waal's surface energy
$\dot{\varepsilon}$	equivalent tensile strain rate
$\dot{\varepsilon}_F, \sigma_F$	material parameter in non-Newtonian flow law $\dot{\varepsilon} = \dot{\varepsilon}_F(\sigma/\sigma_F)^n$
ε_c	critical uniaxial strain for crazing
ζ	amplitude of sinusoidal perturbation of craze front
λ	extension ratio of craze fibrils relative to undeformed polymer glass
λ_{DZ}	extension ratio of shear deformation zone

$(\lambda_{max})_{\|\|}$, $(\lambda_{max})_\perp$	maximum extension ratio of a single entangled chain parallel and perpendicular to the preorientation direction
λ_{max}	maximum extension ratio of a single entangled chain ($= l_e/d$)
λ_{net}	maximum extension ratio of entangled chain network
λ_p	extension ratio achieved in uniaxial preorientation above T_g
λ_x	X-ray wavelength
$\lambda_{\|\|}$, $\lambda_\perp$	extension ratio measured in craze fibrils grown by applying tensile stress parallel, and perpendicular, to the axis of molecular orientation M
ν	Poisson's ratio
ν_E	density of chains between entanglement junction points
ν'_E	density of chains between entanglement junction points after geometrically necessary entanglement loss
ν_E^s, ν_E^d	density of entangled chains lost by scission or disentanglement respectively when forming a fibril surface
ϱ	polymer density
β, β^*	coefficient of proportionality between average hydrostatic stress (σ_0) and tensile stress S or S_t at craze interface or craze tip respectively (i.e. $(\sigma_0)_m = \beta S$; $(\sigma_0)_m = \beta^* S_t$)
σ	equivalent tensile stress
σ_0	hydrostatic stress (negative pressure)
$\delta\sigma_0$	increment of hydrostatic stress
$\nabla\sigma_0$, s^*	gradient of hydrostatic stress and its sign
$(\sigma_0)_m$	average hydrostatic stress ahead of craze tip or craze interface
$(\sigma_0)_s$	hydrostatic stress at the surface of the void "ceiling" between fibrils at craze interface
$(\sigma_0)_t$	hydrostatic stress at craze void finger tip
σ_D	lower yield stress
σ_f	true tensile stress in the craze fibrils
σ_{ij}	components of stress tensor
σ_∞	tensile stress normal to the craze at large distances from the craze plane

1 Introduction

The use of polymer glasses to make structural components has increased significantly over the last decade but is still limited by the tendency of these materials to fail in a macroscopically brittle manner (i.e., with no large scale plastic deformation before cracking). The brittleness can be traced to the formation under tensile stress of small crack-like defects called crazes. Unlike cracks, these are load bearing because their two surfaces are bridged by many small fibrils with diameters in the range 5 to 30 nm. When true cracks form, however, they do so by breakdown of the fibril structure within a craze, a process which is aided by the high local stresses in the fibrils. Hence to control crack nucleation one must stop craze formation or make it more difficult relative to shear deformation. Moreover craze nucleation, growth and fibril breakdown to form cracks may occur at much lower stresses in the presence of certain environments leading to an important link between crazing and the phenomenon of environment-assisted cracking.

Nevertheless under certain conditions crazing can be beneficial. Crazing is a process of plastic deformation and as such is the most important source of *fracture toughness* G_{I_c} in polymer glasses which deform by crazing rather than shear. The plastic work to create crazes ahead of the growing crack appears as the major contribution to G_{I_c}. By arranging that very high densities of crazes nucleate and grow, from small rubber particles for example, before these crazes break down to form cracks, substantial macroscopic plastic strains due to crazing can be achieved even at impact strain rates. The improved impact toughness produced by rubber modification of polystyrene to produce high impact polystyrene is an example of the exploitation of this aspect of crazing.

To understand these sometimes contradictory effects of crazing on fracture properties, some of which will be considered in much more detail in the subsequent chapters of this volume, one needs more information about these crazes themselves. In this chapter only crazes produced in air will be considered; a recent review of environmental crazing and fracture is available for the reader seeking to compare the properties of air crazes with those of environmental crazes [1]. In addition no attempt will be made to completely cover all the older work on crazing. Excellent reviews of this work by Kambour [2], Rabinowitz and Beardmore [3] and Gent [4] mean that this chapter can concentrate on the more recent developments in this field. Craze nucleation and growth, the micromechanics and microstructure of crazes, and the influence of molecular entanglements in the glass on these aspects of crazing will be reviewed in detail.

2 Craze Nucleation

Early in the study of crazing [5,6] it was recognized that there was usually a time delay between the application of stress and the visual appearance of crazes. This delay time is evidence of a barrier to craze nucleation. The experimental situation may be summarized by Figure 1, which shows the density of crazes as a function of time under various states of stress in polystyrene (PS) from the work of Argon and

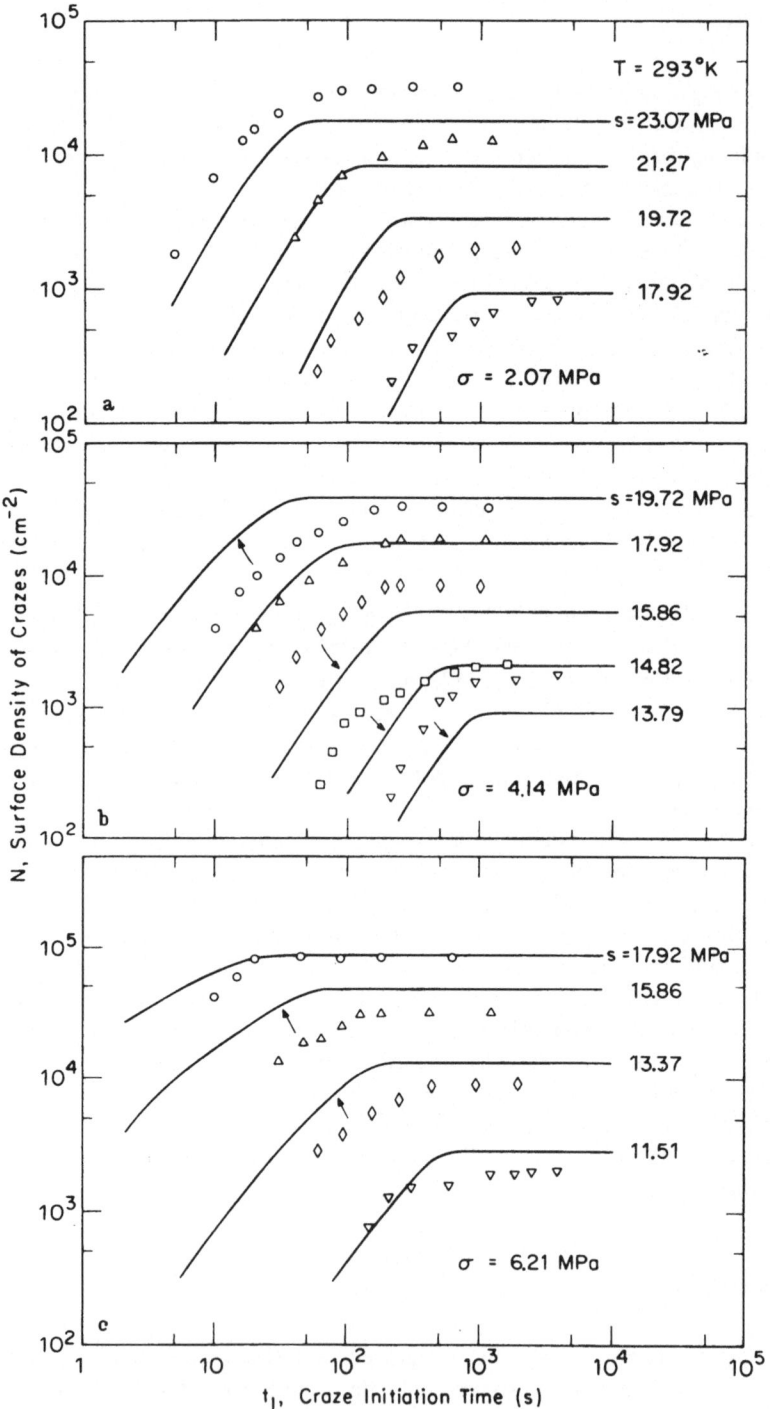

Fig. 1. Surface density of crazes versus time in polystyrene for several levels of hydrostatic tension σ and deviatoric stress s. From Ref. [6], courtesy Taylor and Francis, Ltd.

Hannoosh [6]. Crazes are usually observed to nucleate at the surface of the sample beneath defects such as surface grooves (which were deliberately introduced and quantitatively characterized by Argon and Hannoosh in order to provide reproducible defects), steps or dust particles. Preferential craze nucleation may occur in the interior of samples at the surface of second phase particles, [sometimes these are deliberately added for this purpose, e.g., rubber particles in high impact polystyrene (HIPS)] or if the crazing stress is higher at the surface, e.g. due to molecular orientation there. The importance of these defects for craze nucleation may be inferred, both from direct observation (for example it is easy to identify dust particles in thin film of PS by TEM that acted as sites for craze nuclei) and from the saturation in nucleated craze density observed at long times (Fig. 1). The site saturation is a prominent feature of heterogeneous nucleation, where the nucleation sites are sparse or vary widely in potency. Argon and Hannoosh also found that if they produced small highly perfect samples of PS from single molding pellets that the stress required for craze initiation increased substantially and in several instances in fact these samples underwent shear yielding and necking in tension before crazing.

Also shown on Fig. 1 is the important effect of the state of stress which may be characterized by the *negative pressure* σ_0 and the *deviatoric stress* s_0. These are given by

$$\sigma_0 = (\sigma_{11} + \sigma_{22} + \sigma_{33})/3 \tag{1}$$

and

$$s_0 = \frac{1}{6}[(\sigma_{11} - \sigma_{22})^2 + (\sigma_{22} - \sigma_{33})^2 + (\sigma_{33} - \sigma_{11})^2 + \sigma_{23}^2 + \sigma_{13}^2 + \sigma_{12}^2]^{1/2} \tag{2}$$

where the σ_{ij}'s are the *components of the stress tensor*. From the data in Figure 1 it is clear that craze nucleation kinetics and the saturation craze density increase both with σ_0 and s_0; these results are in qualitative accord with other experiments [7-11] on PS and polymethylmethacrylate (PMMA). There is still considerable controversy about whether air crazes can nucleate when σ_0 is zero or even negative. Sternstein [7], Oxborough and Bowden [8] and Kawagoe and Kitagawa [11] have proposed different air crazing criteria which imply that s_0 for craze nucleation becomes infinite for $\sigma_0 \leqq 0$ (e.g., in pure torsion). More recent tension-superposed hydrostatic pressure [9, 10] and unpublished pure torsion experiments [12] indicate that craze nucleation is possible for $\sigma_0 \leqq 0$. A serious problem in resolving these discrepancies is that the local σ_0 near a stress concentrating defect may be positive while the global σ_0 is negative or zero. The differences in results from one laboratory to another simply may be due to a difference in surface perfection or dust content of the different specimens. The heterogeneous nature of craze nucleation makes the detailed study of this phenomenon quite difficult.

Despite this lack of reproducible experimental evidence on kinetics, one may postulate several microscopic steps involved in craze nucleation. Imagine a polymer surface under simple tension as shown in Fig. 2a. The first is logically plastic deformation by shear at the surface in the region of stress concentration of the defect. The strain softening characteristic of most polymer glasses leads both to an acceleration of the strain rate and a localization of strain as the plastic strain increases. Regions which have locally deformed, but not crazed, have been observed by Wellinghoff and Baer [13] prior to crazing using TEM replicas of gold decorated films stretched on polyethylene terephthalate substrates. At Cornell we have frequently observed incompletely fibrillated plastically deformed zones near craze initiation sites in PS and other polymer films using TEM; further along their length these zones turn into crazes.

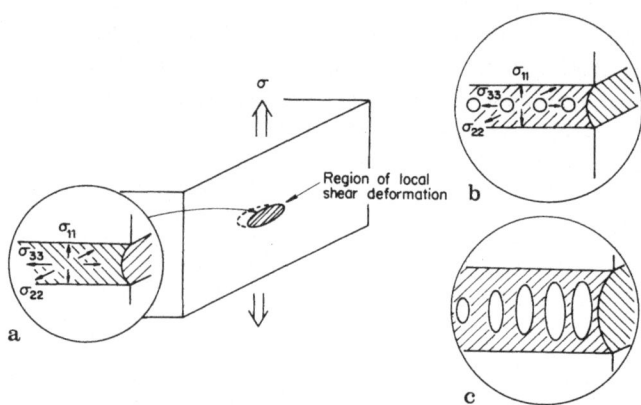

Fig. 2a—c. Schematic drawing of several postulated microscopic steps in craze nucleation: **a** Formation of a localized surface plastic zone and buildup of significant lateral stresses. **b** Nucleation of voids in the zone to relieve the triaxial constraints. **c** Further deformation of polymer ligaments between voids and coalescence of individual voids to form a void network

The localization of plastic strain also leads to the *development of lateral stresses* $(\sigma_{22}, \sigma_{33})$ in the plastic zone. At this stage one of two things can happen. If shear deformation is to continue, the applied tensile stress must be increased since the lateral stresses have reduced the local deviatoric stress s_0 which is driving the shear deformation. The increase in stress may be enough to cause global plastic shear deformation outside the initial localized zone in which case the polymer glass sample, if thin enough, will probably neck and fail only after large macroscopic plastic deformation. As the local lateral stresses increase however it becomes increasingly probable that voids nucleate in the zone due to the high local σ_0 there as shown in Fig. 2b. If this happens the voids can subsequently grow and interconnect with the remaining polymer ligaments between voids thus strain hardening and stabilizing the void structure as molecules within them become highly oriented, Fig. 2c. Finally the periphery of the voided sheet structure begins to extend within the zone and simultaneously this structure begins to thicken. Craze nucleation has been accomplished and craze growth has begun.

It is not so easy to decide which of the steps in this sequence is the critical one for craze nucleation. The existing microscopic models of this process postulate different critical steps. The model of Argon and Hannoosh [6], the most detailed and the most thoroughly tested experimentally of these, assumes that the critical event in craze nucleation is the formation of a certain level of porosity by the thermally activated formation of shear patches which form voids at their ends when blocked. The critical porosity level is supposed to decrease exponentially with the level of global hydrostatic tension σ_0 normalized by the tensile yield stress Y. The treatment bypasses the question of surface energy of the pores that are created [14], assuming that this energy can be supplied by the unstable shear patch, and that furthermore the only surface energy necessary in pore formation is separation of molecules against their intermolecular (van der Waals) attraction. In light of the highly entangled state of polymer chains in high molecular weight glassy polymers this assumption seems particularly dubious. As will be discussed at length below entanglements appear to have a strong effect on crazing.

Kawagoe and Kitagawa [11] proposed a somewhat different microscopic model in which two shear bands or patches are assumed to interact at their intersection to form a microcrack which expands elastically to a cylinderical pore. In this treatment although the surface energy of the pore is considered, it is supposed that no thermal activation of pore formation is possible.

Gent [15] proposed a quite different model for craze nucleation. The local hydrostatic stress σ_0 (concentrated by the presence of flaws) was supposed to decrease T_g of the glass to the ambient temperature. Upon reaching the rubbery state, the polymer will cavitate easily to form voids. The main problem with this mechanism is the large stress concentration factors that are necessary for it to operate at room temperature; it also cannot easily account for the time dependence of craze nucleation under constant stress. It provides a possible explanation, however, for the nucleation of crazes II [16]. In view of the fact that the work of Kausch and Dettenmaier on craze nucleation and intrinsic crazing will be treated in detail in Chapter 2, it will not be discussed further here [16].

If craze initiation is truly a nucleation problem (albeit a heterogeneous one) there should be a critical size associated with the critical craze nucleus. Such critical sizes are implied or explicitly determined in the models reviewed above. There is evidence that such a critical size exists. It has been known for some time that rubber modified PS with small rubber particles ($< 1\ \mu m$) is less tough than HIPS with larger particles. Recent experiments have traced the reason for this decrease in toughness to the poor craze nucleation efficiency of the small rubber particles [17]. At first glance this result seems hard to justify theoretically since it is well established [18,19] that the stress concentration at the surface of a spherical rubber particle is about the same as that of a spherical void, i.e., it is roughly 2, independent of the size of the particle. However one can easily show that this stress concentration falls to less than 1.5 within one tenth of the particle diameter from the particle surface, hence the extent of the high stress region around the particle decreases with the particle size. When the size of this highly stressed region becomes comparable to, or smaller than, the size of the critical craze nucleus the rubber particles should cease to be good craze nucleation sites, as observed. From the rubber particle size which ceases to be effective in HIPS, (diameter $< 1\ \mu m$), the lateral size of a critical craze nucleus in PS can be estimated to be ca. 75 nm, or 3 fibril spacings, which seems very reasonable.

3 Craze Growth

There are two important questions about craze growth, namely what are the mechanisms of craze tip advance (expansion of the craze periphery generating more fibrils) and craze thickening (normal separation of craze surfaces lengthening the craze fibrils). Unlike the cloudy experimental situation regarding craze nucleation, that regarding craze growth now seems quite clear.

Crazes in isotropic amorphous polymers grow on the plane of maximum principal stress [20,11]. This result may be predicted [1] from the mechanical principle of maximum plastic resistance [21] and the craze fibril geometry. Since craze fibrils can support significant tensile forces across the craze but not shear forces the plastic resistance during craze growth is maximized by maximizing the normal stress on the plane of the craze. In oriented (non-isotropic) polymers however crazes no longer grow on the plane of maximum principal stress [22,23] probably because the craze fibril structure and craze surface stress for thickening depends on craze orientation. In light of these craze growth trajectories, it is perhaps not surprising that Argon and Salama [24] have shown that air craze growth kinetics in PS (unlike craze nucleation kinetics) are only sensitive to the magnitude of the maximum principal stress. Craze growth can occur in stress fields where σ_0 is zero or even negative.

3.1 Craze-tip advance mechanisms

It is observed that the normal craze fibril structure can be observed just behind the craze tip where the craze is as thin as 5—10 nm [25]. This observation was difficult to reconcile with early models [14] of craze tip advance which postulated that this occurred by repeated nucleation and expansion of isolated voids in advance of the tip. One problem was to explain how the void phase became interconnected while the craze was still so thin. Another was that the predicted kinetics of craze growth appeared to be incorrectly predicted; indeed since this mechanism almost involves the same steps as the original craze nucleation, it is hard to understand how craze growth could be so much faster than craze nucleation as observed experimentally.

A more recent hypothesis [29] is that the craze tip breaks up into a series of void fingers by the Taylor meniscus instability [26-28]. Such instabilities are commonly observed when two flat plates with a layer of liquid between them are forced apart or when adhesive tape is peeled from a solid substrate [26,27]. The hypothesis in the case of a craze is that a wedge-shaped zone of plastically deformed and strain softened polymer is formed ahead of the craze tip (Fig. 3a); this deformed polymer constitutes the "fluid" layer into which the craze tip "meniscus" propagates whereas the undeformed polymer outside the zone serves as the rigid "plates" which constrain the fluid. As the finger-like craze tip structure propagates, fibrils

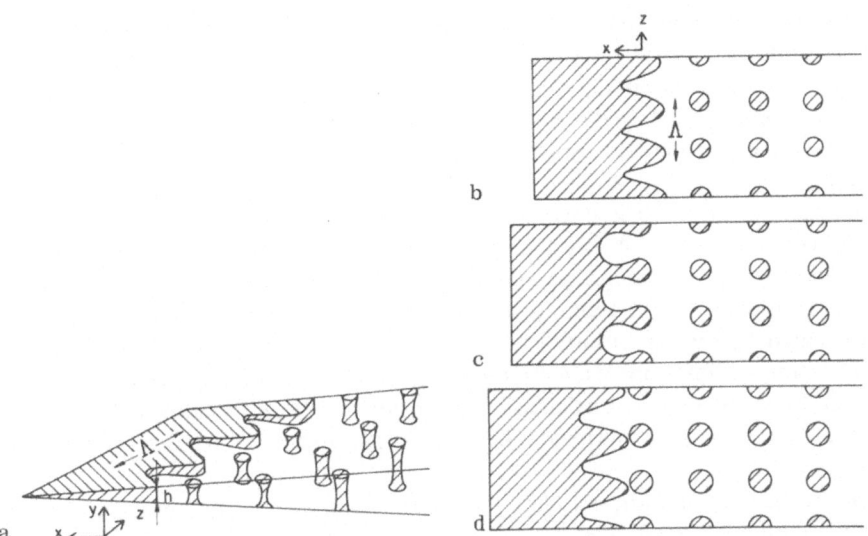

Fig. 3a—d. Schematic drawing of craze tip advance by the meniscus instability mechanism. **a)** Drawing showing wedge of deformed polymer ahead of the void fingers and trailing fibrils. **b)** to **d)** xz sections through the craze showing the sequence of events as the craze tip advances by one fibril spacing. The void finger and fibril spacing in this drawing shows a much more regular structure than observed experimentally

develop by deformation of the polymer webs between fingers, as shown in Fig. 3b—c. The interconnected void network develops naturally right at the craze tip.

Stereo-transmission electron microscopy of craze tips has shown that the meniscus instability is the operative craze tip advance mechanism in a wide variety of glassy polymers [30,31]. Figure 4 shows a craze tip in a thin film of a styrene-acrylonitrile copolymer (PSAN). The void fingers are clearly visible. No isolated voids can be

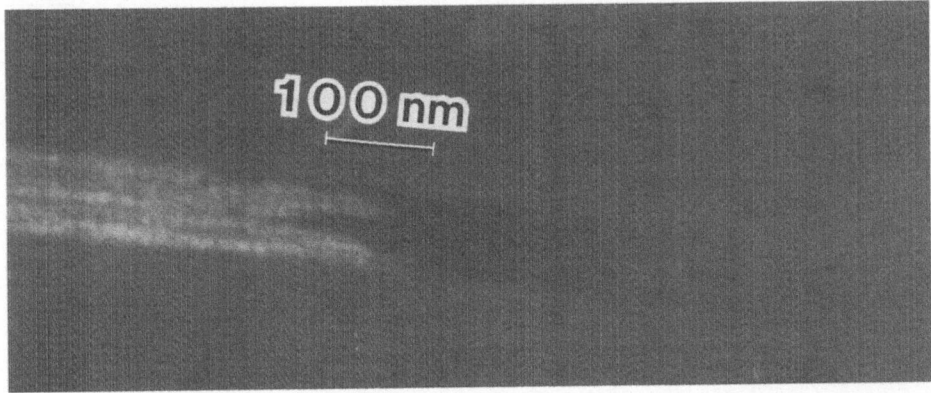

Fig. 4. TEM micrograph of a craze tip in poly(styrene-acrylonitrile) (stained with OsO_4). The film is tilted so the craze front (normal to the film surface) can be seen in projection. Note the stained wedge of plastically deformed polymer ahead of the craze tip. Courtesy of Dr. A. M. Donald

detected ahead of the craze tip. This particular sample has been exposed to a 2% aqueous solution of OsO_4 which, while it does not alter the void finger structure, apparently stains preferentially the polymer in the plastic zone ahead of the craze tip. This zone appears as a wedge-like extension of the craze, an observation that is confirmed by experiments in which one surface of a film was gold decorated prior to crazing [30].

The craze tip advance is conveniently modelled by the advance of a meniscus between two rigid plates a distance h apart where h is the thickness of the craze just behind the craze tip (Fig. 3a). Argon and Salama have approached the problem somewhat differently by assuming the plastic zone in advance of the craze tip is approximated by the logarithmic spiral slip line field of an elastic-perfectly plastic medium; this assumption leads to much larger values of h than those actually observed. In lieu of a realistic calculation of h using the strain softening characteristics of the polymer glass h is treated here as a parameter to be determined from experiment.

Suppose that an initially straight meniscus, lying at $x = x_0$, is perturbed into a small amplitude sinusoid, i. e.,

$$x = x_0 + \zeta \sin \frac{2\pi z}{\Lambda} \tag{3}$$

thus introducing a new radius of curvature $R = [-(4\pi^2/\Lambda^2)\zeta]^{-1}$. There will also be a gradient of local hydrostatic stress $d\sigma_0/dx$ ahead of the craze tip which will tend to increase the amplitude of the perturbation. The perturbation will not grow unless the increment of hydrostatic stress $\delta\sigma_0 = (d\sigma_0/dx)\zeta$ exceeds that required to produce the new radius of curvature, i.e., Γ/R, where Γ is the *surface energy*. There is thus a minimum wavelength for a perturbation to grow

$$\Lambda_m = 2\pi \sqrt{\frac{\Gamma}{d\sigma_0/dx}} . \tag{4}$$

Such a wavelength will neither grow nor shrink. The fastest growing wavelength is just $\Lambda_c = \sqrt{3}\,\Lambda_m$ [28,29]. For a material with a non-Newtonian flow law of the form $\dot{\varepsilon} = \dot{\varepsilon}_F(\sigma/\sigma_F)^n$, where σ and $\dot{\varepsilon}$ are the *equivalent stress* and *equivalent strain rate* in tension and σ_F and $\dot{\varepsilon}_F$ are material parameters, $d\sigma_0/dx$ is computed to be [28,29]

$$\frac{d\sigma_0}{dx} = \frac{2\sigma_F}{\sqrt{3}h}\left(\frac{2(n+2)\,v_0}{\sqrt{3}\dot{\varepsilon}_F h}\right)^{1/n} \tag{5}$$

and

$$\Lambda_c = 2\pi\sqrt{3}\sqrt{\frac{\sqrt{3}\Gamma h}{2\sigma_F}}\left(\frac{\sqrt{3}\dot{\varepsilon}_F h}{2(n+2)\,v_0}\right)^{1/2n} . \tag{6}$$

Following Fields and Ashby [28] as illustrated in Appendix I, one can also estimate the steady state craze tip velocity to be

$$v_0 \simeq \frac{\sqrt{3}}{2} \frac{\dot{\varepsilon}_F h}{(n+2)} \left[\frac{\sqrt{3}h(\beta^*S_t)^2}{8\sigma_F\Gamma} \right]^n \left(1 - \frac{2\Gamma}{\beta^*S_t h} \right)^{2n} \tag{7}$$

where β^*S_t is the *hydrostatic stress* midway between void fingers (taken to be proportional to the average tensile stress S_t at the craze tip). For PS the wavelength of the instability is predicted to be in the range 15 to 35 nm, roughly the spacing of the fingers and the spacing between fibrils just behind the craze tip observed by TEM. It is also roughly the average spacing between fibrils in PS crazes determined by SAXS [32], although since most of the fibril length is formed at the craze surfaces (by craze thickening) rather than by advance of the craze tip the connection here is much more tenuous. (It seems unlikely that the spacing between fibrils set at the craze tip when the fibrils are as short as 5 nm should necessarily persist into the mature craze where the fibrils are ~ 500 nm long [33]).

Note that the velocity of craze tip advance is very sensitive to the energy of the surface being created (v_0 decreases strongly as Γ increases since $10 < n < 20$ for most polymer glasses [34,35]). This feature will be invoked later to explain the important effect of entanglements on the stresses required for crazing.

3.2 Craze Thickening

As the craze expands in area by advance of the craze tip it must also increase in thickness in a direction normal to the craze surfaces. There are two quite different mechanisms by which this thickening might occur. One might imagine that once fibrils are created at the craze tip that they extend in length by creep with no new polymer being drawn into the fibrils at the craze surfaces. Such a fibril creep mechanism would produce a craze that would become weaker and weaker as the craze grew longer and thicker due to the fact that the volume (area) fraction of fibrillar material in the craze would decrease rapidly with craze thickness and correspondingly the true stresses in the fibrils would rise dramatically. This mechanism could be invoked implicitly to account for the apparent existence of a critical crack tip opening displacement in fracture experiments [36,37] and as a reason for the supposed desireability of limiting craze lengths in rubber toughened polymers [38]. It has also been used to model the kinetics of air craze growth [39].

A second thickening mechanism is possible however. New polymer could be drawn into the fibrils from the craze interfaces maintaining the extension ratio of the fibrils constant for a given stress on the craze surfaces. This surface drawing mechanism [25] has as its analog the cold drawing of macroscopic polymer fibers. Such drawing occurs usually by formation of a neck with a certain "natural" draw ratio and then an increase in the length of this neck by continued drawing of new polymer into the neck at the "shoulders" which separate it from the undrawn polymer outside. Crazes which thicken by the surface drawing mechanism would not grow automatically weaker as they grow longer.

Clearly the way to decide between these two mechanisms is to measure either the *volume fraction* v_f of fibrils or its inverse the *fibril extension ratio* λ, at various positions along the craze. If fibril creep is the mechanism of craze thickening λ should be low in thin regions of the craze, i.e., near the craze tip where the fibrils have just been created, and high in thick regions of the craze, i.e., where the fibrils are "older". Three separate types of experiments have recently examined this question and come to the identical conclusion that air craze thickening occurs predominantly by the surface drawing mechanism. In the highest resolution experiments, the mass thickness contrast on TEM micrographs of crazes in thin films have been used to deduce the v_f and λ profile along the craze. Figure 5 shows the

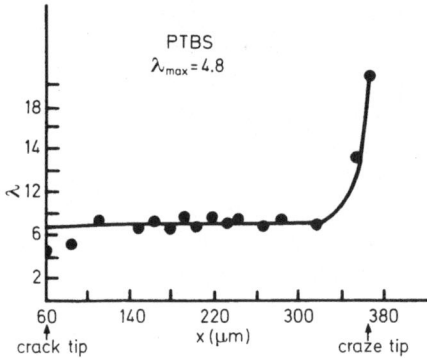

Fig. 5. Craze fibril extension ratio vs. distance from crack tip in a craze grown in a poly(tert-butylstyrene) film (From Ref. [40], courtesy of John Wiley & Sons.)

results of such measurements on a craze in poly(tertbutylstyrene) that was grown from a crack tip. The fibril extension ratio is constant along most of the craze (even near the crack tip where the craze thickens dramatically) and actually increases near the craze tip. Fibril creep cannot explain these results whereas surface drawing can. (Discussion of the increase in λ behind the craze tip which depends on the local stress field of the craze is deferred to a later section). Similar TEM results are obtained on both isolated crazes and crazes grown from crack tips in a wide variety of glassy polymers [25,40].

Verheulpen-Heymans [41] has measured the fibril volume fraction profile along isolated crazes in polycarbonate using an optical technique whereas Trent, Palley and Baer [42] have measured it in isolated polystyrene crazes in thin films by comparing craze displacements measured from the displacement of bars of an evaporated metal grid intersecting the craze thicknesses. They use TEM of the unstressed film to make the measurements. Both groups find that v_f is independent of craze thickness.

Final pieces of evidence are TEM micrographs of crazes grown from crack tips under step straining. Figure 6 shows an example of an air craze grown from a crack tip in PS by increasing the strain from 1 % to 1.5 % to 2 % to 2.5 % to 3 % at 1 minute intervals [43]. Note that the craze contains "ridges" of slightly lower extension ratio material which mark the position of the craze-bulk polymer interface at the time the steps in strain were made. (The material at these interfaces apparently stress ages and becomes more difficult to draw during the holding period between step changes in strain.) It is clear that new polymer is drawn into the fibrils during each increment in craze thickening caused by the strain steps. Furthermore since the extension ratio of the oldest (inner) region of the craze is about the same as the youngest (outer) region, c.f., the craze contrast in these regions, any contribution of creep to the thickening must be small. Thus craze thickening by surface drawing seems to be the dominant mechanism.

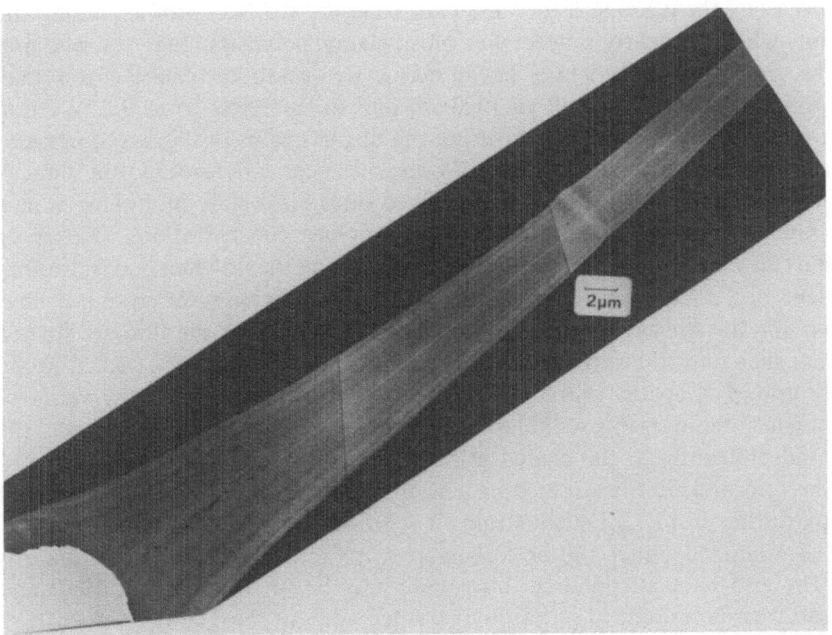

Fig. 6. TEM image of a craze in polystyrene, strained in 5 stages at one minute intervals (From Ref. [43], courtesy Butterworth & Co., Ltd.)

The reader should however beware of extrapolating these results on air crazes created by applying uniaxial tension with those on crazes produced by other means. For example environmental crazes that grow rapidly in highly plasticizing, but slowly diffusing, environments probably thicken primarily by fibril creep since craze thickening by surface drawing will ultimately be limited by diffusion of the environment into the craze-bulk polymer interface (Crazes produced by fast diffusing, highly plasticizing environments, e.g., CO_2 in PS [44], thicken by surface drawing). Crazes at crack tips undergoing fatigue loading may also break down by a fibril creep process. Indeed these results on craze thickening do not imply that fibril failure under increasing tension cannot ultimately occur by a localized creep process somewhere along the fibril. All that can be said is that this creep, if it occurs, is not homogeneous enough to give an appreciable contribution to the total craze thickening.

4 Craze Micromechanics

To understand the details of the craze extension ratio profile one must have knowledge about the mechanics of the craze on a microscopic scale. The most important parameter is the normal tensile stress applied to the craze surface by the craze fibrils. Since the fibril structure is very weak in shear, any shear stresses S(x) on the craze surface must be negligible. The most useful method for determining these stresses experimentally is to measure the craze surface displacement profile w(x) along the craze and then use it to compute the S(x) profile making the assumption, which is nearly correct for most glassy polymers, that the polymer outside the craze only experiences linear elastic deformations. Since crazes may typically range in length from 50 to 1000 μm and in thickness from 0.1 to 2 μm, very high resolution methods for measuring the displacement profile are necessary. The most satisfactory method involves TEM measurements of crazes in thin films. A film, typically 0.5 μm to 1 μm thick is produced on a glass slide by pulling it at a constant rate from a solution of the correct polymer concentration. Thicker or thinner films may be prepared by increasing or decreasing the polymer concentration. The grid bars of a ductile copper grid are coated with the same solution of polymer. After removing the solvent from the film and grid, portions of the film are floated off the glass slide onto the surface of a water bath where they may be picked up on the copper grid. The coated grid and film are now briefly exposed to the vapor of the solvent which removes any wrinkles from the film and at the same time causes the film to bond uniformly to the coated grid. After another solvent removal step in vacuum the grid and film is strained in tension until crazes appear. Since the grid deforms plastically from very small strains it serves to hold the film under tension. The grid is examined under the optical microscope and a suitable grid square is selected. The grid bars surrounding this square may be cut with a razor blade and the film can then be examined in its final strained state in the TEM.

The *stress profile* S(x) along the interface of an isolated craze (one that is not grown from a crack tip) can be considered to be the sum of two terms

$$S(x) = \Delta S(x) + \sigma_\infty ,$$

where σ_∞ is the *applied tensile stress* normal to the craze and $\Delta S(x)$ is the "*self stress*" *of the craze*, the stress that would have to be applied to the craze-bulk interfaces (in the absence of any other stresses) to produce the *displacement* w(x) of these interfaces. There are many different, but still correct, approaches which will solve this problem. One is to use the Fourier transform method of Sneddon [45] in which

$$\Delta S(x) = - \frac{2}{\pi} \int_0^\infty \bar{p}(\xi) \cos (x\xi) \, d\xi \qquad (8)$$

where

$$\bar{p}(\xi) = \frac{\xi E^*}{2} \int_0^a w(x) \cos (\xi x) \, dx .$$

Here E* is an *effective Young's modulus* of the polymer (E* = E for plane stress and E/(1 — v²) for plane strain; v = Poisson's ratio) and a is the half length of the craze. The craze displacement profile is computed from

$$w(x) = \frac{1}{2} T(x) (1 - v_f(x)) \tag{9}$$

so that measurements of the craze thickness $T(x)$ and *fibril volume fraction* $v_f(x)$ profiles from the TEM images of the craze suffice to determine the $w(x)$ and $\Delta S(x)$ profiles as well. In this case the applied stress σ_∞ is perturbed by any defects in the bonding of the films, the formation of other crazes in the same square as the one being analyzed and any stress relaxation in the film. For this reason it is useful to grow crazes from "cracks" centered in each grid square which are produced by "burning" a slot in the film with the intense electron beam of an electron microprobe. Another method of stress analysis, based on the work of Bilby and Eshelby[46] is appropriate in this case. The craze is modelled as an array of continuously distributed dislocations of Burgers vector b (the actual value of b may be set arbitrarily since it eventually cancels out in the final results) which give rise to the same surface displacements as the actual craze[1]. The linear *"dislocation density"* is given by

$$\alpha(x) = -\frac{2}{b} \frac{dw(x)}{dx} . \tag{10}$$

By requiring that the stress singularity at the stationary crack tip must vanish the level of applied stress can be determined to be[47]

$$\sigma_\infty = \frac{E^*b}{2\pi} \int_c^a \frac{dx\,\alpha(x)}{\sqrt{x^2 - c^2}} \tag{11}$$

where c is the *half crack length* and a is the *half length of the craze plus crack*. The craze surface stress $S(x)$ at a point $x > c$ is given by[47]

$$S(x) = \frac{E^*b}{2\pi} x \sqrt{x^2 - c^2} \int_c^a \frac{dx_1\,\alpha(x_1)}{\sqrt{x_1^2 - c^2}(x^2 - x_1^2)} \tag{12}$$

The dislocation method of stress analysis is also useful for determining craze stress fields in anisotropic (e.g., oriented) polymers[48]. All one needs here is the stress field of a single dislocation in a single crystal with the same symmetry as the oriented polymer (the text by Hirth and Lothe[49] provides a number of simple cases plus copious references to more complete treatments in the literature); the craze stress field can be generated by superposition of the stress fields of an array of these dislocations of density $\alpha(x)$. Dislocations may also be used to represent the self-stress fields of curvilinear crazes (produced by craze growth in a non-homogeneous stress field for example). Such a method has been developed by Mills[50]

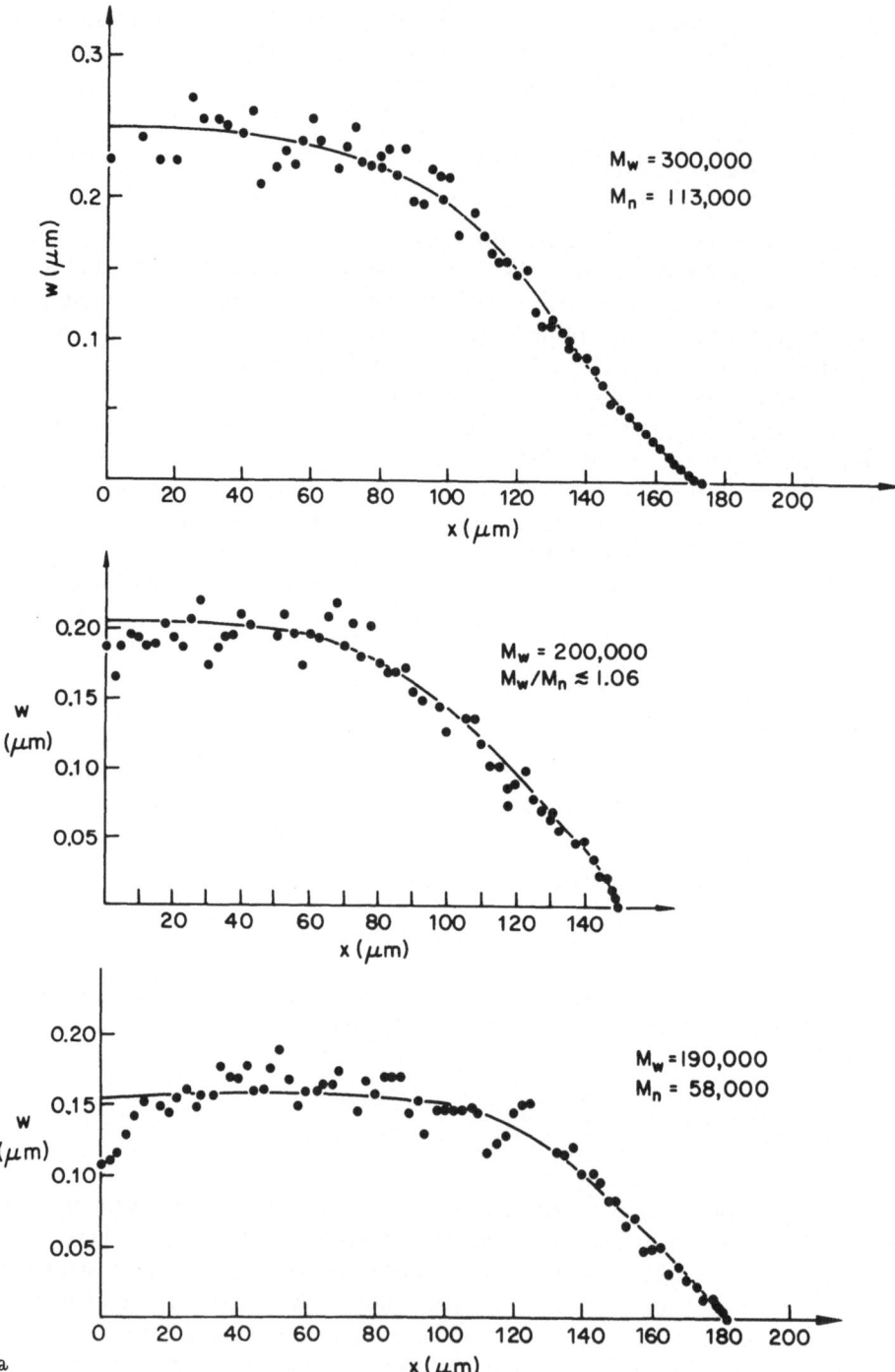

a

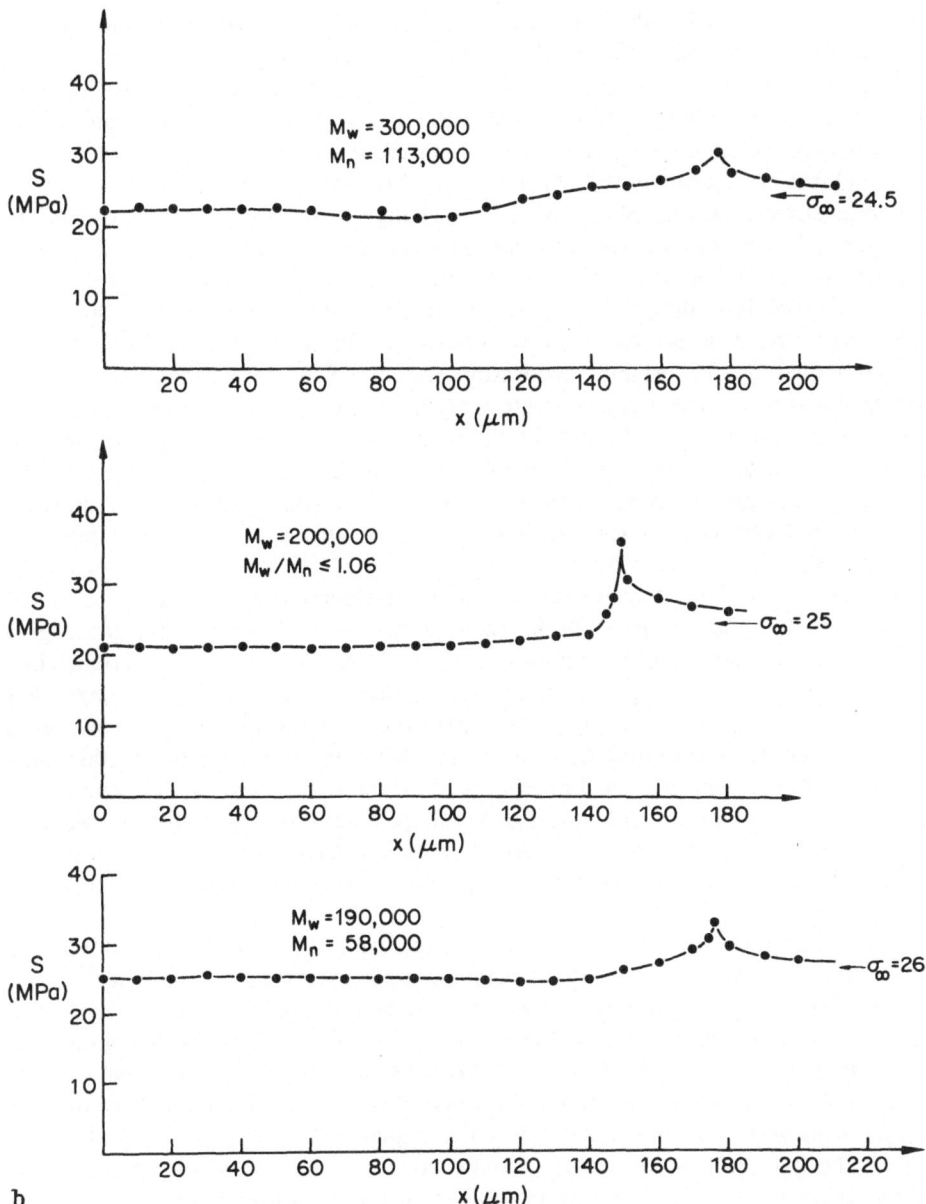

Fig. 7. **a** Crazes displacement profiles and **b** craze surface stress profiles in a commercial polydisperse PS, a monodisperse PS and a blend of a high and a low molecular weight monodisperse PS (5% M_w 4000 in 200,000 M_w)

in his attempt to solve the problem of two interacting crazes. One should note however that these dislocation methods which assume that the craze displacements are always normal to the craze surfaces (w(x) is represented by a set of edge dislocations with Burgers vector normal to the craze surfaces) are not applicable to a case where the state of external stress along the craze surface changes with time. Although the craze initially follows the plane of maximum principal stress and thus the shear stresses on that plane are zero, a change in the stress conditions (due to the approach of a nearby craze or to changes in the external loading) may cause shear stresses to appear on the plane of the craze. Since the craze by virtue of its fibril structure has almost no resistance to shear it will effectively respond to these shear stresses as a shear crack with correspondingly large shear displacements not embodied by the original dislocation model. (These shear displacements of course, could be modelled as a separate array (actually a pileup) of edge dislocations with Burge s vector parallel to the craze surfaces [46]. The importance of this effect has een clearly demonstrated by Takahashi and Hyodo [51], who examined craze and crack growth under superposed static tensile and ultrasonic shear stresses, but current treatments of the interacting craze problem [50,52] ignore it and are therefore incorrect.

Finally the problem of converting craze displacements to stresses may be solved completely numerically using finite element methods [53]. In general these give results that agree well with the methods outlined above but because of computer time constraints they generally give a more coarse-grained stress profile. An important limitation of the Fourier transform and distributed dislocation methods however is that they make the assumption that the craze or the craze plus crack is embedded in a sheet of unlimited extent loaded at infinity. For crazes or cracks whose lengths are a significant fraction of the total grid square width this assumption becomes questionable. The finite element method does not have this limitation and thus potentially can offer important insight into the errors involved in making this assumption.

Figure 7a shows some typical craze surface displacement profiles, and Fig. 7b shows the surface stress profiles, from isolated crazes in polystyrene [54]. The samples represent a monodisperse PS, a typical polydisperse PS (a commercial polymer containing no mineral oil lubricant), and a blend of high and low molecular weight monodisperse PS. All show the same general features, a higher than average stress at the craze tip, which drops over a certain distance behind the craze tip to a roughly constant value not much below the applied stress σ_∞ over the rest of the craze. The load-bearing character of these crazes is clear. Evidently propagation of the craze tip by the meniscus instability is a more difficult process than the craze thickening by surface drawing. But as comparison between these stress profiles shows there are also differences between crazes in these PS's which may be traced to rather subtle differences in the displacement profiles just behind the craze tip. The abrupt rise in craze displacement, the higher stress at, and more rapid drop in stress behind, the craze tip is a characteristic of crazes in all monodisperse PS examined ranging in molecular weight from 100,000 to 2×10^7. The differences thus seem to be traceable to the low molecular weight tail of the polydisperse sample, a view which is reinforced by the w(x) and S(x) profiles of the molecular weight blend which shows intermediate behavior between the monodisperse and polydisperse samples. The physical reason for these differences is not completely clear at present, although in view of the important effects of molecular entanglements on craze micromechanics and microstructure presented in a later section we believe it is an effect of the dilution of the molecular entanglement network of long chains by polymer chains too short to entangle [54].

From the surface stress profile S(x) and fibril volume fraction profile $v_f(x)$ it is also possible to find the true stress in the craze fibrils $\sigma_f(x)$ which is

$$\sigma_f(x) = S(x)/v_f(x) .$$

For the polystyrene crazes in Fig. 7 these values are approximately 4 to 5 times S(x).

This section has emphasized the experimental determination of, rather than theoretical models for, craze micromechanics. At present none of the theoretical models is capable of representing the detail which is observed experimentally. The best isolated craze model is that of Verhuelpen-Heymans and Bauwens [39] which approximates the craze as tip process zone of length r where the stress is high and constant and a body zone where the stress is lower. However the model cannot predict craze growth kinetics (it is based on a fibril creep craze thickening assumption) and gives us no guidance in understanding the differences between one polymer and another (e.g., Fig. 7). A recent paper by Chudnovsky, Baer and Palley [55] adopts the same mechanical model but makes assumptions about craze growth that are incompatible with it. (They assume thickening stops along the body zone at a certain time but that the tip section continues to grow and thicken; the displacement profiles generated by this growth will not produce the assumed step stress profile.) The Dugdale model [56,57] of a craze at a crack tip is widely used and attractively simple; however, its assumption that S(x) is constant along the craze is not always obeyed, even qualitatively, by crazes growing from crack tips. At the moment then experiments continue to lead theory in this area; a more complete discussion of various models may be found in a previous review [1] and in Chapter 3.

5 Craze Microstructure

5.1 Results from Different Methods

The detailed microstructure of the fibrils within the craze offers clues as to the processes occurring at the craze surface and the craze tip as well as having an ultimate bearing on craze instability by fibril breakdown. Craze fibril arrangements and dimensions may be studied by either diffraction methods (X-ray or electron) or directly by TEM. Small angle X-ray scattering offers an important advantage here in being able to give information on the fibril structure averaged over many crazes in bulk samples, conversely the electron microscopic methods are better suited for investigating the fibril structure in small regions of a single craze.

A TEM micrograph of a thick region of a craze (near its origin) is shown in Fig. 8a. While one can detect the fibrillar nature of the craze, the strongly

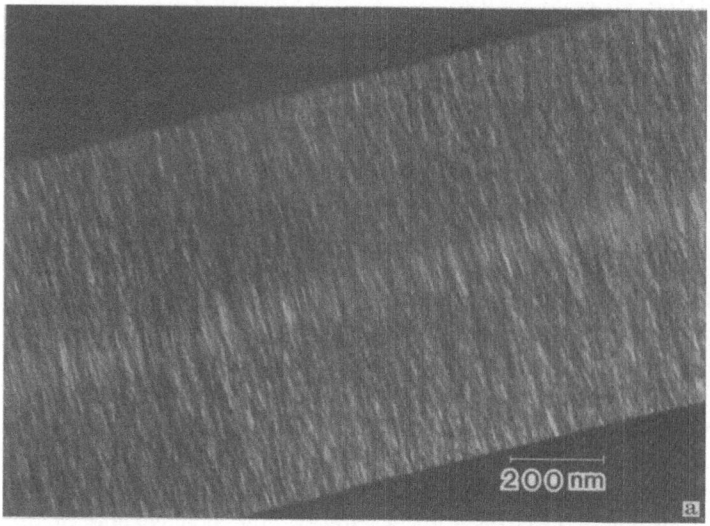

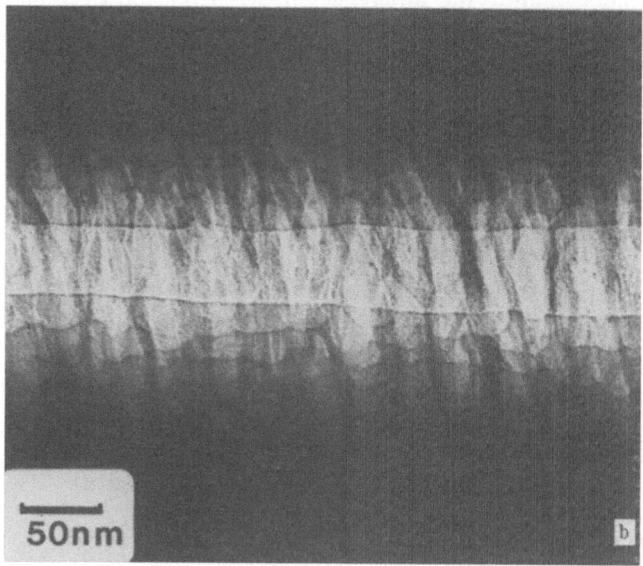

Fig. 8. a TEM image of a wide region of a craze in a thick PS film. **b** TEM image of a craze in a thinner (250 nm) PS film, tilted about an axis parallel to the craze

overlapping nature of projected image of the forest of fibrils makes it difficult to be certain of details, such as average fibril diameters and spacings. One can do somewhat better by working with thinner films (250 nm) to reduce the overlap problem and by tilting to produce stereo TEM images (a single tilted image of an air craze in a thin PS film is shown in Fig. 8 b) but it is very tedious work indeed to measure enough fibrils to get good statistics. For air crazes in PS films a range of fibril diameters from 4 to 10 nm was determined with a mean value of 6 nm [30]. Distances between the centers of nearest-neighbor fibrils were found to fall in the range between 20 and 30 nm. Although the dominant fibril direction is normal to the craze surfaces, there are numerous fibril branches (or tie fibrils [58-60]) which connect one main fibril with another. These demonstrate that the surface drawing process of craze thickening can alter the craze fibril structure.

Nevertheless one can idealize the fibril structure as a forest of isolated, cylinderical fibrils, all oriented perpendicular to the craze surfaces. Such an array of fibrils in the 5 to 10 nm size range should scatter X-rays strongly in the small angle scattering regime [32,33,61,62]. One expects to see the scattering as a disk in reciprocal space (basically the shape transform of a single cylinderical fibril) which is modulated by interference effects caused by any short range ordering of the fibril centers. (It is clear from micrographs such as Fig. 8b and the X-ray evidence itself that no long range order of the fibril centers exists.) That expectation is fulfilled if certain precautions are taken such as measuring the SAXS from crazed samples before unloading [32][1].

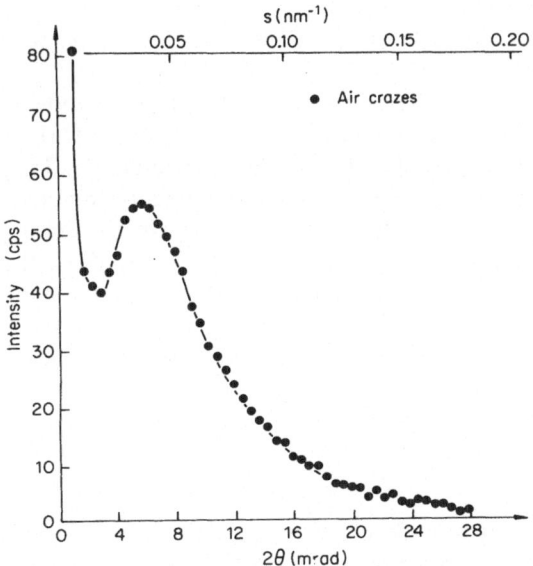

Fig. 9. Small angle X-ray scattered intensity from craze fibrils of air crazes in PS

Figure 9 shows the scattered intensity from the craze fibrils of an air craze in PS. The high angle tail of the curve obeys Porod's law [63],

$$i(s) = k_1/s^3 \tag{11}$$

---

[1] One must also ignore a high intensity scattering spike normal to the craze in reciprocal space. This intensity originates from X-ray reflection from the craze surfaces [32]

where s is the magnitude of the scattering vector $2 \sin \Theta / \lambda_x$ and λ_x is te X-ray wave length. From k_1 and the invarient

$$Q = 2\pi \int ds \; si(s) \; , \tag{12}$$

one may determine a *mean fibril diameter* [32,33]

$$\bar{D} = \frac{\langle D^2 \rangle}{\langle D \rangle} = \frac{1}{\pi^3 (1 - v_f)} \frac{Q}{k_1} \tag{13}$$

where v_f may be taken from the TEM mass thickness contrast measurements. For PS $v_f \simeq 0.25$. The mean fibril diameter for the PS air crazes of Fig. 9 is 6 nm in good agreement with the direct TEM measurements of D in crazes in thin films discussed above.

The SAXS pattern also gives information on the arrangement of fibrils. The maximum in the scattering curve in Fig. 9 represents an interference between X-rays scattered from neighboring fibrils. Detailed analyses [32,33] show however that the packing of fibrils is rather liquid-like, with only a small peak in the fibril axis radial distribution function at a distance 20—25 nm in PS giving rise to the interference effect observed. There is also a large spread in fibril diameters, with many more small fibrils than large ones.

Paredes and Fischer [33,64] have measured the fibril diameters in polycarbonate (PC) and polymethylmethacrylate (PMMA) crazes produced by straining at various rates at temperatures close to, but below, the glass transition temperature T_g. They find that although the lower yield stress σ_D (the stress for deformation by craze thickening at the given strain rate and temperature) decreases with increasing temperature and decreasing strain rate and $\bar{D}$ increases with these variables, the product $\sigma_D \bar{D}$ is a constant. Paredes and Fischer rationalize this observation by assuming that all the work of craze thickening deformation [$S(1 - v_f)dT$] is stored as the new surface free energy of the craze fibrils [$(4v_f \Gamma / \bar{D})dT$]. This assumption[1] leads to the equation

$$S\bar{D} = \frac{4 \, v_f \, \Gamma}{(1 - v_f)} \tag{14}$$

which shows that since $S \simeq \sigma_D$ and v_f is roughly constant during surface drawing, the product $\sigma_D \bar{D}$ is a constant. The experimental products however are about a factor of 4 larger than predicted (e.g., 0.2 for PS vs. 0.05 predicted and 0.58 for PC vs. 0.16 predicted). While some of that discrepancy might be ascribed to the fact that the correct value for Γ should also reflect some covalent bond breaking energy of the highly entangled polymer chains as suggested by Paredes and Fischer, it seems clear that ignoring the viscous work of the plastic deformation process itself is not correct. This frictional dissipation represents another important sink for the work of crazing that is not considered in the Paredes and Fischer energy balance.

---

[1] The equation of Paredes and Fischer [33] is missing the factor v_f since they took σ_D as equal to the true stress in the fibrils during drawing

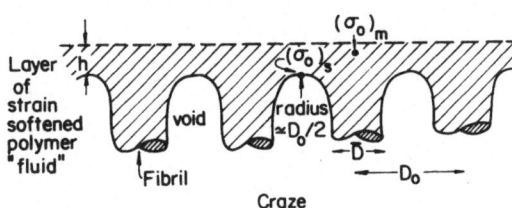

Elastically deformed polymer glass

Layer of strain softened polymer "fluid"

void

Fibril

radius $\approx D_0/2$

$(\sigma_0)_m$

$(\sigma_0)_s$

D

D_0

Craze

Fig. 10. Schematic drawing showing important parameters of the craze-bulk polymer interface

An alternative model is shown schematically in Fig. 10. Just as for the craze tip, the strain-softened polymer being actively deformed at the bulk polymer interface is idealized as a thin layer of non-Newtonion fluid. The velocity v of plastic advance of the craze interface depends on the gradient in hydrostatic tension $\nabla\sigma_0$ as

$$v = s^* \frac{\sqrt{3}\dot{\varepsilon}_F h}{2(n+2)} \left(\frac{\sqrt{3}h}{2\sigma_F}\right)^n |\nabla\sigma_0|^n \tag{15}$$

where s^* is the sign of $\nabla\sigma_0$, h is the thickness of the softened fluid layer and σ_F, $\dot{\varepsilon}_F$ and n are parameters of the non-Newtonian flow law, $\dot{\varepsilon} = \dot{\varepsilon}_F(\sigma/\sigma_F)^n$. From Fig. 10 it seems reasonable to approximate $\nabla\sigma_0$ as

$$\nabla\sigma_0 \simeq \frac{2[(\sigma_0)_m - (\sigma_0)_s]}{D_0} \tag{16}$$

where $(\sigma_0)_m$ and $(\sigma_0)_s$ are the hydrostatic tensions in the layer directly above the fibril and at the surface of the void "ceiling" between fibrils respectively and where D_0 is both the average spacing between fibrils and the fibril diameter before deformation, i.e., $D_0/\bar{D} = \lambda^{1/2}$. The hydrostatic tension at the void surface $(\sigma_0)_s$ is determined by surface tension

$$(\sigma_0)_s \simeq 4\,\Gamma/D_0 . \tag{17}$$

Following the treatment of Fields and Ashby[28] for $(\sigma_0)_m$ we assume it to be proportional to the average tensile stress S on the craze interface

$$(\sigma_0)_m = \beta S \tag{18}$$

The gradient $\nabla\sigma_0$ is then

$$\nabla\sigma_0 = \frac{2\beta S}{D_0}\left(1 - \frac{4\Gamma}{\beta S D_0}\right) . \tag{19 a}$$

It seems reasonable to assume that the fibril structure (value of D_0) with the fastest growing interface will be that ultimately observed; in turn from Eqs. 15 and 19, that D_0 is the one which maximizes $\nabla\sigma_0$. Setting the derivative d $\nabla\sigma_0/dD_0$ equal to zero we find the fastest growing D_0 to be

$$D_0^* = \frac{8\Gamma}{\beta S} \tag{19b}$$

and the maximum $\nabla\sigma_0$ to be

$$\nabla\sigma_0^* = \frac{(\beta S)^2}{8\Gamma} \tag{20}$$

The maximized craze interface velocity from Eq. 15 is

$$v = \frac{\sqrt{3}\dot\varepsilon_F h}{2(n+2)}\left(\frac{\sqrt{3}h(\beta S)^2}{16\sigma_F\Gamma}\right)^n \tag{21}$$

Since the average fibril diameter

$$\bar{D} = D_0^*\lambda^{-1/2} \tag{22}$$

Eq. 19 implies that

$$S\bar{D} = 8\Gamma/(\beta\lambda^{1/2}) \tag{23}$$

With $\beta \leqq 1$ this relation predicts values for the $S\bar{D}$ product that are closer to experiment than Eq. 14 and avoids the unphysical assumption that all deformation energy is converted into craze surface energy.

As implied by the discussion above craze fibril extension ratio or its inverse the fibril volume fraction of the craze is an important parameter of the microstructure. Fibril volume fractions can be measured by several different methods. The refractive index n_c of the craze can be measured by measuring the critical angle for total reflection of light by the craze surface. Using the Lorentz-Lorenz equation v_f then can be computed from n_c [65-67]. The method is difficult because small variations in the plane of the craze produce uncertainly in the angle; in addition it cannot be used to measure v_f for very thin crazes or in very small regions along a craze. An alternative method is to determine v_f from the mass thickness contrast of TEM images of crazes in polymer thin films [25,68,69]. Microdensitometry of the craze image yields values of the optical density of the craze (Φ_{craze}), the film (Φ_{film}) and a hole through the film (Φ_{hole}). The value of v_f is then given by

$$v_f = 1 - \frac{\ln(\Phi_{craze}/\Phi_{hole})}{\ln(\Phi_{film}/\Phi_{hole})}. \tag{24}$$

It is important that either optical or TEM measurements of v_f be made while the crazes are under load. Unloading the craze specimen not only allows retraction of

Fig. 11. TEM image of a craze grown from a crack tip in PS, overlayered by the STEM signal obtained by scanning over the craze on the line indicated by the arrow. The signal which increases to the right corresponds to the electrons that are not scattered out of the detector aperture

the highly drawn craze fibrils but in the case of macroscopic specimens actually puts the craze into compression, producing a buckling of the fibrils that can be detected with SAXS [32].

New scanning transmission electron microscopes (STEM) promise to greatly simplify the measurements on crazes in thin films. In the scanning transmission mode a spot is scanned over the specimen and a signal proportional to the electrons scattered within a certain angular aperture is displayed on a cathode ray tube. This signal is proportional to the optical density on the conventional TEM image. Figure 11 shows such a conventional image of a PS air craze at a crack tip overlayed by a plot of the STEM signal obtained by scanning across the craze at the position indicated. By measuring the height of this signal in the crack, in the film and in the craze v_f may be determined from Eq. (24). This procedure will eliminate the necessity for tedious microdensitometry and also should permit examination of crazes in radiation sensitive polymers since only the line scanned by the electron beam is damaged.

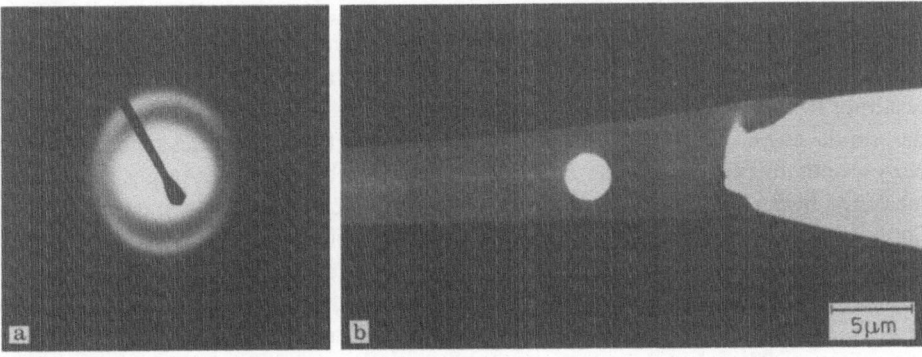

Fig. 12. a Wide angle selected area electron diffraction pattern from the white region of the air craze in PS shown in **b**. From Ref. [52], courtesy of Dr. B. D. Lauterwasser

Selected area electron diffraction (SAED) techniques can also be used to determine other aspects of craze microstructure. The molecular orientation within the craze fibrils in a PS air craze is clearly indicated by the arced SAED pattern shown in Fig. 12 [70,52]. Comparison of similar patterns with those from uniformly drawn films (drawn at 90 °C to a draw ratio of 6) showed that the orientation in the craze fibrils was higher than that in the uniformly drawn film [70]. The fibrils in the atactic PS are still amorphous but very highly oriented. Brown also extracted interference functions from his data which in principle permits comparison with interference functions computed from various structural models of the oriented fibrils. The extraction presents a number of problems [52] among them, correcting for a possibly non-uniform background and for the presence of inelastic scattering which becomes very significant at low angles. Given the fact that at present there is no well established theory that permits one to connect the anisotropic interference function for an amorphous polymer to conventional measures of its orientation[1], it seems more worthwhile to rely on the qualitative comparison discussed above.

A standard way to measure molecular orientation is to measure the optical birefringence. Thick enough crazes may be grown to permit such measurements with the aid of an optical microscope. Under these circumstances one finds that the birefringence is dominated by the form birefringence of the cylindrical fibrils. For example the birefringence of air crazes in PS is large and positive although it is known that the birefringence of highly oriented PS is large and negative [72]. Brown sought to overcome this problem by allowing the craze to imbibe an index of refraction matching liquid and was able to show that under these conditions the birefringence was negative as expected. Uncertainty is introduced however since any liquid which matches the index of refraction is likely to be a crazing agent and plasticizer for the polymer. Thus there is at the moment no truly satisfactory method of quantitatively measuring the molecular orientation in craze fibrils. Given that measurements of this orientation might offer important clues as to the nature of fibril failure process, new methods which can circumvent the difficulties outlined above are needed.

Finally selected area low angle electron diffraction may be used to investigate the microstructure of regions of single crazes in thin films using much the same analysis as that for SAXS of crazes in bulk specimens. Brown [73,74] has also pioneered the use of this technique. He finds low angle electron scattering curves that are qualitatively similar to the SAXS curves with a disk in reciprocal space corresponding to diffraction from the fibrils and a spike normal to the craze resulting from refraction of electrons from the craze surfaces. However the scale of the small angle electron fibril scattering pattern from air crazes in PS films corresponds to fibril diameters and fibrils spacings about 3 times larger than those measured by SAXS in bulk PS [32] or those determined by TEM on somewhat thinner films [30]. Brown has tentatively suggested the possibility of trapped solvent in the film decreasing the craze surface stress S necessary to thicken the craze. From the results of Paredes and Fischer a decrease in S will cause a corresponding increase in $\bar{D}$ and D_0. However to increase $\bar{D}$ by a factor of 3 requires a decrease in S by the same factor. That large a decrease in S is not compatible with craze micro-

---

[1] Recent progress along these lines has been made however [71]

mechanical results on similar PS thin films which yield S's for craze growth close to those observed for crazes in bulk specimens. Another possibility is that the thinness of the film relaxes the plastic constraint in the thickness direction at the craze surfaces so that the fibril drawing conditions are altered. That would not explain the discrepancy between the low angle SAED results and the TEM results on thinner films; moreover the results on film thickness effects reported in the next section show that such a thickness transition does not occur until the PS films are much thinner ($\sim$ 100 nm). The reason for the discrepancy thus remains uncertain.

5.2 Stress Effects on Craze Microstructure

To this point the craze fibril volume fraction v_f and fibril extension ratio λ have discussed as if they were true constants of the craze. While this view is approximately correct, one would expect the draw ratio of the polymer fibrils to depend somewhat on the stress at which they are drawn, since the polymer in the fibrils should have a finite strain hardening rate. Experimental evidence for just such stress effects on λ is discussed below.

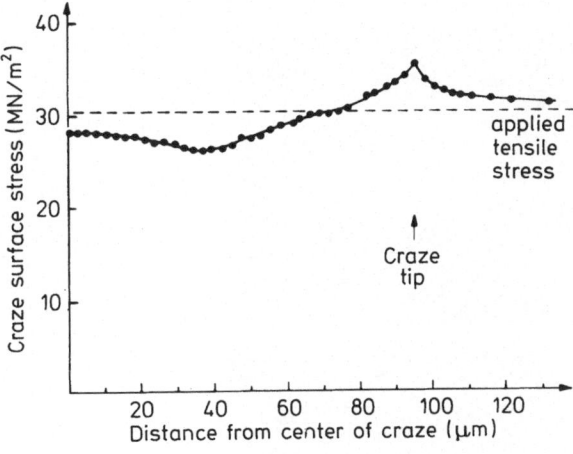

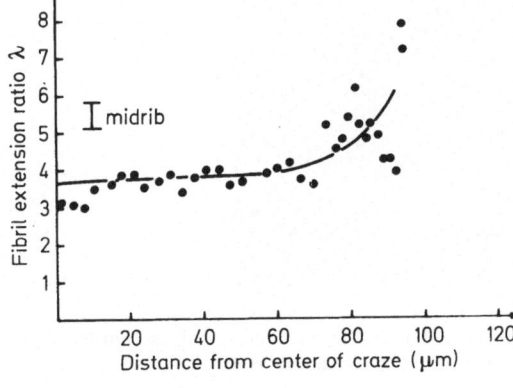

Fig. 13. Fibril extension ratio profile $\lambda(x)$ and surface stress profile S(x) along an air craze in PS. From Ref. [25], courtesy Taylor and Francis, Ltd

Figure 13 shows the craze fibril extension ratio profile $\lambda(x)$ and the surface stress profile $S(x)$ along a craze in PS [25]. It may be seen that the region of high stress just behind the craze tip corresponds to a region where the craze fibrils drawn under these conditions have higher extension ratios. This idea also accounts for the craze "midrib", a layer of higher λ fibrils along the center plane of the craze which may be seen in Fig. 8a. Since the center region of fibrils was drawn just behind the tip of the growing craze (which might be expected to have an even higher stress at its tip than the static craze of Fig. 13) their higher λ seems a natural

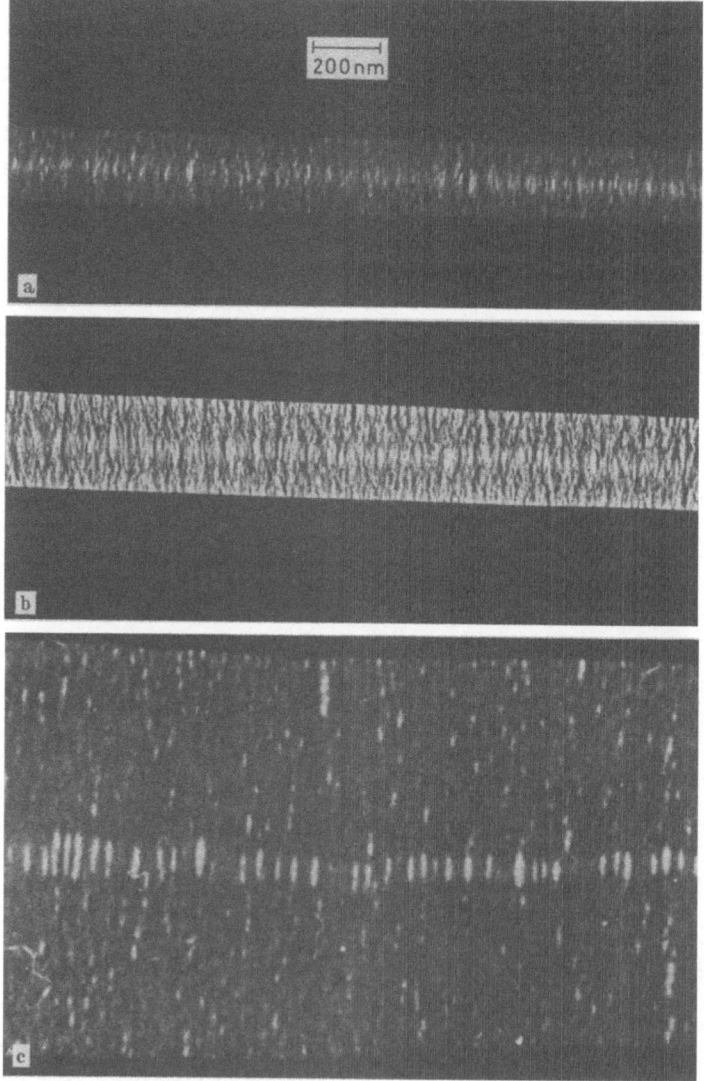

Fig. 14a—c. Craze microstructure in **a** a thick film (1.2 µm), **b** a moderately thick film (0.45 µm) and **c** a very thin film (0.1 µm). From Ref. [76], courtesy Chapman and Hall

result. It is also observed that crazes grown from crack tips have higher average values of λ than isolated crazes in the same polymer [40] and that the midrib expands in thickness as the crack tip is approached [75]. Both results may be qualitatively rationalized if the effects of drawing stress on fibril draw ratio are taken into account.

A quite different kind of stress effect on craze microstructure arises from the presence or absence of lateral stresses parallel to the craze front or the craze-bulk polymer interfaces due to plastic constraint of the deforming polymer there. In the meniscus instability theory of craze tip advance or the model of craze surface drawing outlined above the fibrillation is driven by a hydrostatic tension σ_0 ahead of the craze tip or interface. In a bulk sample or thick film this σ_0 will arise naturally due to the lateral constraint placed on the thin plastically deforming region (wedge in the case of the craze tip, layer in the case of the craze interface) by the adjacent undeformed polymer. If the film becomes thin enough however (so that its thickness approaches the fibril spacing of a craze in a bulk polymer) the plastic constraint normal to the thickness of the film becomes relaxed. Such a thickness transition has been observed in PS to occur at a film thickness of ca. 150 nm [76]. Crazes in films 100 nm or thinner have a very different microstructure from those in thicker films as shown in Fig. 14. While craze images in the thick films show a typical fibrillar microstructure, those in the very thin film show a microstructure more like a perforated sheet. The surface stresses necessary for "craze" thickening in the thinner film are also consistently lower than those necessary for craze thickening in the thicker films. Such decreases in S are expected both from the absence of lateral constraint and from the fact that fibrillation at the interface is not necessary in the thinner films (the voids that do arise appear to be born in the drawn material behind the interface). Unfortunately much of the early TEM investigations [58-60] of crazes in thin PS films used films of 100 nm thickness or less. It also seems possible that early observations of a dramatic change in fibril microstructure along crazes in PS are due to the fact that these crazes were grown in films of non-uniform thickness, with the crazes nucleating in the thinnest regions and growing until their tips stopped in thicker regions [60]. Such structures have never been observed by us in uniformly thick films.

6 Influence of Molecular Entanglements in Crazing

The effects of molecular entanglements on the rheology of polymers above the glass transition temperature are large and have been recognized for a long time [77,78]. One of the most prominent effects is the appearance of a rubbery plateau in the shear modulus G above T_g. For large enough molecular weights the short term shear modulus does not fall to zero at T_g but rather at some higher temperature which increases with molecular weight. Using the simple theory of rubber elasticity one can compute a molecular weight between "entanglement crosslinks" M_e from the level of G_N^o at the plateau, i.e.,

$$M_e = \varrho RT_A/G_N^o \qquad (25)$$

where ϱ is the polymer density, R is the universal gas constant and T_A is the absolute temperature. Even though the modern theories [79-81] of entangled melts show that it is more realistic to view the entanglements as less localized constraints (i.e., slip rings or tubes) than junction points in a truly crosslinked network, it is nevertheless useful for a first approximation to large strain behavior to model the polymer as a network. It is also useful to define several parameters of such an entanglement network. The density v_E of chains between entanglement junction points in the network is

$$v_E = \varrho N_A / M_e . \tag{26}$$

The chain contour length l_e between entanglements is given by

$$l_e = l_0 M_e / M_0 \tag{27}$$

where l_0 is the average projected length of stiff units along the chain and M_0 is the average molecular weight of these units. The entanglement mesh size d, the straight line distance between entanglements is given by the root-mean-square end-to-end distance of a chain of molecular weight M_e, i.e.,

$$d = k(M_e)^{1/2} \tag{28}$$

where k may be determined directly from neutron scattering measurements [82,83] of the radius of gyration of molecular coils in the glass or from light scattering measurements of molecular coil size in dilute solution in a Θ-solvent [84]. For most polymers k lies between 0.05 and 0.09 nm/(mol wt)$^{1/2}$ [84].

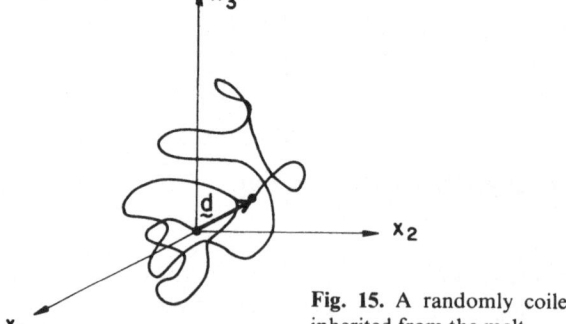

Fig. 15. A randomly coiled chain in the entanglement network inherited from the melt

A single chain in the entanglement network is shown schematically in Fig. 15. If the chain is extended along its end-to-end vector d, it cannot be extended more than a maximum extension ratio λ_{max}, i.e.,

$$\lambda_{max} = l_e / d . \tag{29}$$

For a network however, the average chain will have equal projections on each of three perpendicular axes. Imagine stretching this network along an arbitrary axis, say x_3, by an extension ratio λ_{net} and allowing the network to contract by $\lambda_{net}^{-1/2}$ along both the other axes to conserve volume. If the "average" chain between junction points deforms affinely it will be fully stretched only when λ_{net} satisfies the following equation,

$$\lambda_{net}^2 + \frac{2}{\lambda_{net}} = 3\,\lambda_{max}^2 , \tag{30}$$

where λ_{max} as before is $1/d$. For $\lambda_{max} > 2$ one finds to a very good approximation

$$\lambda_{net} \simeq \sqrt{3}\,\lambda_{max} . \tag{31}$$

The potential importance of molecular entanglements in the uniform deformation of glassy polymers (e.g., in the development of molecular orientation, for instance) was recognized early [85,86] and has been a continuing research theme [87-91] (see also Chapter 2). Since many polymer glasses craze rather than deform uniformly by shear it has been difficult to generalize these results, particularly to see if the entanglement network inferred from orientation studies below T_g has any correspondence to the entanglement network inferred from measurements of melt rheology.

Much less attention has been given to the effects of entanglements on the mechanical properties of crazes and on the competition between crazing and other,

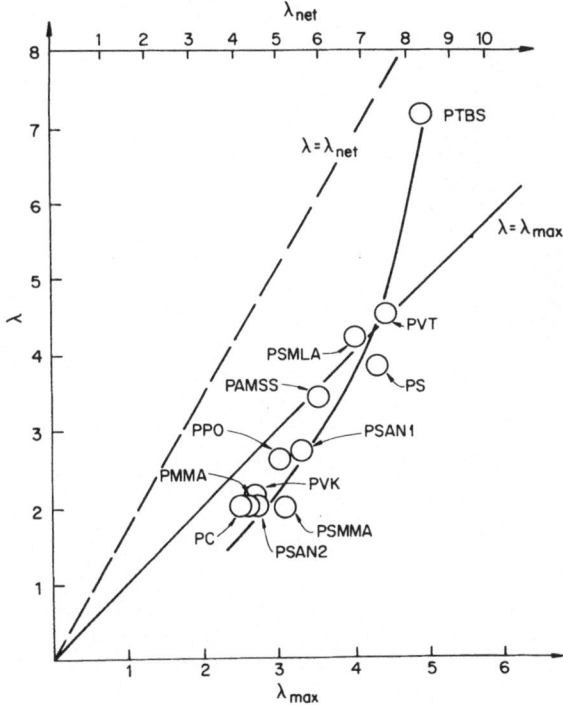

Fig. 16. Experimental extension ratio of crazes in various homopolymers and copolymers plotted against the theoretical maximum extension ratios of a single entangled chain and the entanglement network, λ_{max} and λ_{net}, respectively

more ductile modes of deformation such as shear. An important early piece of evidence was that a critical molecular weight M_c exists below which stable crazes [92-95] do not form. Typically $M_c \simeq 2M_e$; below M_c polymer glasses become very brittle [92,96,97]. In this case the full contribution to the fracture toughness G_{I_c} due to craze deformation at the crack tip is not available [98,99]. These results indicate that entanglements are important in stabilizing craze fibrils. The same conclusions have been derived by Kausch, Dettenmaier and Jud from their studies of crack healing and intrinsic crazing (see Chapter 2). More evidence comes from recent TEM measurements of craze fibril extension ratios λ. If entanglements stabilize craze fibril deformation one would expect λ to be correlated with λ_{max} and λ_{net}; λ_{net} should be an upper limit for λ. Figure 16 shows precisely the expected correlation for a series of homopolymers and copolymers [100] for which λ_{max} could be determined. For the low λ_{max} (high entanglement density) polymers $\lambda \simeq 0.8\lambda_{max}$ but it rises to and above λ_{max} for the higher λ_{max} polymers. The increase in λ above this level at high λ_{max} may be due to increased chain scission caused by the higher fibril true stresses for these polymers. As the entanglement density increases in this series of polymers (decreasing λ_{max}) there is increasing competition between shear deformation in the form of plane stress deformation zones (DZ's) and crazing [101-103]. Figure 17 shows a TEM micrograph of a DZ in poly-(1,4-dimethyl-2,6 phenylene oxide) [PPO] [152]. The zone is not fibrillated as in a craze but consists of polymer drawn to a uniform extension ratio λ_{DZ}. It thickens by drawing more polymer in from the edge of the zone. In this respect the DZ is very similar to a craze [104,105]. Also similar is the fact that λ_{DZ} correlates strongly with λ_{max} (and λ_{net}) of the entanglement network as Fig. 18 shows [104]. The λ_{DZ} values increase with

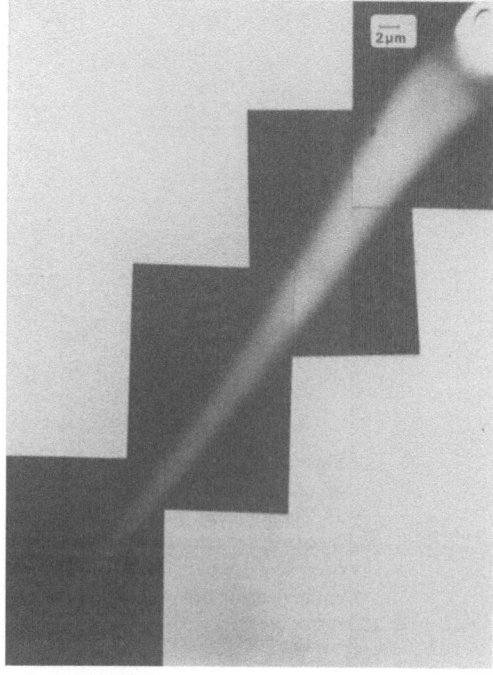

Fig. 17. TEM image of a plane stress shear deformation zone in a thin film of PPO. From Ref. [102], courtesy Butterworth and Co., Ltd.

λ_{max}, clustering around the line $\lambda_{DZ} = 0.6\,\lambda_{max}$. In all cases however λ_{DZ} is less than λ of crazes in the same polymer. This observation indicates that the chain scission/chain slippage necessary for fibrillation in a craze modifies the entanglement network (increases the effective λ_{max} of the polymer fibrils) and so increases λ above λ_{DZ}.

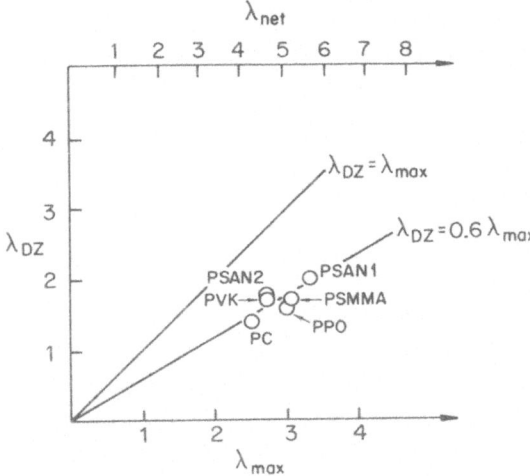

Fig. 18. Experimental extension ratios of deformation zones in various homopolymers and copolymers plotted against the theoretical maximum extension ratios of a single entangled chain and the entanglement network, λ_{max} and λ_{net}, respectively

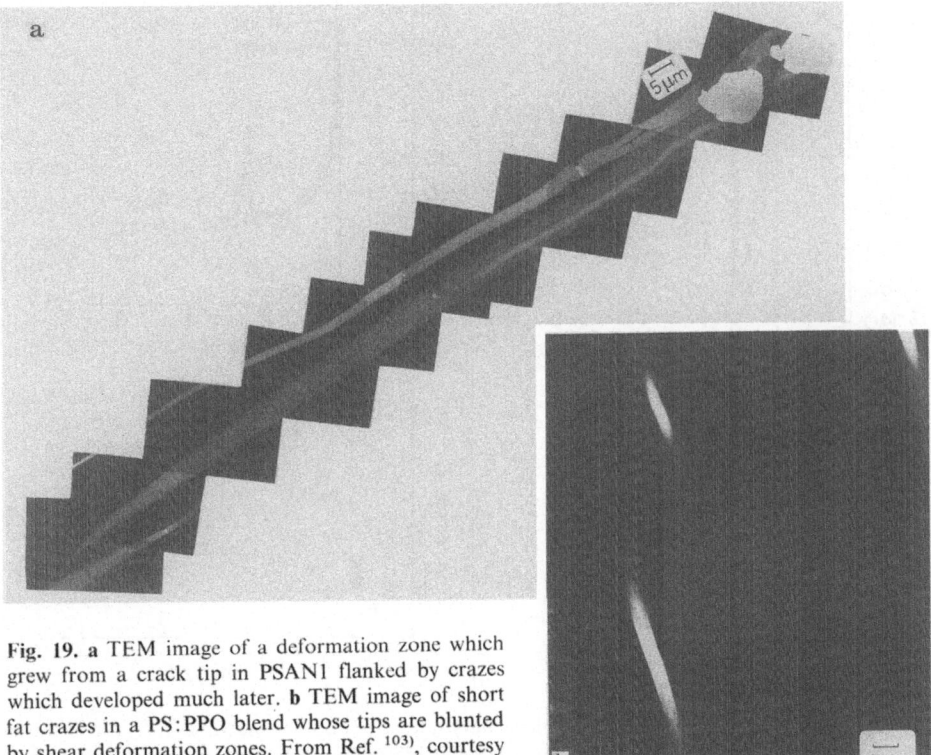

Fig. 19. a TEM image of a deformation zone which grew from a crack tip in PSAN1 flanked by crazes which developed much later. b TEM image of short fat crazes in a PS:PPO blend whose tips are blunted by shear deformation zones. From Ref. [103], courtesy of Chapman and Hall

The competition between shear DZ formation and crazing is strongly dependent on the entanglement density. The highest entanglement density polymer, polycarbonate [PC] forms shear DZ's readily and rarely crazes whereas the intermediate entanglement density polymers such as PPO and poly(styrene acrylonitrile) PSAN copolymers may exhibit both crazing and DZ formation in the same film. It is common to see DZ's and crazes side-by-side as in Fig. 19a (the DZ formed first) as to see short fat crazes whose tips are blunted by shear bands or zones (Fig. 19b) [103]. The lower entanglement density ($\lambda_{max} > 3.5$) polymers never form deformation zones but only

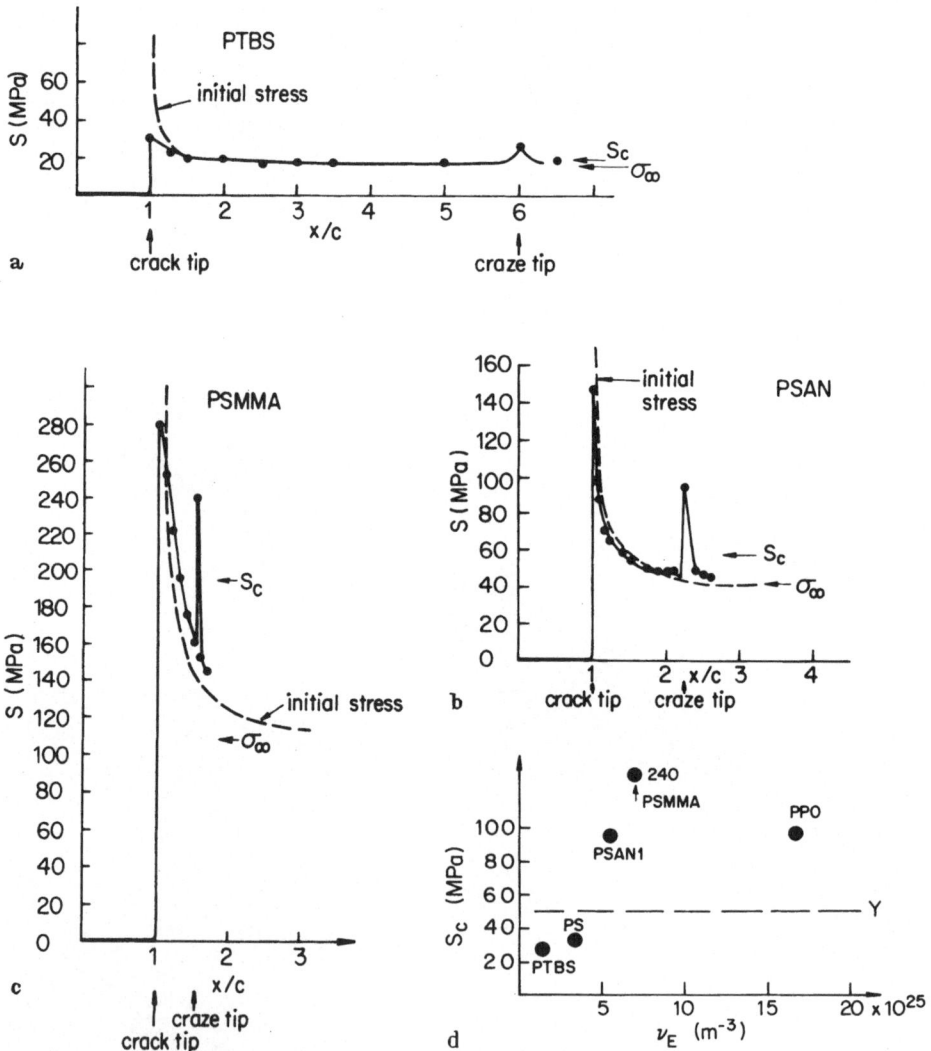

Fig. 20a—d. Surface stress profiles for polytertbutylstyrene [PTBS], poly(styrene-26% acrylonitrile) [PSAN1] and poly(styrene-65% methylmethacrylate) [PSMMA]. The stress at the craze tip S_t is plotted vs. ν_E in **d**. The value of the shear yield stress Y of polycarbonate is indicated

craze. (Polymethylmethacrylate [PMMA] is an exception to these rules in that it normally only crazes although is has a low λ_{max}.)

Part of the reason for this increased tendency toward shear deformation at high entanglement density can be found in the strong increase in the crazing stress with entanglement density. Figure 20 shows the surface stress profiles S(x) for crazes grown from crack tips in films of polytertbutylstyrene [PTBS], poly(styrene 26% acrylonitrile) [PSAN1] and poly(styrene 65% methylmethacrylate) copolymer [PSMMA] [40]. Both the stress at the craze tip S_t and the average level of craze surface stress S_c increases with increasing entanglement density (decreasing λ_{max}); a plot of S_t vs v_E is also shown in Fig. 20d. The shear deformation stress (the yield stress Y) is not expected to depend on v_E since the motion of molecular segments much smaller than l_e are thought to be responsible [106–109] for yield. Although Y will increase with $(T_g—T_a)$ and change with the size of the molecular segment involved in yield [107,108], a single value corresponding to the Y of polycarbonate at 300 K is indicated on Fig. 20. For polymers with v_E above about 5×10^{25} m^{-3} shear yield should occur in preference to crazing. Because of individual variations in Y from polymer to polymer, in isolated instances crazing may be favored over shear even above this value; this may be the reason for the anomalous behavior of PMMA.

Physical aging (annealing the glass at temperatures close to, but below, the glass transition temperature) is known to raise the shear yield stress [110–112]. Its effects on the crazing stress are less sure but what data exist [113] suggest little or no effect. Thus physically aging a polymer, which in the unaged condition would deform mostly by shear deformation, may cause it to deform mainly by crazing. Precisely such transitions are observed in films of PPO, PSAN, PSMMA and PPO-PS blends [103]. On the other hand even prolonged aging of PC films, the highest entanglement density polymer in this series, did not result in crazing competing consistently with shear deformation.

The thin film nature of these experiments should be emphasized. Little, if any, lateral tensile stress develops in the thickness direction of the DZ. In thick sections, particularly beneath notches, one must expect such lateral stresses to develop thereby decreasing s_0 and causing shear deformation to stop. While the decrease in s_0 also decreases the craze nucleation rate this is offset by the increase in σ_0 caused by the lateral constraint stress. Thus as straining is continued, crazing becomes favored over shear in the center of thick specimens. One can approximate the effect of the lateral constraint stress in uniaxial tension by simply adding it to the yield stress in Fig. 20, at the same time maintaining the crazing stress unchanged. Under these conditions even polymers like PC which will not craze readily at room temperature in thin films, will craze. In fact the formation and breakdown of crazes in the center of notched bar impact specimens of PC has an important effect on the measured impact toughness and the so-called ductile-to-brittle transition in this test [112,114].

Compatible polymer blends are interesting systems to test these ideas on entanglement effects. The fact that properties can be changed continuously by alloying has generated considerable practical interest in these systems. Melt rheological properties are no exception; M_e may be varied continuously from its values for either of the two components. A system that is particularly interesting is PS-PPO. These polymers are compatible over the entire composition range. Pure PS has a low entanglement density and normally crazes whereas pure PPO normally forms shear DZ's if not physically aged. The entanglement molecular weight of the blends has been measured as a function of the volume fraction χ of PPO in the blend up to $\chi = 0.5$ [115]. The data follow the relationship

$$M_e(\chi) = \frac{M_e(PS)}{1 + 3.2\chi} \tag{32}$$

which was used to extrapolate the data to the regime between $\chi = 0.5$ and $\chi = 1$. The average chain contour length between entanglement l_e is computed as

$$l_e = \frac{l_0}{M_0}(\chi) \, M_e(\chi) \tag{33}.$$

where $\dfrac{l_0}{M_0}(\chi)$ values are found by linearly interpolating between those of PS and PPO. The mesh size d may be computed as

$$d = k(\chi) \, [M_e(\chi)]^{1/2} \tag{34}$$

where $k(\chi)$ is linearly interpolated between the values 0.70 and 0.84 for pure PS and pure PPO respectively[1]. The single chain extension ratio λ_{max} is taken as l_e/d as before. Figure 21 shows the theoretical curve (solid line) of λ_{max} versus χ. Also given in the figure are experimental values of λ in crazes in selected blends. Note that both λ and λ_{max} fall rapidly between $\chi = 0$ and $\chi = 0.25$; beyond $\chi = 0.25$ there is little change in either quantity. Just as interesting is the observation that a few deformation zones as

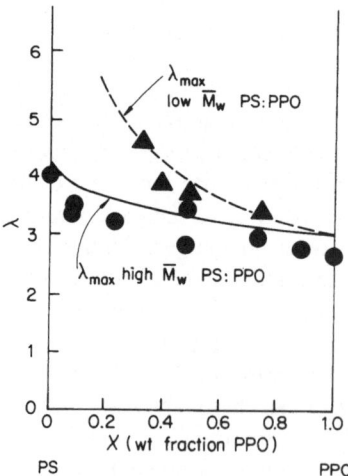

Fig. 21. Craze fibril extension ratio plotted against weight fraction PPO for two types of PS/PPO blends. 0, high M_w PS blend, Δ, low M_w PS blend. The theoretical curves for λ_{max} are also shown as solid lines

---

[1] Originally d was computed [102] by linear interpolation between the d values for PS and PPO. The present method seems more consistent with the picture of an "average" chain between entanglements

well as crazes are observed in the as prepared 10% PPO:PS blend. As χ is increased further DZ's compete increasingly with crazes. Above $\chi = 0.5$ crazes are never observed in films which have not been physically aged to suppress shear deformation. The data for $\chi > 0.5$ are obtained from films that were physically aged. The behavior of the blends is entirely consistent with the behavior of homo- or co-polymers with the same entanglement density.

Perhaps the most convincing test for entanglement effects is to dilute the high molecular network with small molecules incapable of entangling with the chains. Under these conditions M_e should go up as

$$M_e(\chi) = \frac{M_e(\chi = 1)}{\chi} \tag{35}$$

where χ is the volume fraction of high molecular weight polymer in the blend. Such experiments have been carried out with high molecular weight PPO and low M_w (M = 4000) PS acting as a diluent [102]. This PS is well below its entanglement molecular weight ($M_e = 19,000$) and so is not expected to participate in the entanglement network. The chain contour length l_e in these blends is just that of PPO with a

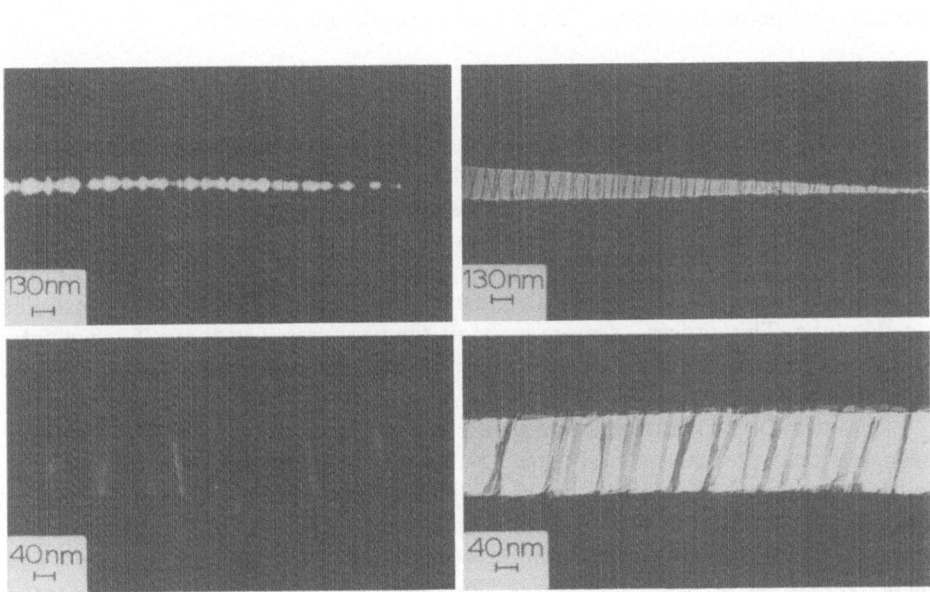

Fig. 22. TEM images of crazes in a uniaxially oriented PS film produced by straining parallel to, and perpendicular to, the orientation direction

                                irradiated   unirradiated

Comparison of irradiated and unirradiated specimens strained to 5.0%

$M_e = M_e(\chi)$. It is expected that the short PS chains will act as a good solvent for PPO thus expanding the PPO coils from their usual θ solvent configurations. A reasonable approximation for d is [102, 116)]

$$d(\chi) = d(PPO)/\chi^{5/8} \tag{36}$$

where d(PPO) is the mesh size in pure PPO. Computing λ_{max} as before results in the dashed line on Fig. 22. The triangles represent experimental points of λ determined on crazes in these blends by the TEM mass thickness contrast method. Again the agreement between the predicted variation of λ and that observed is very good.

Work underway on crazes in blends of high and low M_w PS shows similar trends [117)]. In addition if the high molecular weight chains are long enough the configuration of the coils in the blends will be either those typical of those of the polymer in a good solvent or those typical of the polymer in a θ-solvent, depending on the molecular weight of the low M_w PS [116)]; λ_{max} is predicted to be decreased in a good solvent vs. its value in a θ-solvent.

Prior molecular orientation (either above or just below T_g) should change the starting extension ratios of the entanglement network and therefore λ of any crazes produced. Consider a polymer melt extended in uniaxial extension to $\lambda_3 = \lambda_p$, $\lambda_2 = \lambda_1 = \lambda_p^{-1/2}$. The new λ_{max} in a direction parallel to the molecular orientation should be

$$(\lambda_{max})_{||} = \lambda_{max}/\lambda_p \tag{37}$$

whereas that perpendicular to the molecular orientation should be

$$(\lambda_{max})_\perp = \lambda_{max}(\lambda_p)^{1/2} . \tag{38}$$

Here λ_{max} is the single chain limiting extension ratio in the isotropic, unoriented polymer with the same entanglement weight. A complication is that the process of orientation above T_g may result in a loss of entanglement constraint, effectively increasing M_e and λ_{max}. One of the ways this loss can take place in the current versions of the tube model [79, 80)] is by tube relaxation [118)] whereby the process of tube retraction of molecules neighboring the tube in question can lead to a loss of constraint on this tube. This process is more rapid at temperatures just above T_g than tube reptation [79)] but is still quite dependent on molecular weight and molecular weight distribution.

Experiments in which crazes were grown in a highly oriented PS film qualitatively bear out these predictions [48)]. Figure 22 shows micrographs of crazes in a highly uniaxially oriented PS thin film grown with the applied tensile stress parallel to and perpendicular to the axis of molecular orientation M. It is obvious that the parallel crazes in which the fibril orientation direction is parallel to M have a lower λ and higher v_f than the perpendicular crazes. The parallel crazes have $\lambda = 2.2$ whereas the perpendicular crazes have $\lambda = 20$. For comparison crazes in unoriented PS have $\lambda = 4$. From the equations above assuming $\lambda_{||} \simeq (\lambda_{max})_{||}$ and $\lambda_\perp \simeq (\lambda_{max})_\perp$, one can compute λ_{max} corresponding to the M_e in the oriented PS. This computation yields $\lambda_{max} = 9.6$ implying considerable entanglement loss in the orientation process. An alternative explanation of the high λ in the perpendicular crazes is that there is substantial chain scission in these fibrils as a result of the higher true stresses there. At this time, which, if either, of these explanations is correct is uncertain.

Finally it seems obvious that if increasing the entanglement density has such strong effects, chemically crosslinking the polymer chains together should be equally effective based on the changes this should produce in chain density and λ_{max}. Chemical crosslinking should have little effect at first (when the molecular weight between crosslinks $\gg M_e$) but should decrease λ_{max} and thus λ of the crazes when the crosslink density reaches high enough levels. It should also begin to suppress crazing relative to shear deformation. Preliminary experiments indicate that both these predictions are qualitatively correct. It is possible to crosslink PS thin films by either γ-irradiation or electron irradiation. As crosslinking proceeds there is at first little effect on craze microstructure and the uniaxial strain ε_c to cause crazing [119] ($\varepsilon_c \simeq 0.5\%$), then a marked increase in ε_c (to 3%) where unfibrillated DZ's begin to compete with crazing [120]. At still higher crosslink densities no crazing or DZ formation is observed to strains exceeding 12%. It appears that increasing crosslink density, unlike the entanglement density, increases both the shear yield stress and the crazing stress. The crazing stress however increases much more rapidly. At the highest crosslink densities the films fracture in a brittle manner at $\sim 2\%$ strain with no evidence of crazing or other plastic deformation. A dramatic illustration of the effect of crosslinking is shown in Fig. 23, which shows irradiation crosslinked and uncrosslinked PS films strained to 5%. The uncrosslinked film has crazed and some of the crazes have broken down to form cracks. The crosslinked film has neither crazed nor cracked at this strain.

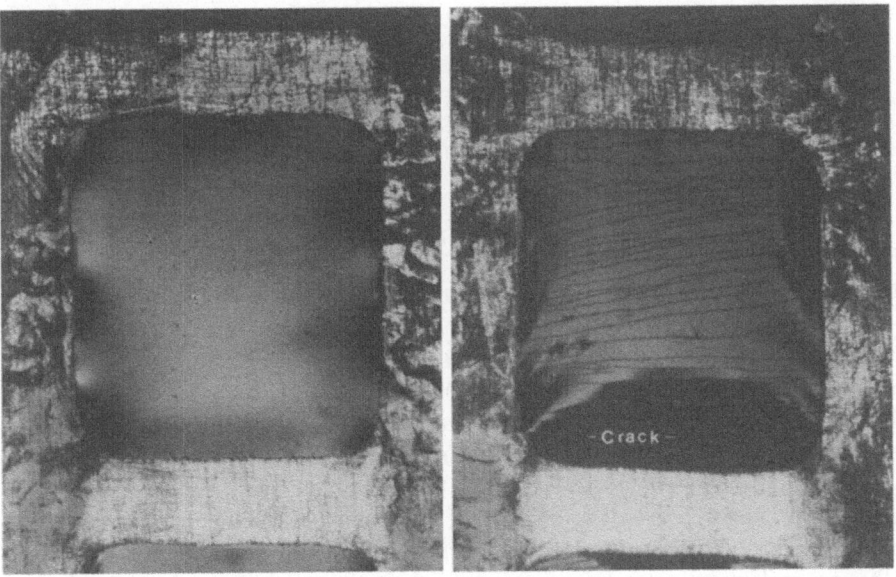

Fig. 23. Optical micrographs of grid squares of PS films strained to 5%. The film on the left has been crosslinked by electron irradiation; the one on the right is uncrosslinked

The results reviewed above clearly demonstrate the importance of entanglement density for crazing. This section will attempt to develop a simple theoretical model to rationalize these effects. In doing so it must be realized that the earlier approximation that the entanglement network is unmodified by crazing cannot be strictly true. Figure 24a shows a schematic view of the finger-like craze tip and one of the long chain molecules of the entanglement network. One end of the molecule is in a web on one side of a void finger whereas the other end is in the web on the other side. Clearly if the craze tip is to advance one of two events must occur. Either one end of the molecule must disentangle itself from the surrounding molecules in its web or the molecule must break somewhere near its center. This breakage or disentanglement of molecules will effectively reduce the number of entanglements in the network that is incorporated into the fibrils, raising λ_{max} and λ_{net}. Very similar events must take place at the craze-bulk polymer interface where polymer is drawn into the fibrils.

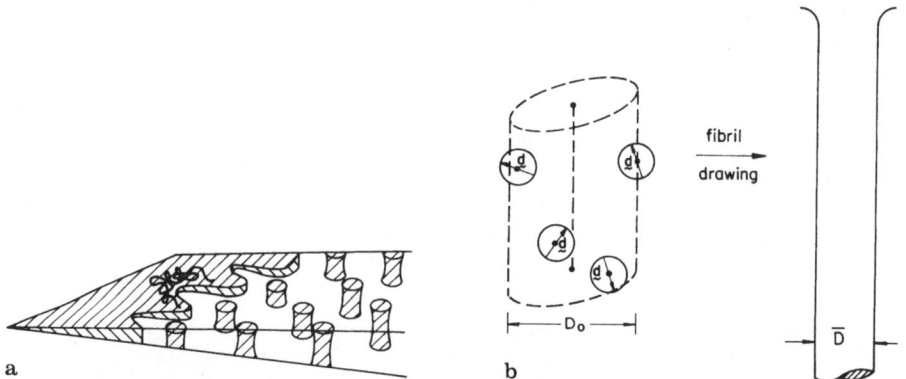

Fig. 24. a Schematic view of the craze tip showing a molecule about to be drawn into two different fibrils. **b** "Phantom" fibril of undeformed polymers, diameter D_0, is ultimately drawn into a fibril of diameter $D = D_0\lambda^{-1/2}$. The entangled chains in the isotropic network, represented by their end-to-end vectors d, are also shown

One can estimate the fraction of the original network chains that remain unbroken after fibrillation. Consider, for example, the cylinder of undeformed polymer of diameter D_0 shown in Fig. 24b. After fibril formation this "phantom" fibril cylinder is drawn into a fibril of diameter $\bar{D} = D_0\lambda^{-1/2}$. The entangled chains in the isotropic polymer before deformation may be represented by their end-to-end vectors d. The center points of these vectors may be placed randomly with a density v_E within the cylinder and the vectors are oriented randomly. If a given vector lies wholly within the cylinder, this chain will contribute to the entanglement network of the drawn fibrils. If the vector breaks the surface of the cylinder this chain must be either broken or disentangled to create the fibril surfaces and so does not contribute to the entanglement network of the drawn fibril which has an entanglement density v_E'. The ratio of v_E'/v_E simply represents the probability that a random vector of length d will lie wholly within the cylinder of diameter $D_0 = \bar{D}\lambda^{1/2}$. The difference $v_E - v_E'$ represents the geometrically necessary loss of entanglement density. The problem of computing v_E'/v_E is difficult to solve exactly in closed form but an approximation (actually an upper bound) for the case when $\frac{d}{D_0} < 1$ is

$$\frac{v_E'}{v_E} = 1 - \frac{d}{D_0} + \frac{1}{3}\left(\frac{d}{D_0}\right)^2. \tag{39}$$

For air crazes in PS where $\bar{D} = 6$ nm [32)], d = 9.6 nm [100)], $\lambda = 4$, and $D_0 = 12$ nm, v_E'/v_E is predicted to be 0.41. Production of small fibrils in PS must involve substantial chain scission or disentanglement simply as a consequence of geometry. On the other hand as the entanglement density increases, d decreases, and fibrils of the same diameter may be produced without decreasing the v_E'/v_E ratio to the same extent as in PS. For example for fibrils of diameter $\bar{D} = 6$ nm in PC where d = 4.4 nm and $\lambda = 2$, v_E'/v_E is predicted to be 0.57.

To estimate experimentally the actual decrease in v_E' in the crazes over v_E in drawn but unfibrillated polymer, one can use the λ and λ_{DZ} data. In addition the perforated sheet crazes in very thin (100 mm thick) PS films [75)] should involve little entanglement network modification and λ in these crazes can be used to estimate λ_{DZ} for this polymer. Since $\lambda_{DZ} \propto (v_E)^{-1/2}$ and $\lambda \propto (v_E')^{-1/2}$

$$\frac{v_E'}{v_E} = \left(\frac{\lambda_{DZ}}{\lambda}\right)^2. \tag{40}$$

For isolated PS crazes $\lambda = 4$, $\lambda_{DZ} = 2.3{-}2.6$ and thus $v_E'/v_E = 0.33{-}0.42$ which is close to that predicted. For the other polymers v_E'/v_E ranges between 0.81 for PSAN2 (a random copolymer with 34 wt. % styrene and 66 wt. % acrylonitrile) to 0.48 for PPO. Although fibril diameters have not been measured in these polymers, the results are consistent with fibril diameters in the range 5–25 nm. For the high molecular weight PS-PPO blends with wt. fraction χ of PPO equal to 0.1 and 0.25, one might expect that the fibril diameter is unchanged from PS. On this basis one can predict that v_E'/v_E should be 0.43 for the $\chi = 0.1$ blend and 0.46 for the $\chi = 0.25$ blend; the v_E'/v_E ratios inferred for the λ_{DZ}/λ ratios measured experimentally for these blends are 0.45 and 0.43 respectively, in good agreement with those computed.

The arguments above show clearly that a considerable loss of entanglement density occurs due to the formation of the fibril surfaces. There are two possible general mechanisms by which this loss can occur, i.e., chain scission or chain disentanglement. Chain disentanglement rates r_d should be strong functions of molecular weight M. For example, if chain disentanglement occurs by reptation [79, 80)] the rate should decrease as M^{-3}; if it occurs by the tube retraction and relaxation processes discussed above r_d should decrease as M^{-2} [118)]. Chain scission, on the other hand, should be independent of M but should depend strongly on the applied stress level. The rates of both disentanglement and chain scission will increase strongly with temperature.

In polystyrene air crazes the predominant mechanism of entanglement loss seems clear. It is observed that both the uniaxial stress for crazing [97)] and the craze fibril extension ratio [121] is essentially constant for molecular weights above 200,000. Furthermore crazes in lightly crosslinked high molecular weight PS (where the molecular weight between crosslinks is greater than 50,000) have similar λ's and fibril microstructures to those in uncrosslinked PS [119, 120]. These results are opposite to what would be expected if chain disentanglement were the dominant mechanism of entanglement loss, but are reasonable if chain scission is the most important mechanism. One must conclude that considerable chain scission accompanies formation of air crazes in PS. These results on PS at room temperatures however cannot necessarily be generalized to all glassy polymers or even to PS at elevated temperatures. One must expect that in polymers of lower molecular weight, with smaller crossectional areas per chain or at temperatures just below the glass transition temperature where crazing stresses are low, that chain disentanglement may play an increasing role in accommodating the geometrically necessary entanglement loss.

Can this geometrically necessary entanglement loss via chain scission or disentanglement account for the strong dependence of craze growth kinetics/craze growth stresses on entanglement density? The evidence favors such an interpretation. Consider the advance of the craze tip represented schematically in Fig. 24a, and imagine that chain scission is the only mechanism of geometrically necessary entanglement loss. The energy to create new void surface area Γ is not just the van der Walls surface energy of intermolecular separation γ but must also contain the energy to break the primary covalent bonds of the chains that must undergo scission. The resulting surface energy Γ is

$$\Gamma = \gamma + \frac{1}{4} d v_E U \tag{41}$$

where U is the energy needed to break a single primary chain[1].

One may estimate the value of Γ for PTBS and PC which represent the two extremes of entanglement density using $U = 6 \times 10^{-19}$ J and $\gamma = 0.04$ J/m² for both. For PC, $\frac{1}{4} d v_E U = 0.20$ J/m² and $\Gamma = 0.24$ J/m²; for PTBS, $\frac{1}{4} d v_E U = 0.026$ J/m² and $\Gamma = 0.066$ J/m². In turn these values of Γ produce substantial predicted differences in craze growth kinetics. Substituting these values into Eq. (7) the craze tip velocity at constant $S_t = 100$ MPa is predicted to decrease by a factor of 10^{17} from PTBS to PC (values for h of 10 nm and for n, the power law exponent, of 17 are assumed for both) or equivalently the value of S_t to give the same craze tip growth rate increases by a factor of 2.8. Since the measured stress S at the craze tip in PTBS is ~ 27 MPa, the craze tip stress in PC is predicted to be 74 MPa, well above its

---

[1] The question of whether U is the energy to break a single bond U_b or whether it is the energy stored in an entire chain in the entanglement network ($= \bar{n} U_b$ where $\bar{n}$, the number of bonds in the chain, is proportional to λ_{max}^2) is not clear for the case of crazing. Lake and Thomas [122] have argued in favor of $\bar{n} U_b$ for crosslinked elastomers under conditions where the stress transfer along the chain is virtually instantaneous. One can only load such a chain through its crosslinked ends. In crazing of polymer glasses one suspects that during rapid loading considerable load transfer to neighboring chains along a given chain occurs. The situation is intermediate between only loading a single bond to failure and an entire chain but closer to the former

shear yield stress at room temperature. The fact that PC thin films normally deform by shear rather than crazing at room temperature is thus a consequence of the extra energy required to cause the geometrically necessary entanglement loss associated with the production of fibril surfaces in this highly entangled polymer. This argument for the craze tip growth may also be extended to craze thickening by the surface drawing mechanism and to the initial nucleation of voids leading to craze nucleation. In both cases the extra surface energy $\frac{1}{4} dv_E U$ markedly decreases the rate.

This extra surface energy hypothesis also has interesting consequences for the temperature (or strain rate — low strain rates correspond to high temperature) dependence of crazing. As temperature is increased the rates for disentanglement of chains of moderate molecular weight become much larger and entanglement loss by this mechanism, either before or after a chain scission event becomes important. The effective value of Γ decreases since only $v_E^s = v_E - v_E^d$ chains per unit volume must now be broken (v_E^d is the density of entanglements lost through disentanglement). Since Γ decreases with temperature, one expects especially in the highly entangled polymers such as PC, that crazing will become an increasingly competitive mechanism with shear deformation at high temperatures. For polymers with a low entanglement density, e.g., PS, the term $\frac{1}{4} v_E dU$ is not nearly as large a fraction of Γ as for PC and the temperature dependence of the stress for craze tip advance should be less pronounced. The increasing tendency toward crazing with increasing temperature in the highly entangled polymers has been observed [123, 103]. No data to check the second prediction are currently available. It would also be interesting to examine the temperature dependence of shear vs. crazing of highly entangled polymers (e.g., PC) of increasing molecular weight. The decrease in crazing stress (relative to shear yield stress) should be pushed to higher temperatures as molecular weight is increased in much the same way the terminal relaxation region is [124]. At the moment no measurements of this type exist.

7 Craze Fibril Breakdown and Craze Fracture

The starting point in any discussion of craze fibril breakdown should be to compute the true fibril stress, $\sigma_t = S\lambda$. Clearly the larger the fibril extension ratio the higher its true stress will be and the more likely it is to break down for a given surface stress, S. Since decreasing the entanglement density of the polymer increases λ, such decreases should produce decreases in craze fibril stability. The effect of decreasing v_E is larger than just the increase in λ, since decreasing v_E also strongly decreases the number of load bearing chains (entangled chains) that survive fibrillation and are incorporated in the fibril. A particularly illuminating case (which can be studied also experimentally) is that of a high molecular weight polymer (for example PS with $M_w = 3 \times 10^5$) diluted with low molecular weight polymer with molecular weight below the entanglement limit (for example PS with $M_w = 19,000$; this PS is a long enough molecule to act as a "theta" solvent). In the section below we will compute the average number of remaining loadbearing (entangled) chains in a fibril, the force per load bearing chain and the probability that a fibril of a given length has at least one entangled chain in every segment along this length, all as a function of χ, the volume fraction of high molecular weight chains.

It is convenient to consider as a reference state the cylinder of isotropic undeformed polymer (diameter D_0) that will subsequently be drawn into the fibril. The number of entangled chains that cross a given crossectional area of the undeformed cylinder is the density of chains crossing unit area $v_e d/2$ [122]) times the area of the cylinder $\pi D_0^2/4$. Only a fraction $q = v_E'/v_E$ of these chains however survives the geometrically necessary entanglement loss that accompanies fibril surface formation. As before the upper limit of this fraction is approximately given by Eq. (39) for $D_0 > d$ and by

$$q \equiv \frac{v_E'}{v_E} \simeq \frac{1}{3} \left(\frac{D_0}{d} \right)^2 \tag{42}$$

for $D_0 < d$. The total number of effectively entangled chains that cross a given section of the deformed fibril (assuming affine deformation) is then just

$$n_e = \frac{1}{8} v_E d \pi D_0^2 q . \tag{43}$$

The force on a given chain is given by

$$f = \pi D_0^2 S / 4 n_e \tag{44}$$

where S is the craze surface stress.

The entangled chains are stretched to a length which we call the entanglement transfer length $l \simeq d\lambda$ along the fibril axis. The distance l represents a natural scale of length along a fibril. If all molecules across a cross section suffer breaks within one length l along the fibril, then the fibril will have no load bearing capacity. If the same number of breaks is pread over a much larger distance than l, the fibril should survive since most of the entangled chains in any one length are still intact and load transfer is possible. Hence a fibril of total length T has $T/l = Z$ entanglement transfer lengths along it.

We first compute the probability p that at least one entangled chain is left unbroken in a given entanglement transfer length. Since for the assumed random chain scission the probability of a certain number of breaks follows a binomial distribution one finds

$$p = 1 - (l - q)^{n_0} , \tag{45}$$

where $n_0 = n_e/q$, the number of chains before deformation and scission in the cylinder. The probability P that there is at least one entangled chain in each entanglement transfer length is given by

$$P = p^Z . \tag{46}$$

These quantities are computed for blends of χ volume fraction of high molecular weight PS in low molecular weight PS. It is assumed that D_0 stays constant at 12 nm, $S = 30$ MPa, $v_E = [v_E] \chi$ and $d = [d]/\chi^{1/2}$ where [ ] denotes the value for pure high molecular weight PS.

The computed values of f and n_e are shown in Figure 25 along with the estimated theoretical bond strength for PS of 3×10^{-9} N [125]. Note that for $\chi < 0.25$, the average force on any unbroken chains exceeds the theoretical bond strength. Note also that the number of unbroken chains in each fibril is quite small even for pure high molecular weight PS and it decreases rapidly as the high M_w PS is diluted. Figure 26 show the probability that at least one entangled chain remains unbroken

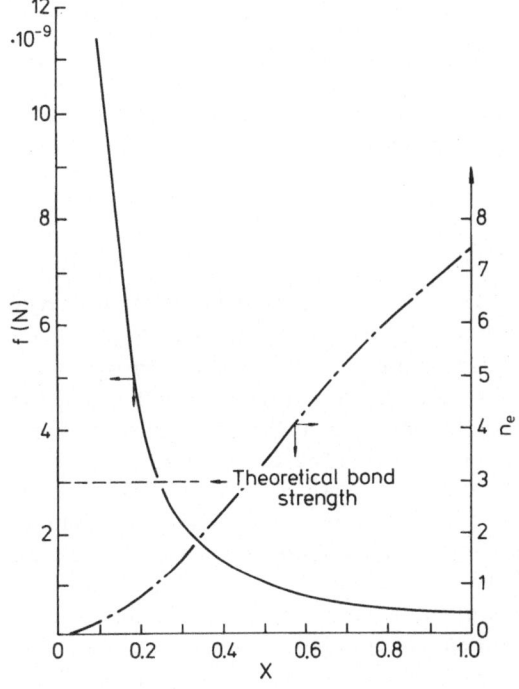

Fig. 25. The computed average number of unbroken entangled chains n_e and the force per chain f in fibrils in crazes in blends of χ wt. fraction high M_w PS with low M_w PS

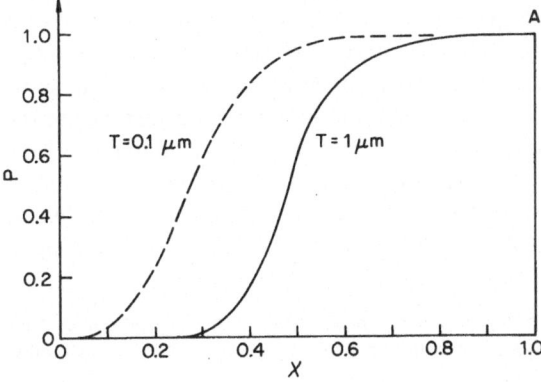

Fig. 26. The calculated probability that at least one entangled chain remains unbroken in each entanglement transfer length along craze fibrils of total length (total craze thickness) T as a function of χ the weight fraction high M_w PS in the blend

in each entanglement transfer length along the fibril for craze thickness (total fibril length) T of 1 μm and 0.1 μm. Normal isolated air crazes in PS usually lie between these two values. For the 1 μm thick craze most fibrils are not load bearing (or at least have weak spots with no entangled chains) below $\chi = 0.5$ whereas the thinner crazes are 50 % load bearing down to $\chi = 0.25$. Experimentally one finds that crazes become very fragile in blends below $\chi \simeq 0.5$ and it is almost impossible to find even a thin intact craze in blends below $\chi = 0.3$ [117]. This behavior is observed even in blends where ultrahigh molecular weight PS (2×10^7) is used as the high molecular weight component.

Although craze fibrils in pure high molecular weight PS are stable in the sense that even very long craze fibrils, e.g., 100 μm, have only a small probability of having less than 1 entangled chain in each entanglement length, not much additional chain scission or chain disentanglement after fibrillation is required to cause these fibrils to be unstable. For example the average force per entangled chain is already $\sim \frac{1}{6}$ the theoretical bond strength at zero K[1]. From reaction rate theory of bond scission one expects that the rate k_b of bond breakage is given by

$$k_b = k_0 \exp \left(- \frac{U^* - l_{act} f}{k_B T_a} \right) \tag{47}$$

where the attempt frequency $k_0 \simeq 10^{12}$ Hz, the bond breaking activation energy $U^* \simeq 6 \times 10^{-19}$ J and the activation distance $l_{act} \simeq 2 \times 10^{-10}$ m. The rate of bond breakage in the average chain ($n_e = 7.4$) is negligibly small at 30 MPa. However the probability that there will be at least one entanglement transfer length along the fibril where there are 2 entangled chains left unbroken is 0.14. The rate of failure of 1 μm long fibrils due to such weak spots is predicted to be 4×10^{-17} sec^{-1} at $S = 30$ MPa and 4×10^{-5} sec^{-1} at a modest craze overload of $S = 40$ MPa. While a more sophisticated calculation of such weak spot effects is necessary for accurate results [126], the present calculation shows that time-dependent chain scission can contribute to initial craze fibril breakdown.

An equally, and perhaps more, important contribution is likely to be due to chain disentanglement by chain reptation or tube relaxation during localized creep of the fibrils. At first sight this statement seems improbable since it was shown above that craze thickening occurs by surface drawing rather than fibril creep and that for high molecular weight polymers the geometrically necessary entanglement loss is probably caused by chain scission. The fact that surface drawing dominates the craze displacement rates however does not preclude local creep of the fibrils in weak spots where there are few entanglements. Such local creep would not contribute much to the overall craze displacement but could still cause fibril failure. Moreover the time available for disentanglement in the fibrillation zone at the craze bulk interface is small, characteristically < 50 sec at typical craze interface displacement rates of 0.1 nm sec^{-1}, and the molecular weight of the polymer is still high. Once these

---

[1] This treatment assumes that the non-entangled chains in the fibril bear no load. Since they probably have some loadbearing capacity, this calculation probably overestimates the instability of fibrils

molecules are incorporated into the craze fibrils, much longer times are available for disentanglement. (Craze failure times of 10^6 secs are not unusual.) More importantly the molecular weight of the chains that are incorporated into the craze fibrils are severely reduced due to the geometrically necessary entanglement loss. If all this loss occurs by chain scission it is easy to show that

$$\frac{1}{M_n'} = \frac{1}{M_n} + \frac{(1-q)}{M_e} \tag{48}$$

where M_n and M_n' are the number average molecular weights before and after fibrillation and $q = v_E'/v_E$ as before. For random chain scission the weight average molecular weight $M_w' \simeq 2M_n'$ [127]. For example consider monodisperse PS of $M_n = M_w = 300,000$. Since $q \simeq 0.41$ and $M_e = 19,000$, the final molecular weights of the polymer in the fibrils will be $M_n' = 29,100$ and $M_w' = 58,200$, a severe reduction.

Since the time for chain disentanglement by tube relaxation decreases as M^2 and that for disentanglement by reptation decreases as M^3, these times are strongly decreased by the decrease in molecular weight. Moreover the presence of nearby fibril surfaces (for 6 nm diameter PS craze fibrils no molecule is further than second nearest neighbor to a fibril surface), means that the normal tube constraints in such fibrils should be strongly relaxed. It would not be surprising at all therefore if local creep leading to final loss of entanglements in some weak spot along the fibril was the major cause of fibril failure. It is most likely that both chain scission and creep disentanglement contribute. Experimental techniques to directly measure the local number of chain scissions, the local entanglement density and the temperature and strain rate dependence of fibril failure in the craze are needed to decide which is most important.

8 Slow Crack Growth

Once a few neighboring fibrils have broken down, a large void is formed in the craze. The presence of such large voids can be inferred from observations of fracture surfaces and TEM micrographs of crazes in thin films [128, 129]. Craze breakdown to form large voids may also occur at unbonded dust or catalyst particles or at poorly bonded rubber particles in rubber modified polymers. If the stress is high enough these voids may grow slowly by slow fibril breakdown at their edges until a crack of critical size has formed within a craze. The crack then propagates rapidly, breaking craze fibrils as it grows [3].

Very slow subcritical crack growth can be approximately[1] characterized by an empirical law [130] of the form

$$V = A'K^{m_c} \tag{49}$$

---

[1] Recent experiments [131] and theory [132] of truly steady state crack growth indicate that there is a threshold K and V and that the exponent m_c changes to become nearly infinite just above the threshold. These complications are ignored in this treatment

where V is the crack velocity and K is the stress intensity factor of the crack. The exponent m_c can be as large as 25. The microscopic process of fibril breakdown at the void (crack) edges is the same process described in detail above. It is sensitive to the molecular force f on a chain in the fibril and thus to the number of remaining entangled chains per fibril n_e, and to the initial area per fibril $\pi D_0^2/4$ and the craze surface stress S since the average force f on a remaining entangled chain is

$$f = \frac{S\pi D_0^2}{4n_e}.$$
(50)

The craze surface stress on a layer of fibrils just next to the void is given approximately by

$$S = K/\sqrt{2\pi D_0}$$
(51)

where D_0 the initial fibril diameter is also the fibril spacing. Thus one obtains

$$V = A'\left(\frac{\pi D_0^2}{4n_e}\frac{K}{\sqrt{2\pi D_0}}\right)^{m_c}$$
(52)

by making the reasonable assumption that the dependence of V on K in the empirical growth law reflects the dependence on the force per entangled molecule f. If an initial penny-shaped crack of radius a_i is assumed ($K_i = 2\sigma_\infty \sqrt{a_i/\pi}$) and $S \simeq \sigma_\infty$ then one can integrate the crack velocity — K relation to obtain the time-to-failure t_f [130,133]

$$t_f = \frac{2}{A'(m_c - 2)}\left(\frac{4n_e}{\sqrt{2}\sigma_\infty D_0^{3/2}}\right)^{m_c}\left(\frac{1}{a_i}\right)^{(m_c-2)/2}$$
(53)

This equation may be recast into a form which reveals the dependence on entanglement mesh size d

$$t_f = \frac{2}{A'(m_c - 2)}\left(\frac{4\pi\varrho k^2 N_A D_0^{1/2}q}{\sqrt{2}\sigma_\infty d}\right)^{m_c}\left(\frac{1}{a_i}\right)^{(m_c-2)/2}$$
(54)

The time-to-failure of the crazed specimen thus depends strongly on the applied stress σ_∞ and on the size of the initial cavity a_i (craze breakdown at large dust or other particles greatly accelerates fracture). The most interesting dependence however is on the material parameters D_0 and d; D_0 at constant σ_∞ increasing with increasing entanglement density through its dependence on Γ and d decreasing with increasing entanglement density. The ratio $q = v_E'/v_E$ also increases strongly with increasing entanglement density. The effects predicted on craze lifetime are profound. Assume for example $m_c = 25$, and D_0 equals a constant 12 nm. Then based on the entanglement mesh size d and $q(D_0, d)$, crazes in PS should have a lifetime 10^6 longer than those in PTBS whereas crazes in PC should have a lifetime 10^{19} times longer than those in PTBS at the same σ_∞. Even if one allows for the factor of 3.5

increase in σ_∞ for constant craze growth velocity predicted for PC over PTBS the PC crazes should live a factor of 2.5×10^6 times longer than those in PTBS (in thin specimens PC will shear yield before this stress is reached). The marked dependence of craze breakdown and crack growth kinetics on entanglement density throws light on an interesting paradox. Those polymers which are most ductile on a microscopic scale, i.e., have the lowest entanglement densities and thus the highest craze λ's in Fig. 18, are most often the most "brittle" in a macroscopic sense. The total volume of craze material is usually too small to make much contribution to the macroscopic strain regardless of the magnitude of λ and increasing λ by decreasing the entanglement density of the polymer also greatly decreases the individual fibril failure time and the total time to craze fracture.

9 Conclusions

1. Craze nucleation appears controlled by the nucleation of voids in localized regions undergoing large unstable plastic deformation. The sensitivity of this process to the nature of the flaw structure of the surface makes a detailed comparison of data between different experimental groups or between experiment and theory very difficult.
2. Craze growth occurs in a lateral direction by advance of a thin finger-like craze tip by the meniscus instability mechanism. Crazes increase in thickness by a surface drawing mechanism in which more polymer is drawn into the craze fibrils at essentially constant extension ratio λ from the craze-bulk polymer interface.
3. Crazes are strongly load bearing. There is a small stress concentration at, and just behind, the craze tip.
4. Craze fibril diameters determined by TEM and SAXS are of the order of 10 nm. Craze fibril volume fractions v_f range from 0.5 to 0.1, depending on the entanglement network of the polymer and the local craze stress.
5. The finite deformability of the molecular entanglement network inherited from the melt governs the extension ratio of the craze fibrils. Logically however creation of small diameter fibrils must result in severe loss of entanglements. This geometrically necessary entanglement loss produces higher λ's in crazes than in unfibrillated deformation zones.
6. For a high molecular weight polymer, well below its glass transition temperature T_g, most of this entanglement loss must occur by chain scission. The extra main chain bond breaking energy thus required to create the fibril surfaces leads to an increasing crazing stress as the entanglement density of the polymer is increased. At high entanglement densities shear deformation can occur at lower stresses than crazing.
7. The loss of entanglements (and the decrease in molecular weight due to chain scission) adversely impacts fibril stability. Fibril breakdown by localized creep should occur more rapidly in polymer crazes with low entanglement densities and small diameter fibrils.

Acknowledgements. I am very grateful for the support and hospitality of Professor H. H. Kausch and the Ecole Polytechnique Federale de Lausanne during my sabbatic leave at which time much of this review was written. The various aspects of the work on crazing at Cornell are supported by grants from the U.S. Army Research Office-Durham and the U.S. National Science Fundation, both directly and through its support of the Materials Science Center at Cornell. Much of that work would have been impossible without the collaboration and advice of Bob Bubeck and Jerry Seitz of Dow Chemical, Roger Kambour of General Electric, and my present and/or former colleagues at Cornell, Athene Donald, David Grubb, Hugh Brown, Bruce Lauterwasser, Terence Chan, Vincent Wang, Nigel Farrar, Chris Henkee and Arnold Yang. I also thank many friends world wide for responding so rapidly to my requests for reprints and preprints of their recent work. I especially appreciate the chance to discuss aspects of this work with M. Dettenmaier, R. A. Duckett, I. Ward, J.-C. Bauwens, M. Bevis, L. C. Struik, W. Döll, R. Schirrer, L. Monnerie, P. DeGennes, L. Bevan, R. N. Haward, L. Nicolais, A. S. Argon, E. L. Thomas, R. S. Porter, D. Pearson and C. Arends.

Appendix I

Steady state craze tip advance

The rate of steady state craze advance may be estimated using a procedure devised by Fields and Ashby [28]. A schematic view of the finger structure of the craze tip is shown in Fig. 3a. The deforming polymer again is represented by a fluid between two rigid plates spaced a distance h apart (normal to the plane of the diagram). The radius of curvature at the tip of a void finger is h/2 in the plane normal to the craze and approximately $r = (\Lambda_s/2)\left(1 - \dfrac{1}{\lambda}\right) = \alpha\Lambda_s$ in the plane of the craze. Here λ is the final extension ratio of the craze fibrils and Λ_s is the steady state void finger spacing (which may differ from Λ_c).

The particle velocity of the non-Newtonian fluid [flow law $\dot{\varepsilon} = \dot{\varepsilon}_F(\sigma/\sigma_F)^n$] at any point, is given by

$$v = s^* \frac{\sqrt{3}\dot{\varepsilon}_F h}{2(n + 2)} \left(\frac{\sqrt{3}h}{2\sigma_F}\right)^n |\nabla\sigma_0|^n \qquad (A1)$$

where $\nabla\sigma_0$ is the gradient of hydrostatic tension and s^*, its sign. At the craze void tip the hydrostatic tension $(\sigma_0)_t$ is determined by surface tension

$$(\sigma_0)_t = \Gamma\left(\frac{1}{\alpha\Lambda_s} + \frac{2}{h}\right) \qquad (A2)$$

Midway between the two fingers it is $(\sigma_0)_m$ which is taken to be proportional to the average tensile stress S_t at the craze tip, i.e.,

$$(\sigma_0)_m = \beta^* S_t . \qquad (A3)$$

Approximating the negative pressure gradient $\nabla\sigma_0$ by

$$\nabla\sigma_0 \approx \frac{(\sigma_0)_m - (\sigma_0)_t}{\Lambda_s} , \qquad (A4)$$

the craze tip velocity v is obtained,

$$v_0 = \frac{\sqrt{3}\dot{\varepsilon}_F h}{2(n+2)} \left(\frac{\sqrt{3}h}{2\sigma_F \Lambda_s}\right)^n (\beta^* S_t)^n \left[1 - \frac{\Gamma}{\beta^* S_t}\left(\frac{1}{\alpha \Lambda_s} + \frac{2}{h}\right)\right]^n \tag{A 5}$$

The steady state finger spacing may be found by finding the Λ_s which maximizes v. This procedure gives

$$\frac{1}{\alpha \Lambda_s} = \frac{\beta^* S_t}{2\Gamma}\left(1 - \frac{2\Gamma}{\beta^* S_t h}\right) \tag{A 6}$$

In turn, substitution of $\alpha \Lambda_s$ into Eq. (A5) yields the craze velocity

$$v_0 = \frac{\sqrt{3}}{2}\frac{\dot{\varepsilon}_F h}{(n+2)}\left[\frac{\sqrt{3}h(\beta^* S_t)^2}{8\sigma_F \Gamma}\right]^n \left(1 - \frac{2\Gamma}{\beta^* S_t h}\right)^{2n} \tag{A 7}$$

10 References

1. Kramer, E. J.: Environmental Cracking of Polymers, Chap. 3 in Developments in Polymer Fracture (ed. E. H. Andrews), Applied Sci. Publishers, Barking, UK, 1979 p. 55
2. Kambour, R. P.: J. Polymer Sci. Macromol. Rev. 7, 1 (1973)
3. Rabinowitz, S., Beardmore, P.: CRC Reviews in Macromol. Sci. 1, 1 (1972)
4. Gent, A. N.: The Mechanics of Fracture, AMD, Vol. 19 (ed. F. Erdogon), New York, ASME, 1976 p. 55
5. Maxwell, B., Rahm, L. F.: J. Soc. of Plastics Eng. 6, 473 (1965)
6. Argon, A. S., Hannoosh, J. G.: Phil. Mag. 36, 1195 (1977)
7. Sternstein, S. S., Myers, F. A.: J. Macromol. Sci.-Phys. B8, 557 (1973)
8. Oxborough, R. J., Bowden, P. B.: Phil. Mag., 28, 547 (1973)
9. Matsushige, K., Radcliffe, S. V., Baer, E.: J. Mat. Sci. 10, 833 (1973)
10. Matsushige, K., Baer, E., Radcliffe, S. V.: J. Macromol. Sci.-Phys. B11, 565 (1975)
11. Kawagoe, M., Kitagawa, M.: J. Polymer Sci.-Polymer Phys. 19, 1423 (1981)
12. Duckett, R. A., Kovacs, A. private communication
13. Wellinghoff, S., Baer, E.: J. Macromol. Sci.-Phys. B11, 367 (1975)
14. Argon, A. S.: Pure Appl. Chem. 43, 247 (1975)
15. Gent, A. N.: J. Materials Sci. 5, 925 (1970)
16. Dettenmaier, M.: Chapter 2 of this volume
17. Donald, A. M., Kramer, E. J.: J. Applied Polymer Sci. 27, 3729 (1982)
18. Ricco, T., Pavan, A., Danusso, F.: Polymer Eng'g. & Sci. 18, 774 (1978)
19. Broutman, L. J., Panizza, G.: Int. J. Polym. Materials 1, 95 (1971)
20. Sternstein, S. S., Onchin, L., Silverman, A.: Appl. Polymer Symp. 7, 175 (1968)
21. McClintock, F. A., Argon, A. S.: Mechanical Behavior of Materials, Addison-Wesley, Reading, MA, 1966 p. 285
22. Beardmore, P., Rabinowitz, S.: J. Materials Sci. 10, 1763 (1975)
23. Kitagawa, M.: J. Polymer Sci.-Polymer Phys. 14, 2095 (1976)
24. Argon, A. S., Salama, M. M.: Phil. Mag. 36, 1217 (1977)
25. Lauterwasser, B. D., Kramer, E. J.: Phil. Mag. A 39, 469 (1979)
26. Taylor, G. I.: Proc. Roy. Soc. A 201, 192 (1950)
27. Saffman, P. G., Taylor, G. I.: Proc. Roy. Soc. A 245, 312 (1958)

28. Fields, R. J., Ashby, M. F.: Phil. Mag. *33*, 33 (1976)
29. Argon, A. S., Salama, M. M.: Materials Science & Eng'rg. *23*, 219 (1977)
30. Donald, A. M., Kramer, E. J.: Phil. Mag. A *43*, 857 (1981)
31. Donald, A. M.: unpublished
32. Brown, H. R., Kramer, E. J.: J. Macromol. Sci.-Phys. *B19*, 487 (1981)
33. Paredes, E., Fischer, E. W.: Makromol. Chem. *180*, 2707 (1979)
34. Bauwens, J. C., Bauwens-Crowet, C., Homès, G.: J. Polymer Sci. A-2, *7*, 1745 (1969)
35. Brady, T. E., Yeh, G. S. Y.: J. Appl. Phys. *42*, 4622 (1971)
36. Morgan, G. P., Ward, I. M.: Polymer *18*, 87 (1977)
37. Döll, W., Schinker, M. G., Könczöl, L.: Int. J. Fracture, *15*, R 145 (1979)
38. Bucknall, C. B.: Toughened Plastics, Applied Science, London, 1977
39. Verhuelpen-Heymans, N., Bauwens, J. C.: J. Materials Sci. *11*, 7 (1976)
40. Donald, A. M., Kramer, E. J., Bubeck, R. A.: J. Polymer Sci.-Polymer Phys. *20*, 1129 (1982)
41. Verhuelpen-Heymans, N.: Polymer *20*, 356 (1979)
42. Trent, J. S., Palley, I. Baer, E.: J. Materials Sci., *16*, 331 (1981)
43. Donald, A. M., Kramer, E. J.: Polymer *23*, 457 (1982)
44. Wang, W.-C. V., Kramer, E. J.: Polymer *23*, 1667 (1982)
45. Sneddon, I. N.: Fourier Transforms, McGraw-Hill, New York, 1951, pp. 395—430
46. Bilby, B. A., Eshelby, J. D.: Dislocations and the Theory of Fracture, Chap. 2, in Fracture, Vol. 1 (ed. H. Liebowitz), Academic Press, New York, 1972. p. 111
47. Wang, W.-C. V., Kramer, E. J.: J. Materials Sci. *17*, 2013 (1982)
48. Farrar, N. R., Kramer, E. J.: Polymer, *22*, 691 (1981)
49. Hirth, J. P., Lothe, J.: Theory of Dislocations, McGraw Hill, New York, 1968 p. 398
50. Mills, N. J.: J. Materials Sci. *16*, 1317 (1981)
51. Takahashi, K., Hyodo, S.: J. Macromol. Sci.-Phys. *B19*, 695 (1981)
52. Lauterwasser, B. D.: Ph. D. Thesis, Cornell Univ., 1979
53. Bevan, L.: J. Polymer Sci.-Polymer Phys. *19*, 1789 (1981)
54. Donald, A. M., Kramer, E. J.: MSC Rep. 4854, Polymer, in press.
55. Chudnovsky, A., Baer, E., Palley, I.: J. Materials Sci. *16*, 35 (1981).
56. Dugdale, D. S.: J. Mechanics of Solids *8*, 100 (1960)
57. Goodier, J. N., Field, F. A.: Proc. Intern. Conf. Fracture of Solids (ed. D. C. Drucker, J. J. Gilman) Met. Soc. Conf., Vol. 20, Interscience, New York, 1963 p. 103
58. Behan, P., Bevis, M., Hull, D.: J. Materials Sci. *8*, 162 (1972)
59. Behan, P., Bevis, M., Hull, D.: Phil. Mag. *24*, 1267 (1971)
60. Behan, P., Bevis, M., Hull, D.: Proc. Roy. Soc. Lond. A 343I, 525 (1975)
61. LeGrand, D. G., Kambour, R. P., Haaf, W. R.: J. Polymer Sci., A2 *10*, 1565 (1972)
62. Steger, T. R., Nielsen, L. E.: J. Polymer Sci.-Polymer Phys. *16*, 613 (1978)
63. Porod, G.: Kolloid Z. *124*, 83 (1951); ibid *125*, 51, 109 (1952)
64. Paredes, E., Fischer, E. W.: J. Polymer Sci.-Polymer Phys. *20*, 929 (1982)
65. Kambour, R. P.: Nature *195*, 1299 (1962)
66. Kambour, R. P.: Polymer *5*, 143 (1964)
67. Kambour, R. P.: J. Polymer Sci. A *2*, 4159 (1964); ibid *2*, 4165 (1964); ibid *4*, 349 (1966)
68. Brown, H. R.: J. Materials Sci., *14*, 237 (1979)
69. Kramer, E. J.: MSC Rep. 4498, to appear in Developments in Instrumental and Physical Characterization of Polymers, Amer. Chem. Soc., — Adv. Chem. Ser.
70. Brown, H. R.: J. Polymer Sci.-Polymer Phys. *17*, 1431 (1979)
71. Biangardi, H. J.: Makromol. Chem. *183*, 1785 (1982)
72. Brown, H. R.: J. Polymer Sci.-Polymer Phys. *17*, 1417 (1979)
73. Brown, H. R.: Ultramicroscopy, *7*, 263 (1982)
74. Brown, H. R.: J. Polymer Sci.-Polymer Physics, in press.
75. Chan, T., Donald, A. M., Kramer, E. J.: J. Materials Sci. *16*, 676 (1981)
76. Donald, A. M., Chan, T., Kramer, E. J., ibid., *16*, 669 (1981)
77. Porter, R. S., Johnson, J. F.: Chem. Rev. *66*, 1 (1966)
78. Graessley, W. W.: Adv. Polymer Sci. *16*, 1 (1974)
79. deGennes, P. G.: J. Chem. Phys. *55*, 572 (1971)
80. Doi, M., Edwards, S. F.: J. Chem. Soc. Faraday Trans. *74*, 1789 (1978)
81. Graessley, W. W.: J. Polymer Sci.-Polymer Phys. *18*, 27 (1980)

82. Wignall, G. D., Ballard, D. G. H., Schelten, J.: J. Macromol. Sci.-Phys. *B12*, 75 (1976)
83. Kirste, R. G., Kruse, W. A., Ibel, K.: Polymer *16*, 120 (1975)
84. Kurata, M., Tsunashima, Y., Iwama, M., Kamada, K. in Polymer Handbook (J. Brandrup, E. Immergut, Eds.), Wiley, New York, 1975, Chap. IV-4
85. Müller, F. H., quoted in P. H. Hermans: The Physics and Chemistry of Cellulose Fibers, Elsevier, New York, 1949, p. 418
86. Haward, R. N.: The Strength of Plastics and Glass, Clever-Hunel, London, 1949, Chap. IV
87. Pinnock, P. R., Ward, I. M.: Trans. Faraday Soc. *62*, 1308 (1966)
88. Haward, R. N.: Chap. VI in The Physics of Glassy Polymers (ed. R. N. Haward), Applied Sci., Barking, UK, 1973
89. Raha, S., Bowden, P. B.: Polymer *13*, 174 (1972)
90. Kahar, N., Duckett, R. A., Ward, I. M.: Polymer, *17*, 136 (1978)
91. Duckett, R. A.: The Natural Draw Ratio, to appear in Proc. Internat. Spring School on Plastic Deformation of Amorphous and Semicrystalline Materials, Les Houches, France, April 19–29, 1982 (ed. B. Escaig, C. G'Sell), (Les Editions de Physique, Les Ulis, 1983) p. 253
92. Gent, A. N., Thomas, A. G.: J. Polymer Sci. A2 *10*, 571 (1972)
93. Brady, T. E., Yeh, G. S. Y.: J. Materials Sci. *8*, 1083 (1973)
94. Wellinghoff, S. Baer, E.: J. Macromol. Sci.-Phys. *B11*, 367 (1975)
95. Lainchbury, D. L. G., Bevis, M.: J. Materials Sci. *11*, 2222 (1976)
96. Kusy, R. P., Turner, D. T.: Polymer, *15*, 394 (1974)
97. Fellers, J. Kee, B. F.: J. Appl. Polymer Sci. *18*, 2355 (1974)
98. Robertson, R. E. in ACS Symp. Toughness and Brittleness of Plastics, Sept. 1974 (R. D. Demin, A. O. Crugnola, Eds.), Adv. Chem. Ser. 154, Amer. Chem. Soc., Washington, 1976 p. 89
99. Kramer, E. J.: J. Materials Sci. *14*, 1381 (1978)
100. Donald, A. M., Kramer, E. J.: J. Polymer Sci.-Polymer Phys. *20*, 899 (1982)
101. Donald, A. M., Kramer, E. J.: J. Materials Sci. *16*, 2967 (1981)
102. Donald, A. M., Kramer, E. J.: Polymer *23*, 461 (1982)
103. Donald, A. M., Kramer, E. J.: J. Materials Sci. *17*, 1871 (1982)
104. Donald, A. M., Kramer, E. J.: Polymer *23*, 1183 (1982)
105. Donald, A. M., Kramer. E. J.: J. Materials Sci., *16*, 2977 (1981)
106. Argon, A. S.: Phil. Mag. *28*, 839 (1973)
107. Argon, A. S., Bessonov, M. I.: Phil. Mag. *35*, 917 (1977)
108. Argon, A. S., Bessonov, M. I.: Polymer Eng. and Sci. *17*, 174 (1977)
109. Argon, A. S., Chap. 3 Inelastic Deformation and Fracture in Oxide, Metallic and Polymeric Glasses, in Glass: Science and Technology (ed. D. R. Uhlmann), Academic Press, NY, 1980, p. 79
110. Allen, G., Morley, D. C. W., Williams, T.: J. Materials Sci. *8*, 1449 (1973)
111. Adam, G. A., Cross, A., Haward, R. N.: ibid. *10*, 1582 (1975)
112. Pitman, G. H., Ward, I. M., Duckett, R. A.: ibid. *13*, 2092 (1978)
113. Verhuelpen-Heymans, N.: ibid. *11*, 1003 (1976)
114. Brown, H. R.: ibid. *17*, 469 (1982)
115. Prest Jr., W. M., Porter, R. S.: J. Polymer Sci.-Polymer Phys. *10*, 1639 (1972)
116. deGennes, P. G.: Scaling Concepts in Polymer Physics, Cornell University Press, Ithaca, 1979
117. Yang, A., Kramer, E. J., unpublished
118. Monnerie, L.: MACRO 82, IUPAC, Amherst MS, July 15, 1982, to be published
119. Farrar, N. R., Kramer, E. J.: Bull. Amer. Phys. Soc. *26*, 463 (1981)
120. Henkee, C., Kramer, E. J.: unpublished
121. Kita, J., Kramer, E. J.: unpublished
122. Lake, G. J., Thomas, A. G.: Proc. Roy. Soc. London Ser. A *300*, 108 (1967)
123. Wellinghoff, S. T., Baer, E.: J. Appl. Polymer Sci. *22*, 2025 (1978)
124. Ferry, J. D.: Viscoelastic Properties of Polymers, 3rd Ed., John Wiley, New York, 1980, p. 366
125. Kausch, H. H.: Polymer Fracture, Springer-Verlag, Heidelberg, 1978
126. Phoenix, S. L.: Internat. J. Fracture *14*, 327 (1978)
127. Flory, P.: Principles of Polymer Chemistry, Cornell Univ. Press, Ithaca, 1953, p. 317
128. Murray, J., Hull, D.: J. Polymer Sci. A2 *8*, 583 (1970)
129. Doyle, M. J., Maranci, A., Orowan, E., Stork, S. T.: Proc. Roy. Soc. *A329*, 135 (1972)

130. Beaumont, P. W. R., Young, R. J.: J. Materials Sci. *10*, 1334 (1975)
131. Aleshin, V. I., Aero, E. L., Lebedeva, M. F., Kuvshinskii, E. V.: Mekh. Komposytn. Mater. *N1*, 15 (1979)
132. Hart, E. W.: Int. J. Solids and Structures *16*, 807 (1980)
133. Kramer, E. J.: Mechanisms of Toughening in Polymer Mixtures, in Polymer Compatibility and Incompatibility — Principles and Practices (ed. K. Solc), MMI Press, Midland, MI, 1982, p. 251

Received December 2, 1982
H. Kausch (editor)

Intrinsic Crazes in Polycarbonate: Phenomenology and Molecular Interpretation of a new Phenomenon

Manfred Dettenmaier
Laboratoire de Polymères, Ecole Polytechnique Fédérale de Lausanne, Chemin de Bellerive 32, CH-1007 Lausanne

List of Symbols

A, A_0	Amplitudes of waves scattered by the sample and by one electron
D	Fibril diameter
E	Young's modulus
EM	Electron microscopy
F	Form factor
$\hat{F}_2, \hat{F}_3$	Two and three dimensional Fourier transformations
$\tilde{I}$	Smeared scattering intensity
I_p	Scattering intensity integrated along the meridian
J_0, J_1	Zero and first order Bessel functions
L	Length of fibrils
LS	Light-scattering
M, M_0	Molecular weights of the polymer and of the monomer unit
M_e	Molecular weight of chains between entanglement points
N	Number of random links between cross-links and entanglement points
N_0	Number of random links between entanglement points for the unstretched polymer
$P_2(\xi)$	Second order Legendre Polynomial in ξ
PC	Polycarbonate
PET	Poly(ethylene terephthalate)
PMMA	Poly(methyl methacrylate)
PS	Polystyrene
R	Distance between sample and detector
SAXS	Small-angle X-ray scattering
SEM	Scanning electron microscopy
T	Temperature
T_g	Glass transition temperature
V	Sample volume
V_{cr}	Volume of the sample which is occupied by crazes
V_0	Initial sample volume
I, II	Refer to extrinsic and intrinsic crazes
c	Constant
$\vec{e}, \vec{e}_0$	Unit vectors in the directions of the incident and scattered waves
f	Area per fibril
g	Interfibrillar correlation function
h_{cr}, h_f, h_v	Volume fractions of the samples which are occupied by crazes, fibrils and voids
k	Parameter
l_e	Length of the chain between entanglement points
l_0	Length of the projection of the monomer unit on the chain axis
Δn	Optical birefringence
Δn^{max}	Maximum birefringence obtained by complete orientation of random links
Δn^{II}	Birefringence at intrinsic craze initiation
r	End-to-end distance of chains

r_e	End-to-end distance of chains between entanglement points
$\vec{s}$	Reciprocal vector
s_{max}	Magnitude of the reciprocal vector at the scattering maximum
s_r, s_3, s_Φ	Cylindrical coordinates of $\vec{s}$
t	Drawing time
v	Volume fraction of a phase
v_f	Volume fraction of fibrils within a craze
$\vec{x}$	Space vector
x_r, x_3, x_Φ	Cylindrical coordinates of the space vector
$\vec{y}$	Space vector
α	Ratio of the incremental volume strain and the incremental elongational strain
β	Constant
γ	Correlation function
ε	Elongational strain
$\dot{\varepsilon}$	Strain rate
ε_{cr}	Craze strain
ε_{el}	Elastic strain
ε_{sh}	Shear strain
ζ	Interfacial length
λ	Extension ratio
λ^e	Maximum extension ratio of chains between entanglement points
λ^{sh}	Extension ratio after shear yielding
λ^I, λ^{II}	Extension ratios at craze I and II formation
λ_f	Extension ratio of craze fibrils
ϱ, ϱ_{cr}	Densities of the uncrazed and crazed material
ϱ_e	Electron density
σ	Stress
σ^{II}	Stress at craze II initiation
σ_D	Lower yield stress
χ	Area

1 Introduction

The considerable interest which has been focused in the past on the phenomenon of cavitational plasticity in polymers (crazing) and on the resulting fibrillar microstructure (craze) is well documented in the large amount of literature published on this subject (see. e.g. [1−4]). Despite the enormous effort which has been invested, crazing is still not completely understood. One basic problem in studying the ordinary type of crazing, which is reported in the literature, arises from the fact that this phenomenon is frequently controlled by foreign particles and surface grooves which both act as stress concentrators (see e.g. Argon et al. [5,6]). As a consequence, the macroscopic state of the matrix at craze initiation may differ considerably from the local one which is, in general, poorly defined. Under these conditions, a small number of extrinsic crazes grow in amorphous single phase polymers and usually cause the premature rupture of the specimen.

A large number of crazes distributed throughout the sample volume are observed in multiphase systems such as rubber-modified polymers where numerous rubber particles act as stress concentrators. The intensive light-scattering from the multitude of crazes gives rise to the phenomenon of stress-whitening. Evidently, crazing in a multiphase system does not only reflect the intrinsic properties of the different phases but also the interfacial properties and the morphology of the system. Thus crazes in multiphase systems are closely related to the extrinsic crazes mentioned above.

Intrinsic craze phenomena have been reported in a few cases for amorphous single phase polymers. For instance, Goldbach and Rehage [7] observed intrinsic crazes in poly(methyl methacrylate) (PMMA), plasticized at the surface to avoid premature fracture by surface crazes. Hull et al. [8,9], Lainchbury and Bevis [10] and Argon and Hannoosh [6] reported on craze yielding of polystyrene (PS) by formation of numerous intrinsic crazes. Like extrinsic crazing, this phenomenon occurred at small strains of the order of a few percent. Unfortunately, the intrinsic craze phenomena reported above have not been analyzed in detail. In particular it is not always clear whether the term "intrinsic" for the observed type of crazing is justified. In fact, it is difficult to distinguish between extrinsic and intrinsic crazes whenever heterogeneities are involved on a molecular scale. For instance chain ends, weak intermolecular coupling points, and also low molecular weight substances such as monomers, oligomers or even additives may generate defects which allow local plastic deformation to occur. The important points is to demonstrate that craze initiation is governed by intrinsic properties of the polymer matrix.

Recently, Dettenmaier and Kausch [11−14] have observed an intrinsic craze phenomenon in bisphenol-A polycarbonate (PC), drawn to high stresses and strains in a temperature region close to the glass transition temperature, T_g. This type of crazing is not only initiated under extremely well defined conditions which reflect specific intrinsic properties of the polymer but also produces numerous crazes of a very regular fibrillar structure. These crazes were called crazes II in order to distinguish them from the extrinsic type of craze, called craze I. As shown by the schematic representation in Figure 1, a detailed quantitative analysis of intrinsic crazes in terms of craze initiation and microstructure was possible. The basis of this analysis and the results obtained are reviewed in this article.

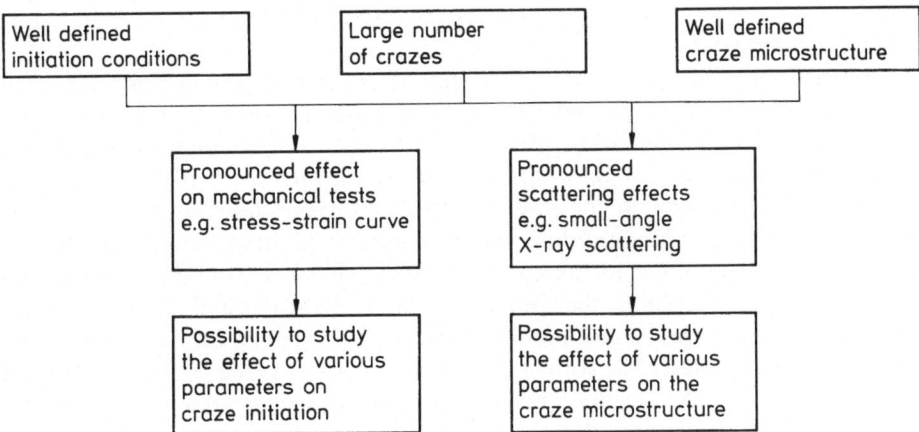

Fig. 1. Scheme for the experimental analysis of intrinsic crazes in polycarbonate

2 Experimental Methods

Sections 4, 5, and 6 are based largely on the investigations from one laboratory. Thus, some information on experimental details may be useful.

2.1 Material

Measurements were conducted on commercially available samples of PC, PMMA and PS. The trade names, the weight average molecular weights and the glass transition temperatures, as measured by DSC, are listed in Table I.

Table 1. Characteristics of Polymers referred to in this Paper

Polymer	Trade name	Weight average molecular weight g/mole	Glass transition temperature °C
polycarbonate	Makrolon 3200	35000	149
poly(methyl methacrylate)	Plexiglas 233	4000000	106
polystyrene	Polystyrol 168 N	360000	103

2.2 Tensile Experiments

PC sheets were pre-oriented above T_g at 160 °C for t = 63 s with a maximal strain-rate of approximately 0.03 s^{-1}. After drawing, the samples were rapidly cooled to room temperature in a stream of cold air whilst keeping their position in the clamps fixed. Each specimen was marked with a series of ink spots. The elongation of the sample was determined by measuring the distance between the ink spots before and after drawing.

For tensile tests below T_g dumb-bell-shaped specimens with a central part of 25 mm length, 4 mm width and various thicknesses between 1 and 3 mm were milled from isotropic and from pre-oriented sheets. Unless otherwise stated, the samples were drawn at 129 °C with a nominal strain rate of 6×10^{-4} s^{-1}. Prior to drawing, they were kept in the heating chamber for one hour in order to reach thermal equilibrium. This procedure, which represents a form of annealing, did not affect the experimental results presented in this paper. Strain transducers were used to measure the elongation of the specimens. Nominal stresses and strain rates were converted into true stresses and strain rates by multiplying and dividing respectively with the extension ratio. The former procedure, which is based on the assumption that the volume of the specimen remains constant during straining, is reasonable for samples before craze growth.

2.3 Birefringence

The birefringence of each sample was determined at a wavelength of 545 nm with a polarizing microscope, equipped with an Ehringhaus compensator. The measurements were carried out by compensating large retardations with a wedge-shaped piece of PC and small ones with the Ehringhaus compensator.

2.4 Density

Density measurements were carried out at a temperature of 23 °C with a density gradient column filled with a solution of NaBr in distilled water. Weighing the samples before and after they were introduced into the solution showed that the liquid did not enter the craze pores.

2.5 Small-angle X-ray Scattering

Small angle X-ray scattering (SAXS) was measured with a Kratky camera using nickel filtered CuKα radiation. The width of the entrance and detector slits were 20 and 50 μm, respectively. No desmearing procedure was applied to the scattering curves. Absolute measurements were carried out by using a Lupolen standard.

The excess scattering due to the presence of crazes was obtained in the following way. From the scattering of the crazed samples, the absorption corrected scattering of the matrix and of the slits was subtracted. In case of extrinsic crazes, the matrix scattering was taken from an undrawn sample. For intrinsic crazes the matrix scattering was evaluated using a sample drawn to an elongation slightly smaller than that necessary for craze initiation. In the latter case the background scattering was usually found to be only a minor part of the total scattering. Measurements on samples with various thicknesses and examination of the primary beam after passage through the sample did not give any evidence for multiple scattering.

For experimental convenience all the scattering curves were taken from samples after stress relief. This should not be a severe limitation. In fact, in subsequent sections experimental evidence will be presented that the fibrillar structure of intrinsic crazes in PC is not strongly affected by stress relief as previously shown for extrinsic crazes by Brown and Kramer [15].

2.6 Scanning Electron Microscopy

Normally the craze microstructure is not directly visible in the scanning electron microscope. Thus, an etching procedure using oxygen ions was employed to remove the plastically deformed layer at the sample surface. Control measurements on uncrazed samples showed that this procedure does not lead to artifacts. The surfaces were coated with gold to reduce surface charging.

3 Structure and Deformation of Amorphous Polycarbonate

3.1 Structure

Any discussion of crazing in molecular terms must take into account the molecular organization in amorphous polymers. Thus, the information available on the structure of amorphous PC will be briefly reviewed in this section. Much of this information results from the enormous efforts that have been invested to resolve the intense controversy concerning the existence of strong intermolecular orientation correlations in conventional polymer melts and glasses (see e.g. the contributions in [16]). A comprehensive review has recently been given by Wendorff [17].

Molecular organization in amorphous polymers can be conveniently considered in terms of conformational, spatial and orientational correlations between chain units.

Recently, Ballard et al. [18] and Gawrisch et al. [19] have investigated the conformation of PC chains in the bulk material by coherent neutron scattering from mixtures of protonated and deuterated molecules. The experimental data of Gawrisch et al. cover a wide range of scattering vectors including the low angle and intermediate scattering region. Both groups come to the same results. From the scattering at small scattering vectors they found, in agreement with the unperturbed chain model [20]

$$\langle r^2 \rangle^{1/2} = \beta M^{1/2} \tag{1}$$

where $\langle r^2 \rangle$ is the mean square end-to-end distance of chains of sufficiently high molecular weight, M is the molecular weight and β denotes a constant which for PC amounts to $\beta = 1.1$ nm $(\text{mole/g})^{1/2}$.

Yoon and Flory [21] have compared the experimental data of Gawrisch et al. to the scattering of unperturbed chains calculated on the basis of the rotational isomeric state theory. They found good agreement up to the highest values of the scattering vector corresponding to distances of approximately one chain diameter

(≈ 0.5 nm). These findings are consistent with previous results for various other poly-mers and oligomers in the melt and in the glassy state [22-27].

Spatial correlations between chain units over small distances, usually called short range order, can be detected by wide-angle X-ray scattering. The radial distribution function of PC calculated from the scattering data contains information on both intra- and intermolecular correlations. The existence of weak intermolecular cor-relations is suggested by two broad, temperature-dependent peaks centered at about 0.5 and 1.0 nm, respectively [28-30]. Wignall and Longman [28] compared the radial distribution functions of samples, showing marked differences in mechanical pro-perties due to different thermal histories. They came to the conclusion that the short range order of the samples was virtually identical.

Density fluctuations resulting from spatial correlations over larger distances have been investigated by small-angle X-ray scattering (SAXS) and light-scattering (LS). In addition to the scattering due to thermal density fluctuations, a strongly angular dependent scattering component is observed for PC and many other amorphous polymers [31-36]. This component must be attributed to large heterogeneities with a broad size distribution. For instance, the angular dependent LS component of PC measured by Dettenmaier and Fischer [35,36] indicates the presence of density fluctua-tions with a correlation length of about 140 nm. The origin of these heterogeneities is not completely understood at present.

Evidence for heterogeneities in PC has also been derived from electron microscopy (EM). Extensive work has been published on the existence of a nodular structure in the range of 3–10 nm (see e.g. [37-39]). According to Yeh et al. [45], craze fibrils are formed by the movement of these nodules. However, both the sample preparation and the resolution of the EM technique have been subject to criticism [41,42]. Kaempff and Orth [43] have studied stretched samples submitted to an ion etching procedure. They observed heterogeneities of about the same size as those detected by LS. Grosskurth has made similar observations for PS [44] and recently also for PC [45]. He assumes that these heterogeneities may be regarded as precursors to crazes [46].

Depolarized light-scattering has proved to be a very powerful method of character-izing the orientation correlations between chain segments in amorphous polymers [33,35]. In fact, the intensity of the depolarized scattered light is directly related to anisotropy fluctuations of the polarizability. Detailed investigations on the orientation correlation between PC chain segments in untreated and thermally pre-treated samples have been performed by Dettenmaier and Kausch [47]. They have demon-strated that appreciable intermolecular orientation correlations do not exist in PC nor are they created if the material is annealed below the glass transition temperature. It is well known that PC can be crystallized thermally or by means of solvents and plasticizers. However, even at a temperature of 190 °C, where the rate of crystallization attains a maximum, prolonged heating is necessary to get some spurious amounts of crystallinity in the form of distorted spherulitic structures [47]. This result clearly demonstrates the strong hindrance to crystallization in PC.

To summarize, it must be assumed that amorphous PC does not exhibit strong intermolecular spatial and orientational correlations between chain segments. This result is consistent with the fact that PC chains are highly entangled (see Sect. 4.2). It is believed that, up to the present time, insufficient experimental evidence has been provided to prove the existence of some sort of molecular organization in glassy polymers which may be regarded as the precursor to crazes. In fact, it will be demonstrated that the intrinsic craze initiation in PC can be understood on the basis of a random phase model.

3.2 Deformation

In the previous section the structure of amorphous polymers has been discussed in terms of intra- and intermolecular correlations between chain segments. The complexity of these correlations both in space and in time is reflected in the variety of deformation mechanisms which may be active. In fact, glassy polymers exhibit mechanical properties which have the attributes of energy and entropy elasticity, viscosity and plasticity to various degrees depending on the time scale of observation, the stress-field, temperature, pre-treatment of the sample and environment. PC displays all these properties under very common experimental conditions and has, therefore, been the subject of numerous investigations. A comprehensive review of these results is beyond the scope of this paper. Instead, this section focuses on some general mechanisms which determine the deformation behavior of glassy PC.

At small stresses and strains, glassy PC exhibits linear viscoelastic behavior. The limit of applicability of the theory of linear viscoelasticity has been investigated by Yannas et al. [48-52] over the temperature range 23 °C–130 °C. The critical strain at which, within the precision of their measurement, deviations from the linear theory occur has been found to diminish from about 1.2% at 23 °C to about 0.7% at 130 °C. According to Jansson and Yannas [52] the transition from linear to nonlinear viscoelastic behavior is marked by the onset of significant rotation around backbone bonds.

In passing through the range of nonlinear viscoelasticity two distinct plastic deformation mechanisms, namely crazing and shear yielding, may be initiated. Before focusing on the crazing of PC some considerations on shear yielding may be useful. The shear yielding of PC has been extensively investigated over wide ranges of temperature and strain-rate [53-61]. The experimental results are generally described by theories based on a stress-biased temperature activated rate mechanism of the type treated in the absolute reaction rate theory of Glasstone, Leidler and Eyring [62]. According to these theories shear strain is produced by thermal fluctuations of the Gibb's free energy which enable a molecular system to overcome free energy barriers and to adopt adjacent positions of stable equilibrium on the Helmholtz free energy contour. Attempts to specify the molecular origin of these energy barriers have been made by Argon et al. [63-66] and Robertson [67,68]. According to Argon and Bessonov [64-66], the activated configuration in shear yielding of PC consists of approximately four molecular segments.

In general, the activation of shear yielding in a glassy polymer reduces its plastic resistance to further deformation. When strain softening occurs the deformation becomes unstable to small perturbations of the stress field. This instability results in the formation and growth of deformation zones, the shape of which are controlled by the strain softening and strain hardening characteristics of the material [69-71].

Deformation zones in PC have been analyzed in terms of shape, size, strain and birefringence [72-75]. The fact that well defined zones exhibit a constant strain has been associated with the existence of a natural draw ratio. Donald and Kramer [76] have provided strong evidence that the extension ratio of the material within these zones is governed by the maximum extensibility of chains between entanglement points.

Since the transformation of unyielded to yielded material is accompanied by appreciable energy consumption, deformation zones are an important source of fracture toughness. The high ductility of PC is directly related to the ease with which these zones may be grown. If shear yielding becomes highly localized or even inactive, for example at low temperatures, high strain rate under plane strain conditions, or by annealing PC, a ductile-brittle transition is observed [74,77-86]. Straightforward interpretations of this transition in terms of basic molecular relaxation mechanisms such as the β-process have raised substantial objections (see e.g. [87-89]).

The remaining part of this section briefly focuses on the ordinary type of crazes in PC, previously referred to as extrinsic crazes or crazes I. More detailed information on this type of craze is given in the other contributions of this volume. The extensive studies on these crazes have revealed clearly the important role of crazing as a plastic deformation process competitive to shear yielding. The molecular mechanism of craze initiation and its relation to shear yielding as discussed above is not completely understood at present. According to Argon [63], crazing results from the combined effect of the deviatoric and the dilatational part of the applied stress tensor. He assumes that the former activates a coarse shear process and that the latter produces the voided craze matter from the yielded polymer. The fact that the first process occurs while the polymer is at stress levels below general yield is attributed to the existence of stress concentrations. The second process results in the formation of a well defined fibrillar craze structure by the mechanism of meniscus instability [90,91].

Since crazing represents a cavitational form of plasticity, it is clear that crazes play an important role in the fracture of polymers. Crazing is, generally, involved when PC fails under plane strain conditions, e.g. in fracture mechanics tests on thick samples with sharp notches [79,80,82,83,93-96], where high triaxial stresses are built up at the notch tip. In samples loaded under plane stress conditions, Haward et al. [97,98] observed a failure mode associated with the formation of diamond shaped cavities. The formation of crazes in PC and the break down of craze fibrils to form a crack may be greatly enhanced by environmental effects. Various gases [99-101], organic solvents [102-109] and even "finger prints" [105] are known to act as crazing agents. The plasticization of the material seems to be the most important mechanism in environmental crazing (see e.g. [2-4]).

Much attention has been focused on the microstructure of crazes in PC [102,105-112] in order to understand basic craze mechanisms such as craze initiation, growth and break down. Crazes I in PC, which are frequently produced in the presence of crazing agents, consist of approximately 50% voids and 50% fibrils, with fibril diameters generally in the range of 20–50 nm. Since the plastic deformation of virtually undeformed matrix material into the fibrillar craze structure occurs at approximately constant volume, the extension ratio of craze I fibrils, λ_f^I, is given by

$$\lambda_f^I = \frac{1}{v_f^I} \tag{2}$$

where v_f^I denotes the volume fraction of fibrillar material in a craze. For PC, equation (2) yields $\lambda_f^I = 2.0$. As in the case of deformation zones, Donald and Kramer [111] found good correlation between λ_f^I and the maximum extension ratio of chains between entanglement points (see also Sect. 4.2). Paredes and Fischer [110,112] investigated the effect of temperature and strain rate on the microstructure of PC crazes produced in tensile tests. From their results it can be concluded, in accordance with the concept of Donald and Kramer, that v_f^I and, hence, λ_f^I are little affected by these parameters. However, the average fibril diameter, $\bar{D}$, rises with increasing temperature and decreasing strain rate. This behaviour has been attributed to the temperature and strain rate dependence of the lower yield stress, σ_{DI}, since $\bar{D}$ is found to remain constant. The observation that $\bar{D}$ remains constant during craze growth indicates that craze thickening occurs primarily by drawing out of more material at the craze boundary and not by fibril creep. This is in agreement with results obtained by Lauterwasser and Kramer [113] and Verheulpen-Heymans [114].

4 Intrinsic Crazing

4.1 Phenomenology

The intrinsic crazing of PC is basically a post-yield phenomenon and must, therefore, be distinguished from the phenomena discussed in the previous section.

In tensile tests at strain-rates of $\dot{\varepsilon} = 2 \times 10^{-5}$–$2 \times 10^{-3}$ s^{-1} intrinsic crazing of PC has been observed in a temperature range of 120–135 °C, i.e. about 15–30 °C below the glass transition temperature [11-14]. Figure 2 shows a nominal stress-strain curve

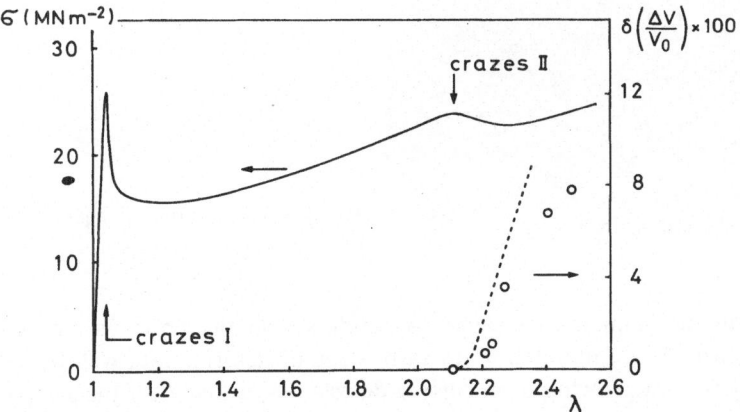

Fig. 2. Nominal stress, σ, and incremental volume strain, $\delta\left(\dfrac{\Delta V}{V_0}\right)$, versus the extension ratio, λ. $\delta\left(\dfrac{\Delta V}{V_0}\right)$ was determined from the strains parallel and perpendicular to the draw axis (straight line) and from density measurements (circles)

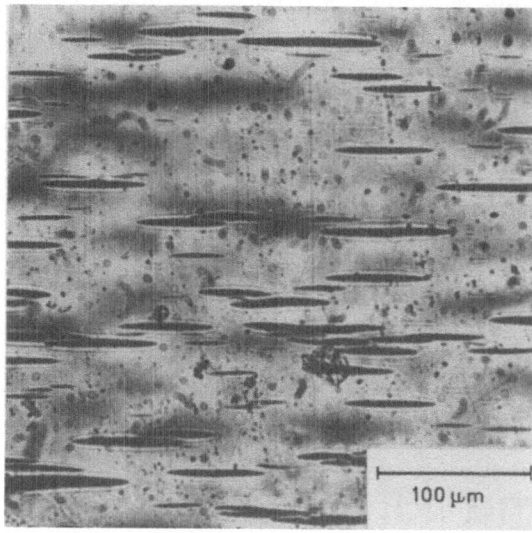

Fig. 3. Optical micrograph of crazes I

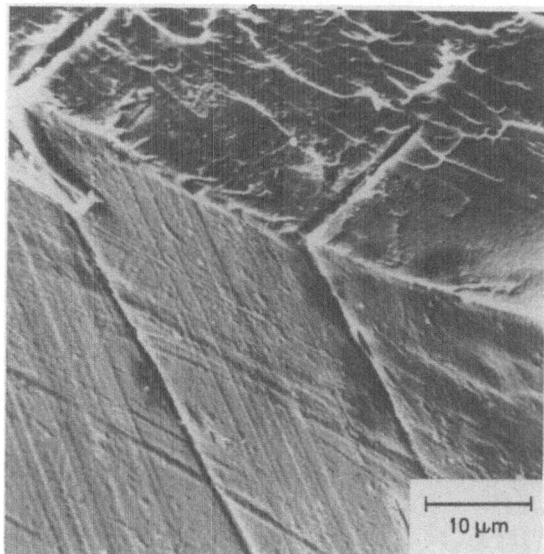

Fig. 4. Scanning electron micrograph
of crazes I

of PC measured in this temperature range. As discussed in the previous section, extrinsic crazes (crazes I) are initiated at an early stage of deformation, well below the yield point. They start growing from surface defects, in particular, if the sample has come into contact with some crazing agent. Crazes I are shown in the optical and scanning electron micrographs in Figures 3 and 4, respectively. These crazes are largely separated from each other and are situated at the surface of the specimen (Fig. 4). Since they occupy only a small fraction of the sample volume, their initiation and growth has no notable effect on the stress-strain curve.

However, at high stresses and strains, well above the yield point, numerous crazes (crazes II) are initiated throughout the sample volume. Their initiation is preceded by some strain hardening of the material and is followed by a strain softening mode leading to the second peak in the stress-strain curve of Figure 2. This peak clearly defines the stress-strain state of the matrix at craze II initiation. The marked influence of the initiation and growth of these intrinsic crazes on the tensile behaviour of the material is clear in view of their large number as shown by the optical and scanning electron micrographs in Figures 5 and 6, respectively. In

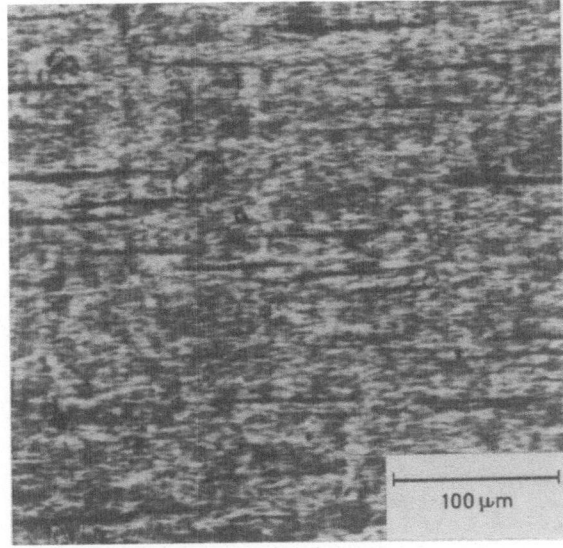

Fig. 5. Optical micrograph of crazes I and II

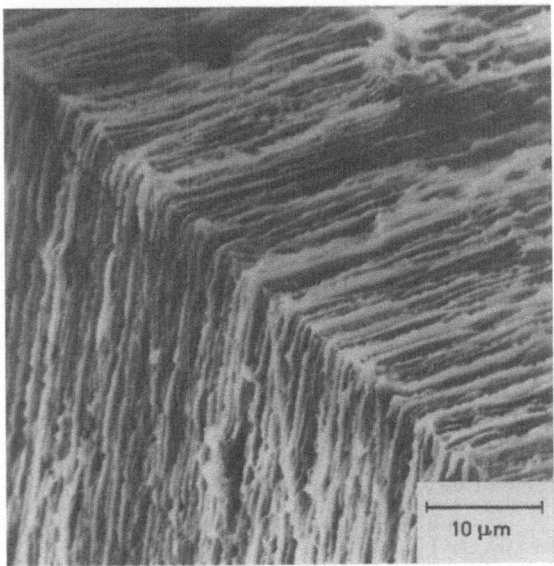

Fig. 6. Scanning electron micrograph of crazes II

Figure 5, crazes I and II can clearly be distinguished, the large and isolated crazes I and the dense pattern of very fine crazes II. The axes of both coincide with the draw axis. Other examples of the coexistence of two distinct types of crazes, whose axes however do not coincide, may be found in the literature [109,115,116]. Unfortunately, the phenomena reported in these publications are not very well understood.

The regular fibrillar microstructure of some large intrinsic crazes and their dense arrangement are shown in the scanning electron micrograph in Figure 7. A quantitative analysis will be given in Section 5.2 by means of SAXS.

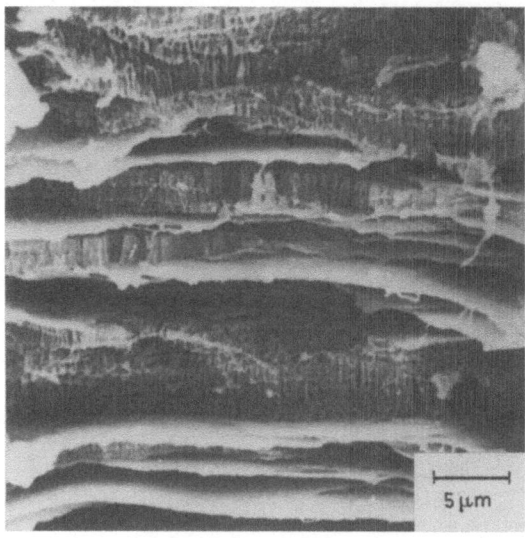

Fig. 7. Scanning electron micrograph of the fibrillar microstructure of crazes II

In order to explain the distinct size of extrinsic and intrinsic crazes in PC, as seen in Figure 5, several factors must be considered. Firstly, the number of intrinsic craze nuclei is very large, consequently, crazes interact at an early stage of growth. Secondly, intrinsic crazes are initiated during large elongations of the specimen when the matrix material has undergone appreciable orientation. Studies [116–118] on crazing in oriented polymers have shown that craze growth is hindered when the growth direction is perpendicular to the orientation axis. Finally, it will be demonstrated in Section 5.2 that intrinsic crazes have a higher volume fraction of fibrils than extrinsic crazes. Thus, stress concentrations are more easily transmitted from one craze boundary to the other and new crazes can be initiated in the immediate neighborhood of already existing crazes.

The formation of intrinsic crazes generally causes stress-whitening of the samples. Under certain conditions of pre-orientation and drawing even color phenomena may be observed. Irradiation by white light causes these samples to appear blue in the direction of backward scattering and yellow in the direction of forward scattering. This behavior has, for example, been observed for samples preoriented above T_g to an optical birefringence of $\Delta n = 0.02$, and afterwards stretched at 120 °C with a strain-rate of $\dot{\varepsilon} = 4 \times 10^{-5}$ s^{-1} [11]. In section E II some evidence will be given that these color phenomena have their origin in interfibrillar interferences of the scattered light.

As demonstrated in Figure 2, the specific volume of PC increases notably when crazes II are initiated, resulting in a loss in density of approximately 8 % at the rupture of the specimen. This is comparable to that found for high impact polymers where the number of crazes is increased artificially by the incorporation of rubber particles.

The question arises as to whether intrinsic craze formation is sufficient to account for the observed plastic strain. Following the approach made by Bucknall and Clayton [119] and Heikens et al. [120] in their studies on multiphase systems, the total elongational strain, ε, must be assumed to have contributions from elasticity, ε_{el}, crazing, ε_{cr}, and shear deformation, ε_{sh}. Assuming the additivity of these components gives:

$$\varepsilon = \varepsilon_e + \varepsilon_{cr} + \varepsilon_{sh} .\tag{3}$$

An analogous equation may be formulated for the volume strain:

$$\frac{\Delta V}{V_0} = \left(\frac{\Delta V}{V_0}\right)_{el} + \left(\frac{\Delta V}{V_0}\right)_{cr} + \left(\frac{\Delta V}{V_0}\right)_{sh} .\tag{4}$$

The contribution of crazing to the volume strain is given by

$$\left(\frac{\Delta V}{V_0}\right)_{cr} = \varepsilon_{cr}\tag{5}$$

while shear deformation does not contribute, so that

$$\left(\frac{\Delta V}{V_0}\right)_{sh} = 0 .\tag{6}$$

The relative contributions of crazing and shear deformation after intrinsic craze formation are of particular interest. They may be derived from the ratio of the incremental volume strain, $\delta\left(\dfrac{\Delta V}{V_0}\right)$, and the incremental elongational strain, $\delta(\varepsilon)$

$$\alpha = \frac{\delta\left(\dfrac{\Delta V}{V_0}\right)}{\delta\varepsilon}\tag{7}$$

where

$$\delta\left(\frac{\Delta V}{V_0}\right) = \frac{\Delta V}{V_0} - \left(\frac{\Delta V}{V_0}\right)^{II} \quad \text{and} \quad \delta\varepsilon = \varepsilon - \varepsilon^{II}\tag{8}$$

$\left(\dfrac{\Delta V}{V_0}\right)^{II}$ and ε^{II} are the values taken at intrinsic craze initiation. To a first approximation $\left(\dfrac{\Delta V}{V_0}\right)_{el}$ and ε_{el} may be considered constant during craze formation. Within this approximation, verified by measuring the strains before and after stress-relaxation, Equations (3–7) yield:

$$\alpha = \frac{1}{1 + \dfrac{\delta\varepsilon_{sh}}{\delta\varepsilon_{cr}}} \quad \text{or} \tag{9}$$

$$\frac{\delta\varepsilon_{sh}}{\delta\varepsilon_{cr}} = \frac{1}{\alpha} - 1 \tag{10}$$

where $\delta\varepsilon_{sh}$ and $\delta\varepsilon_{cr}$ are defined corresponding to Equations (8). Evidently, $\delta\varepsilon_{cr} = \varepsilon_{cr}$ since the small number of extrinsic crazes does not appreciably contribute to the measured strain. As seen from Equation (9), α can take values between 0 (only shear contribution) and 1 (only craze contribution). α has been determined from Equation (7) by recording the strains parallel and perpendicular to the draw axis during a tensile test. In the deformation range, considered here, the incremental volume strain, $\delta\left(\dfrac{\Delta V}{V_0}\right)$, calculated from these strains, is a linear function of $\varepsilon = \lambda - 1$, as shown in Figure 2. Therefore, it must be assumed that the relative contributions of crazing and shear deformation remain constant. The slope of the straight line amounts to $\alpha = 0.46$, which yields by using Equation (10) $\dfrac{\delta\varepsilon_{sh}}{\delta\varepsilon_{cr}} = 1.17$. This result indicates that intrinsic crazing and shear deformation contribute equally to the elongational strain.

It is interesting to compare $\delta\left(\dfrac{\Delta V}{V_0}\right)$ to the change in specific volume, measured for unloaded samples by means of a density gradient column. The data obtained from these measurements are represented as circles in Figure 2. Except for larger values of λ they follow close to the straight line, indicating that the fibrillar and voided microstructure of intrinsic crazes is not strongly affected by unloading. This result is also reflected in the cycling experiment of Figure 8, which does not show a strong effect of intrinsic crazes on the elastically recoverable strains. This behavior is different from the pronounced hard-elastic type of behavior with large recoverable elastic strains as previously reported for other crazes [121–125]. To explain the different behavior, it should be noted that previous measurements were conducted on crazes grown well below T_g. Under these conditions the surroundings of a craze deform elastically and, hence, after stress relief, crazes are forced to contract. The situation is different in the case of intrinsic crazes which grow in highly stressed material close to T_g, where sufficiently fast relaxation processes prevent the build-up of large elastic constraints after craze formation. The molecular origin of these processes will be discussed in Section 4.2.

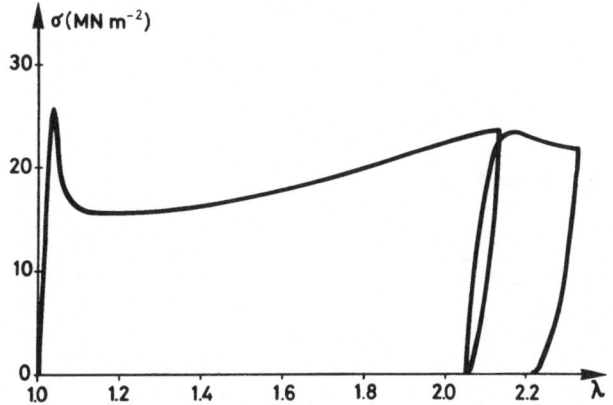

Fig. 8. Cyclic stress-strain behavior of polycarbonate

Intrinsic craze formation is also reflected in the elastic properties of the material. The values of Young's modulus, E, listed in Table 2, clearly reveal both the effect of matrix orientation before craze initiation ($\lambda < \lambda^{II}$) and, the effect of voids after craze formation ($\lambda > \lambda^{II}$). The former causes E to increase, the latter results in a decrease of E.

Table 2. Young's Modulus, E, for Samples stretched at 129 °C with $\dot{\varepsilon} = 6 \times 10^{-4} \, s^{-1}$ to Different Extension Ratios, λ (E was measured under the same drawing conditions)

λ	1	2.1 ($\equiv \lambda^{II}$)	2.5
$E/GN \, m^{-2}$	1.8	2.5	1.8

The following part of this section is concerned with the effect of the drawing temperature on the tensile properties of PC and, in particular, intrinsic craze initiation. It should be mentioned that the formation of extrinsic crazes, e.g. enhanced by weak crazing agents, had no measurable effect on intrinsic craze initiation. Under the experimental conditions used in these investigations, intrinsic crazing has not been observed below 120 °C. From room temperature to 120 °C samples can be stretched to rupture without showing any sign of opacity or cloudiness. In this temperature range shear yielding results in neck formation. The extension ratio, λ^{sh}, of the material within the neck is independent of the drawing temperature (see also Sect. 4.2). The measured value of $\lambda^{sh} = 1.7$ is in good agreement with values reported previously for necked material [53, 126] and for shear bands [72]. After the neck has propagated through the entire gauge length of the specimen, the stress increases continuously until rupture occurs.

In the temperature range of 120–135 °C PC deforms plastically, firstly, by the formation of shear bands and, then, by the formation of intrinsic crazes. The true stress, σ^{II}, at craze initiation was determined from the post-yield peak in the stress-strain curve as shown in Figure 2. As demonstrated in Figure 9 and 10, σ^{II}

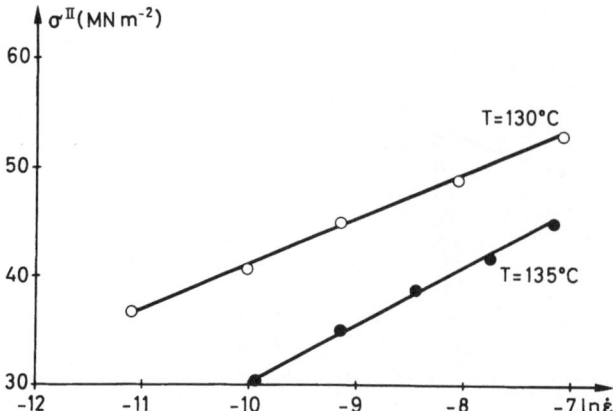

Fig. 9. True stress, σ^{II}, at intrinsic craze initiation as a function of the logarithm of the true strain rate ε. Parameter is the drawing temperature, T

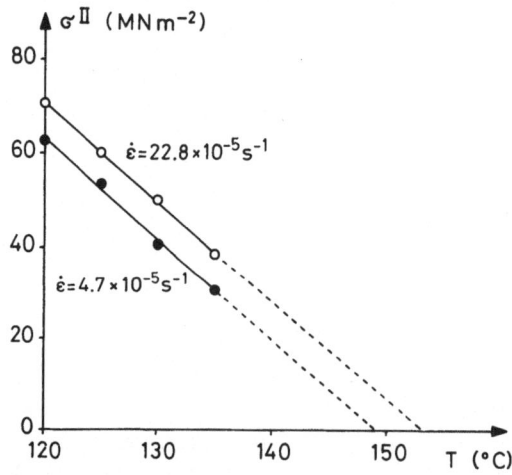

Fig. 10. True stress, σ^{II}, at intrinsic craze initiation as a function of temperature, T. Parameter is the true strain rate, $\dot{\varepsilon}$

strongly decreases with decreasing strain rate and increasing temperature. By extrapolation, σ^{II} is found to become zero close to the glass transition temperature. This behavior may be qualitatively understood if intrinsic crazing is considered as a temperature and stress activated rate mechanism. The quantitative treatment of this mechanism as a rate process requires detailed information on its free energy contour. This information is not available at present. A simple approach based on the Eyring theory of viscous flow may lead to incorrect results as clearly pointed out by Argon and Bessonov[66] and Haussy et al.[127]. The extension ratio, λ^{II}, at intrinsic craze initiation is much less affected by the drawing temperature than σ^{II} and may be considered constant to a first approximation. The measured value of $\lambda^{II} = 2.1$ clearly indicates that, in contrast to extrinsic crazes, intrinsic crazes are initiated in a highly deformed matrix.

As the drawing temperature increases from 135 °C to T_g, stress-whitening decreases in intensity and crazes lose their regular fibrillar microstructure. This temperature range may, therefore, be considered as a transition region between cavitation by crazing and cavitation by irregular void formation. It is interesting to note that Argon and Salama [128] observed a voided cellular structure even at 160 °C when PC was rapidly fractured between two glass plates.

Valuable information on the intrinsic craze phenomenon has been obtained from studies on pre-oriented material [13]. The effect of pre-orientation on craze initiation is central to the discussion of the molecular craze mechanism in Section 4.2. To facilitate the presentation of the experimental results, the following nomenclature is introduced. Quantities which refer to the pre-orientation and to the crazing experiment are labelled with indices one and two, respectively. If no index is used, the quantity refers to the result of both experiments.

Measurements were conducted on PC pre-oriented at 160 °C, i.e. about 10 °C above T_g. The birefringence, Δn_1, of the material was used to characterize the degree of orientation. The nominal stress-strain curves of samples elongated at 130 °C parallel and perpendicular to the orientation axis are presented in Figure 11. Many features of these curves are consistent with reports in the literature (see e.g. [129]) on the deformation behavior of oriented polymers, namely the effect of orientation on Young's modulus, shear yielding and strain-hardening. In addition, each stress-strain curve exhibits a pronounced peak or an inflection above the yield point, indicating the formation of intrinsic crazes. Inspection of the samples shows that at the point where the slope of the curve changes abruptly, the originally transparent samples become opaque. The effect of pre-orientation on the extension ratio, λ_2^{II},

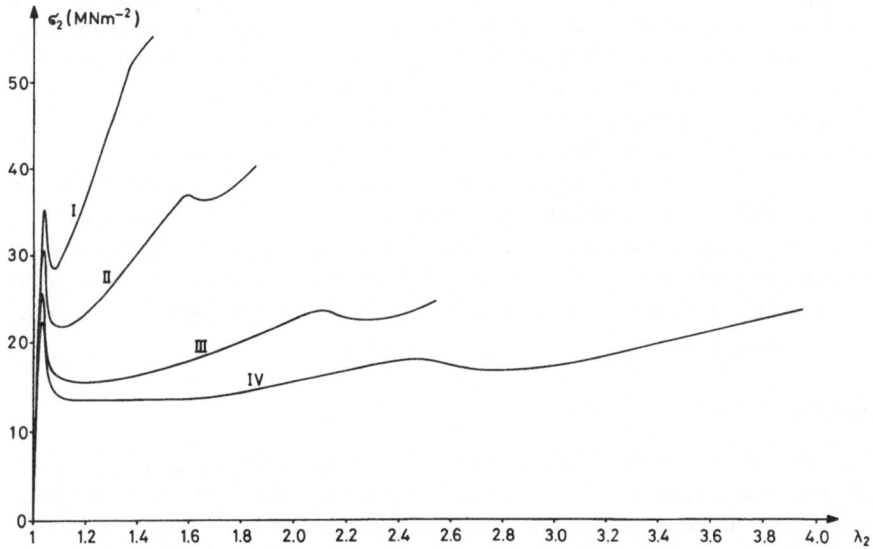

Fig. 11. Nominal stress, σ_2, versus extension ratio, λ_2, for polycarbonate with different values of pre-orientation given by Δn_1. Samples drawn parallel to the orientation axis: $\Delta n_1 = 0.029$ (I), $\Delta n_1 = 0.016$ (II); unoriented sample: $\Delta n_1 = 0$ (III); sample drawn perpendicular to the orientation axis: $\Delta n_1 = 0.017$ (IV)

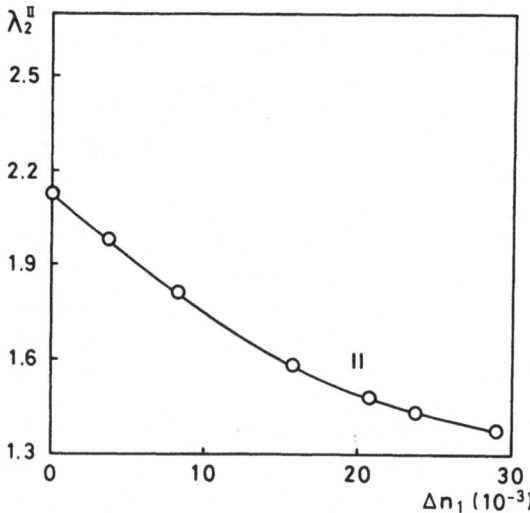

Fig. 12. Extension ratio, $\lambda_2{}^{II}$, at intrinsic craze initiation as a function of Δn_1 for polycarbonate stretched parallel to the orientation axis

at craze initiation is shown in Figure 12 for samples stretched parallel to the orientation axis. $\lambda_2{}^{II}$ strongly decreases as a function of Δn_1. Qualitatively, this behavior may easily be understood by assuming that a certain total deformation, λ^{II}, of the matrix is required to initiate craze II. Pre-stretching by an amount λ_1 will decrease the remaining part, $\lambda_2{}^{II}$. However, the situation is complicated further since $\lambda^{II} = \lambda_1\lambda_2{}^{II}$ does not remain constant but increases with Δn_1. In Section 4.2, this behavior will be quantitatively analyzed on a molecular scale.

The question may arise as to whether, under given drawing conditions, intrinsic craze initiation is a unique function of orientation. The answer must be negative since orientation below and above T_g acts differently. This may be demonstrated by comparing experimental results of Figures 8 and 11 after converting nominal into true stresses. To a first approximation, the orientation of PC during the first cycle in Figure 8 does not affect the remaining part of the stress-strain curve (second cycle). However, as seen in Figure 11, orientation above T_g has a marked effect on the post-yield behavior and hence on intrinsic craze formation. This result is not surprising in view of the behavior of crystalline polymers. In fact, it is well known that their physical properties are not uniquely related to chain orientation but depend also on the morphology of the material. In the subsequent section, some evidence will be given that molecular entanglements are an important factor influencing the mechanical properties of glassy polymers.

4.2 Molecular Interpretation

It has become apparent in the previous section that intrinsic crazing of PC is a post-yield phenomenon which occurs at high deformations of the material. There is much evidence that such phenomena are strongly affected by the existence of an entanglement network frozen-in during the glass transition [130–153]. For example, it is well known that the ultimate mechanical properties such as the critical energy release rate, G_{Ic}, strongly depend on the presence of chain entanglements. This may be demonstrated either by the strong increase of G_{Ic} above a critical molecular weight [135,138–140,151] or by the increase of G_{Ic} with chain interpenetration at interfaces [148,149,153].

The concept of chain entanglement was originally developed in order to explain the deformation of chemically uncrosslinked amorphous polymers above their glass transition temperature (see e.g. [154,157]). The increase in molecular weight dependence of melt viscosity above a critical value M_c and the rubber like response at intermediate times and frequencies may serve as examples which demonstrate the existence of a temporary network of junction points in polymer melts of sufficiently high molecular weight. The entanglement model, the tube model of de Gennes [158] or the Doi-Edwards' model [159], which is essentially a combination of both, represent alternative descriptions of intermolecular coupling effects. In the cited models, the molecular weight of strands between entanglement points and the tube diameter, respectively, are characteristic parameters for the amount of intermolecular coupling whose precise nature is not completely understood at present. The entanglement molecular weight of PC has been measured to $M_e = 2500$ g/mole by Seitz [160]. This rather low value indicates that PC has to be considered as a highly entangled polymer.

In cooling polymer melts below T_g the overall chain conformation and hence the entanglement network is frozen-in. At small deformations hardly any molecular weight effects and consequently no entanglement effects are observed in glassy polymers. However, at large deformations the presence of entanglements imposes considerable restrictions on a further conformational rearrangement of chain segments. It is important to recognize that this effect can play an important role even for small elongations of the sample. In fact, large deformations may only occur in localized plastic zones such as craze fibrils or shear bands.

In the subsequent part of this section an attempt will be made to correlate intrinsic craze initiation with the limiting extensibility of chains between entanglement points. Even though this approach is entirely based on geometrical arguments and ignores time and temperature dependent relaxation processes which precede crazing, it provides a useful basis for a molecular description of craze initiation. The limits of extensibility of the entanglement network can be estimated by considering the maximal extension ratio, λ^e, of strands between entanglement points. λ^e is given by

$$\lambda^e = \frac{l_e}{\langle r_e^2 \rangle^{1/2}} = \frac{M_e}{M_0} l_0 \frac{1}{\langle r_e^2 \rangle^{1/2}} \tag{11}$$

where $\langle r_e^2 \rangle$ is the mean square end-to-end distance for a chain with entanglement molecular weight, M_e, and l_e is the length of the completely elongated chain. M_0 denotes the molecular weight of the monomer unit and l_0 is the length of the projection of the monomer unit on the chain axis. l_0 and $\langle r_e^2 \rangle$ can either be calculated from the chain conformation in the elongated and unperturbed state respectively, or may be evaluated from crystallographic and small-angle neutron scattering data, respectively. If $\langle r_e^2 \rangle^{1/2}$ is calculated from Equation (1) and inserted into Equation (11), λ^e can be written in the form

$$\lambda^e = \frac{l_0}{M_0 \beta} M_e^{1/2} . \tag{12}$$

Because of the very low entanglement molecular weight of PC, Equation (1) and, therefore, also Equation (12) can only be regarded as approximations. With the data for PC listed in Table 3 (for β see also Section 3.1), Equation (12) yields $\lambda^e = 2.0$. Data for PMMA and PS and the resulting values of λ^e are incled in Table 3. They will be discussed in Section 6.

Table 3. Instability Phenomena in Amorphous Polymers: Comparison between the Extension Ratio, λ^{II}, at the Initiation of Stress-Whitening and the Maximum Extension Ratio, λ^e, of Chains between Entanglement Points

Polymer	l_0 nm	M_e $\dfrac{g}{mole}$	β $nm\left(\dfrac{mole}{g}\right)^{\frac{1}{2}}$	λ^e Eq. [12]	λ^{II}
Polycarbonate	1.1	2490 [160]	0.11 [18, 19]	2.0	2.1 [14]
Poly(methyl methacrylate)	0.21	9150 [160]	0.076 [23]	2.7	2.7 [182]
Polystyrene	0.21	19100 [160]	0.067 [22]	4.1	—

In the discussion of the shear yielding and crazing behavior of PC (in Sect. 3.2 and 4.1), the existence of characteristic extension ratios has become apparent: (1) The extension ratio, λ^{sh}, after shear yielding referred to as natural draw ratio, (2) the extension ratio, λ_f^I, of craze I fibrils and (3) the extension ratio, λ^{II}, at craze II initiation.

Table 4. Comparison between the Extension Ratio, λ^{sh}, after Shear Yielding, the Extension Ratio, λ_f^I, of Craze I fibrils, the Extension Ratio, λ^{II}, at Craze II Initiation and the Maximum Extension Ratio, λ^e, of Chains between Entanglement Points

λ^{sh}	1.7	λ^{II}	2.1
λ_f^I	2.0	λ^e	2.0

The values of λ^{sh}, λ_f^I and λ^{II} are listed in Table 4 and compared with λ^e. In accordance with the findings of Donald and Kramer [76,111], λ_f^I compares well with λ^e, and λ^{sh} falls slightly below these values. In addition, Table 4 shows a very good agreement between λ^e and λ^{II} indicating that λ^e must be assumed to govern craze II initiation. It is tempting to conclude that intrinsic crazing in PC results from an instability of the entanglement network at large strains.

In order to test the model of intrinsic craze initiation it would be desirable to produce entanglement networks of different structures and study λ^{II} as a function λ^e. This may be achieved by pre-orientation of PC above T_g. In fact, there is much evidence that the entanglement network in polymer melts is modified increasingly with the magnitude of deformation [132,150,159,161]. The modification must be assumed to arise from the operation of two major mechanisms. The first mechanism involves chain slippage, the second one break-down of entanglement points. It should be noted that under high stresses, e.g. as observed at drawing temperatures close to T_g, these mechanisms may differ from those discussed in the slip-link model of Doi and Edwards [159]. If chain slippage or break-down of entanglement points causes λ^e to increase with the extension ratio, the same should be valid for λ^{II}. The measurements of craze initiation in pre-oriented PC samples, discussed in Section 4.1, are at least qualitatively in accordance with this assumption. In order to derive some quantitative agreement, the function $\lambda^e(\lambda_1)$ has to be known. This, in principle, requires a detailed molecular description of the deformed network which has not yet been given. In view of this difficulty some simple approaches have been made by modifying the basic relationships of rubber elasticity theory in order to fit in conformity with the experimental findings on uncrosslinked polymers above T_g. The birefringence, Δn_1, of uniaxially stretched rubber may be written in the form [129]:

$$\Delta n_1 = \Delta n_1^{max} \langle P_2(\xi) \rangle , \tag{13}$$

where Δn_1^{max} denotes the maximum birefringence resulting from complete orientation of the random links between cross-links. The orientation distribution function of the random links is approximated by the ensemble average of the second order Legendre polynomial in ξ, which is the cosine of the angle between the draw axis and the axis of a random link. In the Gaussian approximation region of rubber elasticity, $\langle P_2(\xi) \rangle$ is given by

$$\langle P_2(\xi) \rangle = \frac{1}{5N} (\lambda_1^2 - \lambda_1^{-1}) \tag{14}$$

where N denotes the number of random links between cross-links. Equations (13) and (14) yield the well-known formula

$$\Delta n_1 = \frac{\Delta n_1^{max}}{5N} (\lambda_1^2 - \lambda_1^{-1}) \tag{15}$$

According to Raha and Bowden [132], Ward et al. [162] and Zanker and Bonart [150] the birefringence of stretched PMMA as a function of the extension ratio can be put into a form very similar to Equation (15):

$$\Delta n_1 = \frac{\Delta n_1^{max}}{5N_0} e^{-k(\lambda_1 - 1)} (\lambda_1^2 - \lambda_1^{-1}) . \tag{16}$$

A comparison of Equations (15) and (16) suggests two distinct interpretations. In retaining the rubber strain function, it may be assumed that N increases with λ_1 according to

$$N(T, t, \lambda_1) = N_0(T, t) e^{k(T, t)(\lambda_1 - 1)} \tag{17}$$

where $N_0(T, t)$ denotes the number of random links in the unstretched polymer as a function of temperature T and drawing time t, and $k(T, t)$ is a parameter which does not depend on λ_1. Another way of interpreting Equation (16) consists in associating the exponential term with a modified rubber strain function [163]. Supposing the first interpretation were true λ^e may easily be calculated. Since

$$\frac{M_e(\lambda_1)}{M_e(1)} = \frac{N(\lambda_1)}{N_0} .$$

(18)

Equation (12) gives

$$\lambda^e(\lambda_1) = \lambda^e(1)\ e^{\frac{k}{2}(\lambda_1 - 1)}$$

(19)

According to the craze II concept outlined above the same Equation should hold for λ^{II}. Before comparing the theoretical and experimental values, the validity of Equation (16) for the deformation of PC above T_g has to be examined. It should be recognized that in the case of PC the assumption of Gaussian chains between entanglement points, inherent to Equation (14), can only be regarded as a rough approximation because of two reasons: Firstly, the number of random links between entanglement points must be considered to be very small. Secondly, the deformation range under investigation extends to values of λ_1 for which λ_1/λ^e is no longer small ($\lambda_1/\lambda^e \geq 0.5$). For this deformation range, a modified rubber strain function, derived from a series expansion of the inverse Langevin function, has been given by Treloar [164]. However, the distribution function of the end-to-end vectors of short real chains is not assumable to a simple mathematical representation as illustrated by an analysis of polymethylene chains of finite length by Jernigan and Flory [165]. In comparing various existing approximations, they found that the Gaussian one leads to a rather satisfying description of the real chains even up to values of λ_1/λ_e close to one. In view of these results and for the sake of simplicity the Gaussian approximation will be adopted and the data obtained for oriented PC will be analyzed according to Equation (16). It will further be assumed that Δn_1^{max} is independent of λ_1. This

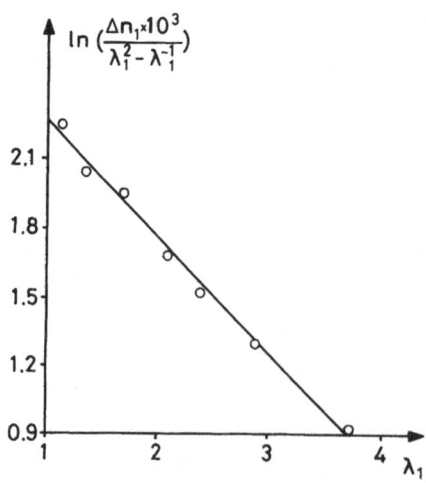

Fig. 13. Plot of $\ln\left(\dfrac{\Delta n_1 \times 10^3}{\lambda_1^2 - \lambda_1^{-1}}\right)$ against λ_1 for pre-oriented polycarbonate in accordance with Equation (16)

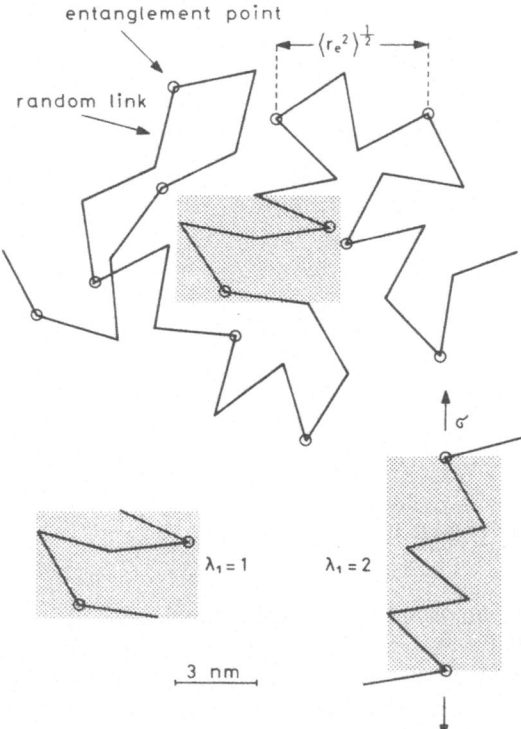

entanglement point

random link

$\langle r_e^2 \rangle^{\frac{1}{2}}$

$\lambda_1 = 1$ $\lambda_1 = 2$

3 nm

σ

Fig. 14. Schematic representation of the deformation of an average polycarbonate chain above T_g

amounts to neglecting the reduction of Δn_1^{max} due to the presence of terminal chains, i.e. chains attached to the entanglement network at one end only.

In decribing the deformation of PC by Equation (16), it is important to note that only the reversible part of deformation has to be taken for λ_1. In general, uncrosslinked polymers have a viscous component as well. The plot presented in Figure 13 shows that the birefringence data of PC stretched above T_g are well described by Equation (16). The constants appearing in this Equation are calculated to $\dfrac{\Delta n_1^{max}}{N_0} = 0.079$ and $k = 0.50$. In view of the approximations involved in Equation (16) both values should be handled with some caution. A schematic representation of the deformation of PC, based on the above results, is given in Figure 14.

The validity of Equation (16) being established, the craze initiation data may be analyzed according to Equation (19). The plot, $\ln \lambda^{II}$ versus λ_1, presented in Figure 15, demonstrates that Equation (19) gives a satisfactory description of intrinsic craze initiation. In view of the assumptions made above, the value of $k = 0.66$ calculated from the slope of the straight line is in sufficient agreement with the value of $k = 0.50$ determined from Equation (16). A possible reason for the fact that both values are not identical may be that a modified rubber strain function should be introduced. If this function causes Δn_1 to increase more rapidly with λ_1 as compared to the normal type of rubber strain function, the fit of Equation (16) to the experimental data will yield a higher value of k. The rubber strain function of Treloar, for example, is of that type.

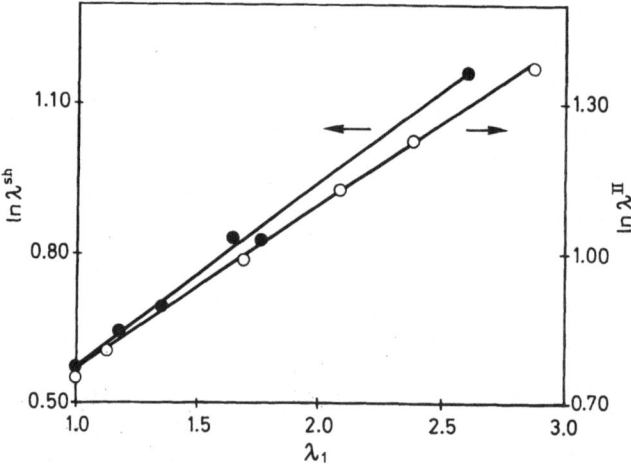

Fig. 15. Logarithmic plot of the total extension ratios at intrinsic craze initiation (T = 129 °C) and after shear yielding (T = 23 °C) λ^{II} and λ^{sh}, respectively versus the extension ratio, λ_1, after pre-orientation. Samples have been stretched parallel to the orientation axis

As previously pointed out, another important role of the entanglement network consists in determining shear yield arrest. Therefore, any modification of the entanglement network affecting λ^e should also be reflected in shear yielding. To assess the effect of pre-orientation on the shear yield arrest in PC, samples were stretched at room temperature under neck formation. The total extension ratio, λ^{sh}, of the material within the neck was measured as a function of λ_1. In agreement with the entanglement concept outlined above, λ^{sh} is found to be roughly proportional to λ^{II}. This is demonstrated in Figure 15, where the plot $\ln \lambda^{sh}$ versus λ_1 has been shifted on the ordinate to superimpose on the plot $\ln \lambda^{II}$ versus λ_1. The fact that $\ln \lambda^{sh}$ has to be shifted to higher values is not surprising since intrinsic craze initiation requires the build-up of large stresses, $\sigma^{II} > \sigma_D$, which necessarily result in further deformation of the material.

It is interesting to compare the shear yield behavior of PC with that of poly(ethylene terephthalate) (PET). For PET with different degress of pre-orientation, Ward et al. [130, 131, 147] measured a constant value for λ^{sh} which they associated with the existence of an entanglement network that remained unaffected during deformation. Thus, the behavior of PET seems to be different from that of PC in Figure 15, where λ^{sh} increases with λ_1. However, recent investigations of Engelaere et al. [166] have shown that λ^{sh} of PET may also increase with λ_1, the increase being more pronounced for samples with low molecular weight than with high molecular weight. This result is plausible since the relaxation of the deforming network must be assumed to occur by molecular weight dependent processes such as chain slippage and disintegration of entanglement points. For high molecular weight material, these processes will be shifted to large time scales or to higher temperatures. In the modified rubber model, the molecular weight dependence of the network relaxation can be taken into account by assuming k = k(t, T, M) in Equations (16) and (17).

An important consequence of the above results with respect to crazes I is obvious. In fact, provided the craze I concept of Donald and Kramer [111] is also applicable to pre-oriented material, the extension ratio and hence the volume fraction of craze I fibrils may be predicted either from Equation (16) together with Figure 13 or from experimental data such as presented in Figure 15.

It must be assumed that the existence of an entanglement network is not only reflected in a limiting extension but also in a limiting orientation of chains between entanglement points. Birefringence measurements which were conducted on un-oriented and pre-oriented PC stretched below T_g are qualitatively in accordance with this assumption. In fact, the birefringence, Δn^{II}, at intrinsic craze initiation varies little with drawing temperature, strain rate and pre-orientation. The effect of the latter parameter is illustrated in Figure 16 where the birefringence is plotted as a function of the extension ratio for unoriented and pre-oriented samples. It is clearly seen that samples with different degrees of pre-orientation exhibit approximately the same birefringence of $\Delta n^{II} = 0.045$ at intrinsic craze initiation. However, further investigations are necessary to substantiate these results since the measured birefringence may be a complicated function of chain orientation. In fact, if PC is stretched below T_g the anisotropy of the polarizability of the chain segments, and hence Δn_I^{max} in Equation (13), may be affected by the deformation process.

In this section, much evidence has been derived that intrinsic craze initiation results from an instability of the entanglement network when stretched to its limiting extension. The activation of this instability depends on a large number of parameters. In addition to those which have already been discussed in this paper, e.g. temperature, strain rate and pre-orientation, other parameters such as the molecular weight and the molecular weight distribution may be important. For instance by decreasing the molecular weight, PC becomes brittle and fails at $\lambda < \lambda^{II}$. The instability of the entanglement network observed in PC may arise from the activation of several mechanisms such as chain slippage, disintegration of entanglement points or chain rupture. Provided the first two mechanisms are involved they cannot be considered identical to those treated in the reptation model of de Gennes [158] and in the sliplink model of Doi and Edwards [159]. In fact, these models do not take into account

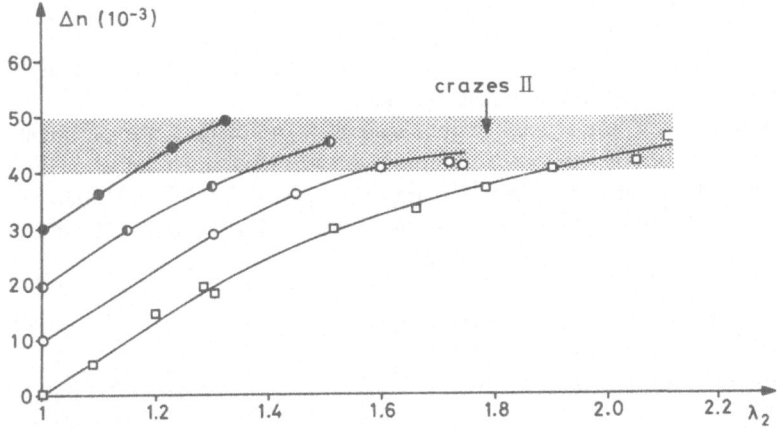

Fig. 16. Birefringence, Δn, against the extension ratio, λ_2, for unoriented and pre-oriented poly-carbonate stretched parallel to the orientation axis

plastic deformation processes initiated by stress-activated conformational rearrangement of chain segments. However, these processes must be assumed to occur under the conditions of intrinsic craze formation.

Since intrinsic crazes are observed in PC stretched under high stresses in a temperature region close to T_g, the model of Gent [167] in which crazing is associated with a stress-induced glass transition seems attractive. However, even for intrinsic crazing there is no conclusive evidence in support of this model at present. To examine the relaxation mechanism of chains during intrinsic craze formation dynamic mechanical measurements on highly stressed samples may be extremely useful.

5 Craze Microstructure as Revealed by Small-Angle X-Ray Scattering

5.1 Scattering Theory

SAXS has proved to be a very powerful tool for a quantitative analysis of the craze microstructure. This is not surprising since characteristic craze parameters such as the fibril diameters and the interfibrillar spacings frequently fall into the range of $1–5 \times 10^2$ nm covered by SAXS. The general theory of SAXS has extensively been treated in the literature (see e.g. [168–171]). Some basic elements of the SAXS theory, which are important in understanding the scattering from crazes, will be outlined in this section. For further details, the reader is referred to several publications [172–178] on the scattering from systems with fibre symmetry and, in particular, to recent publications of Paredes and Fischer [110] and Brown and Kramer [15] on the scattering from crazes.

The scatering of X-rays arises from local fluctuations of the electron density. The scattering theory, which relates these fluctuations to the measured scattering intensity, is generally derived under the following assumptions:
1) incoherent (Compton-) scattering can be neglected
2) reflections do not contribute to the measured intensity
3) multiple scattering does not occur (Born-approximation)
4) the sample dimensions are small as compared to the distance between the sample and the detector system (Fraunhofer-approximation).
Assumptions 1) and 4) are not crucial in the ordinary type of SAXS experiment. However, the justification of the two remaining assumptions must be examined carefully, especially if the scattering arises from voids (see e.g. [15, 177]). Under the assumptions 1)–4), the solution of Maxwell's Equations gives for the amplitude of the scattered waves:

$$A(\vec{s}) = A_e \iiint_V \Delta\varrho_e(\vec{x}) \exp(-2\pi i \vec{s}\vec{x}) \, d\vec{x} = A_e \hat{F}_3\{\Delta\varrho_e(\vec{x})\} \tag{20}$$

where the reciprocal vector $\vec{s}$ is defined by

$$\vec{s} = \frac{1}{\lambda}(\vec{e} - \vec{e}_0) \tag{21}$$

λ is the wavelength of radiation and $\vec{e}$ and $\vec{e}_0$ are unit vectors in the direction of the incident and scattered waves, respectively. s is then equal to

$$s = |\vec{s}| = \frac{2}{\lambda} \sin \theta \qquad (22)$$

if 2θ denotes the scattering angle. (Note: The definition of s does not follow a previous publication [11]). A_e represents the scattering amplitude of one electron. $\Delta\varrho_e(\vec{x})$ is the difference between the electron density $\varrho_e(\vec{x})$ at a point defined by the vector $\vec{x}$ and the average electron density, the average being taken over the whole sample volume, i.e.

$$\Delta\varrho_e(\vec{x}) = \varrho_e(\vec{x}) - \langle\varrho_e(\vec{x})\rangle_v . \qquad (23)$$

In Equation (20) $\Delta\varrho_e(\vec{x})$ is related to $A(\vec{s})$ by a three dimensional Fourier transformation designated by the symbol $\hat{F}_3$. Therefore, the vectors $\vec{s}$ form a three dimensional Fourier or reciprocal space with respect to the real-space vectors $\vec{x}$. The scattering intensity in electron units is calculated from Equation (20) as

$$I(\vec{s}) = \frac{|A(\vec{s})|^2}{|A_e|^2} = |\hat{F}_3\{\Delta\varrho_e(\vec{x})\}|^2 = \hat{F}_3\{\Delta\varrho_e^{*2}(\vec{x})\} \qquad (24)$$

$\Delta\varrho_e^{*2}(\vec{x})$ designates the self convolution or auto correlation of $\Delta\varrho_e(\vec{x})$ which is by definition

$$\Delta\varrho_e^{*2}(\vec{x}) = \underset{v}{\iiint} \Delta\varrho_e(\vec{x} + \vec{y}) \, \Delta\varrho_e(\vec{y}) \, d\vec{y} \qquad (25)$$

According to Equation (24) $\Delta\varrho_e^{*2}(\vec{x})$ is related to $I(\vec{s})$ by a three dimensional Fourier transformation. Since the phase of the scattered waves is lost in passing from $A(\vec{s})$ over to $I(\vec{s})$ only the self convolution of $\varrho_e(\vec{x})$ and not $\varrho_e(\vec{x})$ itself as in Equation (20) is obtained from the scattering experiment. Therefore, additional information is needed for a unique structural description.

By introducing a function

$$\gamma(\vec{x}) = \frac{\Delta\varrho_e^{*2}(\vec{x})}{\Delta\varrho_e^{*2}(0)} = \frac{\Delta\varrho_e^{*2}(\vec{x})}{V\langle\Delta\varrho_e^2(\vec{x})\rangle_v} \qquad (26)$$

which is normalized at the origin, i.e. $\gamma(0) = 1$, Equation (24) may be written in the form

$$I(\vec{s}) = V\langle\Delta\varrho_e^2\rangle_v \, \hat{F}_3\{\gamma(\vec{x})\} = V\langle\Delta\varrho_e^2\rangle_v \underset{v}{\iiint} \gamma(\vec{x}) \exp(-2\pi i \vec{s}\vec{x}) \, d\vec{x} \qquad (27)$$

$\gamma(\vec{x})$ is called characteristic or correlation function and describes the probability of finding a fluctuation of the electron density $\Delta\varrho_e(\vec{x})$ at the point $\vec{x}$ if there is a fluctuation $\Delta\varrho_e(0)$ at the origin.

Provided the sample volume is sufficiently large, $\gamma(\vec{x})$ is obtained from $I(\vec{s})$ by inverse Fourier transformation, the integration being carried out over the whole reciprocal space:

$$\gamma(\vec{x}) = (V\langle\Delta\varrho_e^2\rangle_v)^{-1} \, \hat{F}_3^{-1}\{I(\vec{s})\} = (V\langle\Delta\varrho_e^2\rangle_v)^{-1}x$$
$$\iiint I(\vec{s}) \exp(2\pi i\vec{s}\vec{x}) \, d\vec{s} . \tag{28}$$

In studying uniaxially deformed polymers, we can confine ourselves to systems with cylindrically symmetric correlation functions. It should be recognized that this does not impose any symmetry on the electron density distribution $\varrho_e(\vec{x})$. Introducing cylindrical coordinates (x_r, x_3, x_φ) and (s_r, s_3, s_φ) and integrating over the azimutal angles x_φ and s_φ Equations (27) and (28) become

$$I(s_r, s_3) = V\langle\Delta\varrho_e^2\rangle_v \, 2\pi \int_0^\infty \int_{-\infty}^\infty x_r\gamma(x_r, x_3) \, J_0(2\pi x_r s_r) \exp(-2\pi i x_3 s_3) \, dx_r \, dx_3$$

$$\gamma(x_r, x_3) = (V\langle\Delta\varrho_e^2\rangle_v)^{-1} \, 2\pi \int_0^\infty \int_{-\infty}^\infty s_r I(s_r, s_3) \, J_0(2\pi x_r s_r) \exp(2\pi i x_3 s_3) \, ds_r \, ds_3 \tag{29}$$

where J_0 is the zero-order Bessel Function. The correlation function $\gamma(x_r)$ defined by $\gamma(x_r) \equiv \gamma(x_r, 0)$ is of particular concern in analyzing the craze microstructure. In fact, since crazes form perpendicular to the draw axis, $\gamma(x_r)$ is related to density fluctuations within the craze plane. Therefore, $\gamma(x_r)$ contains information on intra- and interfibrillar correlations, i.e. on the fibril diameter and interfibrillar distances. $\gamma(x_r)$ is calculated from Equation (29) as

$$\gamma(x_r) = (V\langle\Delta\varrho_e^2\rangle_v)^{-1} \, 2\pi \int_0^\infty s_r I_p(s_r) \, J_0(2\pi x_r s_r) \, ds_r$$
$$= (V\langle\Delta\varrho_e^2\rangle_v)^{-1} \, \hat{F}_2^{-1}\{I_p(s_r)\} \tag{30}$$

where

$$I_p(s_r) \equiv \int_{-\infty}^\infty I(s_r, s_3) \, ds_3 \tag{31}$$

is generally called the equatorial scattering component. $I_p(s_r)$ is related to $\gamma(x_r)$ by a two dimensional Fourier transformation. Inversing the Fourier transformation gives

$$I_p(s_r) = V\langle\Delta\varrho_e^2\rangle_v \, \hat{F}_2\{\gamma(x_r)\} = V\langle\Delta\varrho_e^2\rangle_v \, 2\pi \int_0^\infty x_r\gamma(x_r) \, J_0(2\pi x_r s_r) \, dx_r . \tag{32}$$

If both sides of Equation (30) are evaluated for $x_r = 0$, it follows

$$2\pi \int_0^\infty s_r I_p(s_r) \, ds_r = V\langle\Delta\varrho_e^2\rangle_v \tag{33}$$

The integral in the left side of Equation (33) is called invariant since it does not depend on structural details of the specimen. In the case of a two-phase system $\langle \Delta \varrho_e^2 \rangle_v$ is given by

$$\langle \varrho_e^2 \rangle_v = (\delta \varrho_e)^2 \, v(1 - v) \tag{34}$$

where v is the volume fraction of one phase and $\delta \varrho_e$ the difference in electron density between the two phases. Equations (33) and (34) may be used to derive some quantitative information on the craze microstructure. In fact, in the SAXS region, the equatorial scattering component of crazes primarily results from electron density fluctuations associated with the internal craze structure. Because of the large craze dimensions, the scattering from the overall craze is situated at much smaller angles. With this assumption Equations (33) and (34) yield:

$$2\pi \int_0^\infty s_r I_p(s_r) \, ds_r = V_{cr} \varrho_e^2 v_f (1 - v_f) \tag{35}$$

where V_{cr} denotes that part of the sample volume, V, which is occupied by crazes and ϱ_e is the electron density within the craze fibrils. Taking to a very good approximation, for the densities of the craze fibrils and of the matrix, the density, ϱ, of the sample at craze initiation yields for V_{cr}:

$$V_{cr} = \frac{1 - \dfrac{\varrho_{cr}}{\varrho}}{1 - v_f} \, V \tag{36}$$

where ϱ_{cr} denotes the density of the crazed sample. Equations (35) and (36) may be combined to evaluate the volume fraction of craze fibrils, v_f, from the invariant:

$$2\pi \int_0^\infty s_r I_p(s_r) \, ds_r = V \varrho_e^2 \left(1 - \frac{\varrho_{cr}}{\varrho} \right) v_f \tag{37}$$

Evidently, the application of Equation (37) requires ϱ_{cr} to be sufficiently different from ϱ, i.e. the number of crazes must be large. Extrinsic crazes do, generally, not fulfill this requirement. However, as shown in Figure 2, intrinsic crazes in PC cause a marked decrease in density and, hence v_f may be evaluated from the above Equation.

Frequently, SAXS experiments are performed with slit collimated systems. We assume that the conditions are such that the incident beam can be regarded as infinite and of constant intensity in the direction of the x_3-axis and inifinitely narrow in the direction of the x_2-axis of a cartesian coordinate system (x_1, x_2, x_3). The cylinder axis is supposed to be parallel to the x_3-axis and the intensity measurements are performed parallel to the x_2-axis. In the small-angle region we have, by replacing $\sin \theta$ by its argument in Equation (22)

$$s_r \approx s_2 \approx \frac{x_2}{\lambda R} \quad \text{and} \quad s_3 \approx \frac{x_3}{\lambda R} \tag{38}$$

where R is the distance between the sample and the detector. Under the slit collimation, defined above, the "smeared" scattering intensity designated by a tilde is given by

$$\tilde{I}(x_2) = \int_{-\infty}^{\infty} I(x_2, x_3) \, dx_3 = (\lambda R) \int_{-\infty}^{\infty} I(s_r, s_3) \, ds_3 = (\lambda R) \, I_p(s_r). \tag{39}$$

We note that $\tilde{I}(x_2)$ is proportional to $I_p(s_r)$, therefore, $I_p(s_r)$ is directly obtained by measuring with a slit collimated system. It should be realized that an equivalent relationship does not exist if the intensity is measured along the cylinder-axis with the incident beam being infinitely perpendicular to this axis.

Valuable information on crazes can be derived from their scattering behavior at large values of s. Following a derivation given by Porod [168] it may be shown that a two-phase system with sharp phase boundaries and a cylindrically symmetric correlation function obeys the asymptotic law

$$\lim_{s_r \to \infty} I_p(s_r) = \frac{1}{4\pi^3} \, (\delta\varrho_e)^2 \, V \, \frac{\zeta}{\chi} \, \frac{1}{s_r^3} \tag{40}$$

which may be written together with Equations (33) and (34) in another form

$$\frac{\lim\limits_{s_r \to \infty} s_r^3 I_p(s_r)}{2\pi \int\limits_0^{\infty} I_p(s_r) \, s_r \, ds_r} = \frac{1}{4\pi^3} \, \frac{\zeta}{\chi} \, \frac{1}{v(1-v)} . \tag{41}$$

In the above Equations χ is the total area created by the intersection of a plane perpendicular to the cylinder axis and passing through the origin. ζ denotes the interfacial length of the two dimensional cells in this plane. Crazes may be modeled by a system of parallel cylinders. For a realistic description of the craze microstructure, a distribution in fibril diameter, D, must be taken into account. This yields

$$\frac{\zeta}{\chi} = 4v_f \, \frac{\langle D \rangle}{\langle D^2 \rangle} \tag{42}$$

where the average is taken over all fibril diameters. With the definition

$$\bar{D} \equiv \frac{\langle D^2 \rangle}{\langle D \rangle} \tag{43}$$

Equations (41) and (42) give

$$\frac{\lim\limits_{s_r \to \infty} s_r^3 I_p(s_r)}{2\pi \int\limits_0^{\infty} I_p(s_r) \, s_r \, ds_r} = \frac{1}{\pi^3(1-v_f)} \, \frac{1}{\bar{D}} . \tag{44}$$

Since v_f is known from Equation (37), $\bar{D}$ may be evaluated from Equation (44).

The above approach is very general in the sense that the correlation function $\gamma(x_r)$ contains both intra- and interfibrillar correlations. To discuss the scattering pattern of crazes in more detail, another approach is instructive. The existence of a well defined scattering element (craze fibril) suggests distinguishing intra- from interfibrillar correlations and writing the scattering intensity in the form [169]:

$$I(s_r, s_3) = (\langle |F(s_r, s_3)|^2 \rangle - |\langle F(s_r, s_3)\rangle|^2)$$
$$+ |\langle F(s_r, s_3)\rangle|^2 \, G(s_r) \tag{45}$$

where $F(s_r, s_3)$ is the form factor of a craze fibril and $G(s_r)$ is a function which describes the interfibrillar interferences. $G(s_r)$ is related to the interfibrillar correlation function $g(x_r)$ by the Equation

$$G(s_r) = 1 + \frac{2\pi}{f} \int_0^\infty [g(x_r) - 1] \, J_0(2\pi s_r x_r) \, x_r \, dx_r \tag{46}$$

where f is the average area per fibril in the craze plane. If craze fibrils are modeled by cylinders with diameter D and length L, the form factor is given by

$$F(s_r, s_3) = \varrho_e \frac{D^2 \pi}{2} \frac{J_1(\pi D s_r)}{\pi D s_r} \frac{\sin(\pi L s_3)}{\pi s_3} \tag{47}$$

where ϱ_e is the electron density within a cylinder and J_1 the first order Bessel function. The existence of a distribution in fibril dimensions, taken into account in Equation (45) by averaging over all form factors, has several consequences for the scattering pattern of crazes: 1. The interfibrillar correlation function, $g(x_r)$ cannot be derived by Fourier inversion of Equation (45) without any assumption on the distribution function of the fibril dimensions. 2. The fluctuation of the form factor causes some background scattering corresponding to the first term on the right side of Equation (45). 3. The oscillations of the form factor, predicted by Equation (47) are smeared by averaging and, hence, crazes generally do not exhibit intrafibrillar scattering peaks. On the other hand, the craze microstructure necessarily produces interfibrillar correlations which may be sufficiently strong to yield a peak maximum in the scattering intensity. For PC crazes, this maximum has been observed by Paredes and Fischer [110], Dettenmaier and Kausch [11] and Brown and Kramer [15]. Paredes and Fischer have shown that its position, s_{max}, is related to the volume fraction of craze fibrils and to the average fibril diameter by a very simple Equation:

$$s_{max} = \frac{c \sqrt{v_f}}{\bar{D}} \tag{48}$$

where c is a constant that is virtually independent of the precise geometrical arrangement of the craze fibrils. For instance, if the fibrils are supposed to occupy the positions on a hexagonal lattice, it follows from Bragg's law that $c = 1.212$ which gives, inserted into Equation (48)

$$s_{max} = 1.212 \frac{\sqrt{v_f}}{\bar{D}} . \tag{49}$$

The fact that for PC crazes, Equations (44) and (49) gave consistent results ([110, 15], see also Sect. 5.2) provides strong support for the scattering model outlined above.

5.2 Experimental Results

The equatorial SAXS curve of extrinsic crazes in PC is shown in Figure 17 for a sample which has been stretched at $T = 129\ °C$ to $\lambda = 1.8$, i.e. to $\lambda < \lambda^{II}$. The scattering curve exhibits the well pronounced interfibrillar interference maximum discussed in the previous section. From the position, s_{max}, of this maximum the average fibril diameter, D, may be evaluated. If the volume fraction of craze I fibrils is assumed to be $v_f^1 = 0.5$ (see Section 3.2), Equation (49) yields $D = 37$ nm, which compares well with results obtained under approximately the same drawing conditions by Paredes and Fischer [110] and Brown and Kramer [15].

In the presence of intrinsic crazes the scattering intensity increases drastically as shown in Figure 17 for a sample which has been stretched to $\lambda = 2.5$, i.e. to $\lambda > \lambda^{II}$. It is clearly seen that intrinsic crazes also exhibit an interference maximum which is, however, shifted to very small values of the reciprocal vector. The question arises as to whether the SAXS behavior of crazes II may be interpreted, in some analogy to the scattering behavior of crazes I, on the basis of the craze model outlined in the previous section. In this model it has been assumed that the craze microstructure can be described as a two-phase system with a cylindrically symmetric correlation function and with sharp phase boundaries. The existence of a cylindrically symmetric correlation function follows from the fact that the scattering pattern does not change when the specimen is tilted around the draw axis provided correction is made for the changing sample volume. The existence of sharp phase boundaries is confirmed by the Porod plot of the scattering curve. As predicted by Equation (40) the plot $\tilde{I}s^3$ versus s^3,

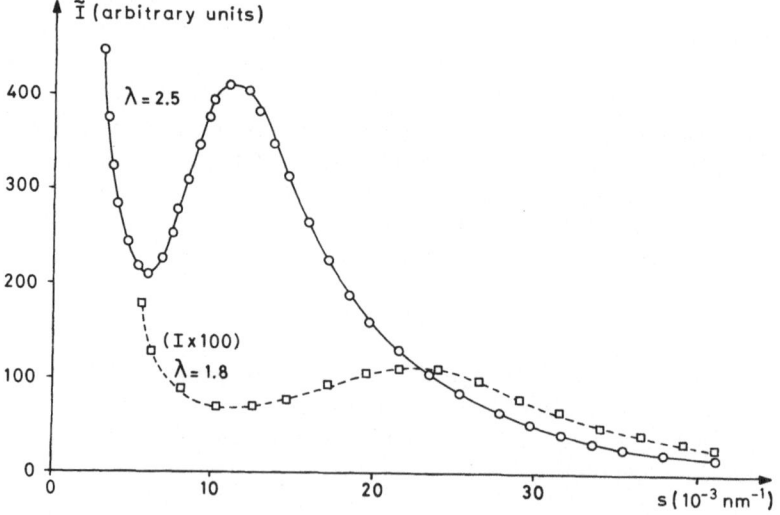

Fig. 17. Equatorial SAXS curves of crazes I (- - - -) and crazes II (———) in polycarbonate. Samples have been stretched to $\lambda = 1.8$ and to $\lambda = 2.5$, respectively

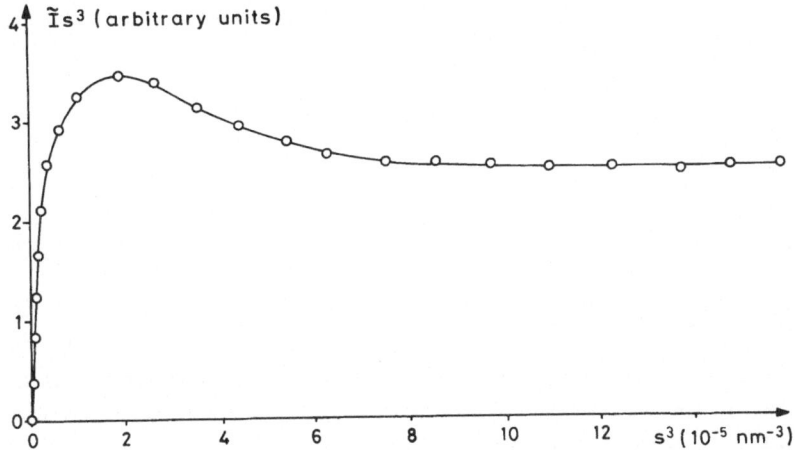

Fig. 18. Porod plot $\tilde{I}s^3$ versus s^3 of the equatorial SAXS curve of crazes II in polycarbonate

shown in Figure 18, attains a constant value for large reciprocal vectors. On these grounds, Equations (37), (44) and (49) may be used to derive some quantitative information on the craze microstructure. From the invariant (Equation (37)), the volume fraction of fibrils within a craze is calculated to $v_f^{II} = 0.76$. Since the craze fibrils contract after stress relief the value measured for v_f may be overestimated. However, this effect should not be very strong as discussed in Section 4.1. The value of v_f^{II} is surprisingly high and must be interpreted in terms of a random close package of craze fibrils. In fact, Stillinger et al. [179] found experimentally $v = 0.81$ for a two-dimensional random close packing of uniform hard disks. A hexagonal close packing yields the maximum value of $v = 0.91$. It is clear that the very dense packing of craze fibrils causes pronounced interfibrillar correlations. On the other hand, it should be realized that the distribution in fibril diameter smears these correlations and, therefore, the equatorial scattering curve of crazes II exhibits only one broad peak (Figure 17). v_f^{II} may be used to evaluate the extension ratio, λ_f^{II}, of craze II fibrils, which is given by

$$\lambda_f^{II} = \frac{\lambda^{II}}{v_f^{II}} . \tag{50}$$

With $v_f^{II} = 0.76$ and $\lambda^{II} = 2.1$, Equation (50) yields $\lambda_f^{II} = 2.76$. The result that

$$\frac{\lambda_f^{II}}{\lambda^{II}} < \frac{\lambda_f^{I}}{\lambda^{I}} \approx \lambda_f^{I} \tag{51}$$

seems plausible since intrinsic crazes, contrary to extrinsic crazes, are initiated in a highly deformed matrix. Therefore, considerable constraints are imposed on the further deformation of the material. The value of v_f^{II} may also be used to derive the volume fractions of crazes, craze fibrils and voids within the sample, namely h_{cr}, h_f and h_v, respectively. These quantities are given by

$$h_{cr} = \frac{1 - \dfrac{\varrho_{cr}}{\varrho}}{1 - v_f}, \tag{52}$$

$$h_f = h_{cr} v_f \tag{53}$$

$$h_v = h_{cr}(1 - v_f) \tag{54}$$

The values of $h_{cr} = 30\%$, $h_f = 23\%$ and $h_v = 7\%$ obtained from the above Equations demonstrate that intrinsic crazes occupy a considerable part of the sample volume, and that most of this volume fraction consists of craze fibrils.

Since v_f^{II} is known, either of the Equations (44) and (49) may be used to evaluate the fibril diameter. The fact that both Equations yield exactly the same value of $\bar{D} = 94$ nm strongly supports the scattering model outlined in the previous section. For crazes II, the fibril diameter turns out to be much larger than for crazes I. Some evidence will be presented subsequently that the difference in fibril diameter primarily reflects the distinct stress-srain state of the matrix at craze I and II initiation.

The meridional scattering curve of intrinsic crazes has not been analyzed in detail. However, it may be worth mentioning that this scattering component continuously increase up to very high intensities at small reciprocal vectors. The same type of scattering behavior has previously been reported for extrinsic crazes [15, 110]. Brown and Kramer [15] assumed that internal reflections at the craze-matrix boundary strongly contribute to the intensity measured along the meridian.

In discussing the structure of intrinsic crazes in PC it should be pointed out that the material within the fibrils is completely amorphous. Both wide-angle X-ray scattering and DSC did not give any evidence for the existence of crystalline regions in

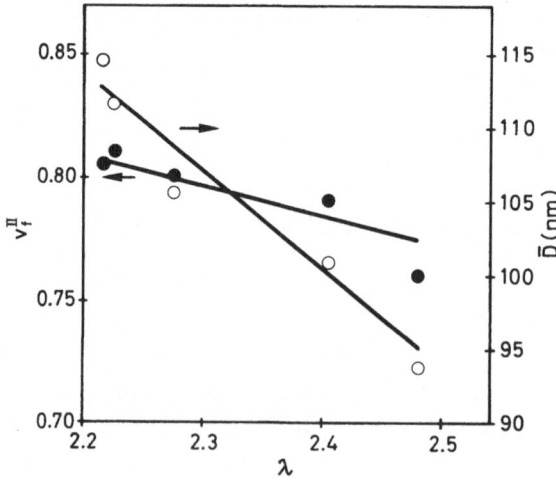

Fig. 19. Volume fraction, v_f^{II}, of fibrils within crazes II and average fibril diameter, $\bar{D}$, as a function of the extension ratio, λ

crazed PC. This result is not surprising in view of the fact that PC is extremely difficult to crystallize (see Sect. 3.1).

The objective of the following part of this section is to assess the effect of different parameters on the craze microstructure. In these studies the SAXS curve of each sample has been subjected to the data analysis outlined above.

The craze microstructure of samples stretched to different extension ratios $\lambda > \lambda^{II}$ has been investigated in Figure 19. It may easily be shown that the observed decrease of $\bar{D}$ with λ cannot be attributed to fibril creep. Otherwise, the relation

$$v_f^{II}(\lambda) \sim \bar{D}^2(\lambda) \tag{55}$$

should hold and, consequently, v_f^{II} should decrease more rapidly with increasing λ than $\bar{D}$. Instead, v_f^{II} decreases only slightly indicating that fibril creep plays only a minor role in craze growth. A possible explanation which accounts for the measured decrease in fibril diameter would be to assume that with increasing extension ratio fibrils split into smaller ones. Observations in support of this mechanism have been reported by Beavan et al. [180, 181].

To discuss the effect of other parameters on the craze microstructure an arbitrary reference state with respect to λ must be defined. This has been done by taking $\lambda = 1.05 \lambda^{II}$.

Because of the very narrow temperature range in which intrinsic crazes are observed it is experimentally more convenient to study the volume fraction of craze fibrils and the fibril diameter as a function of strain rate rather than a function of temperature. The experimental results are presented in Figures 20 and 21. Figure 20 shows that the volume fraction of fibrils is independent of the strain rate and amounts to $v_f^{II} = 0.79$. As noted previously, this value seems to suggest that v_f^{II} is governed by the formation of a random close packing of fibrils which results in a minimum extension of the matrix material when transformed into the fibrillar craze structure. Contrary to the volume fraction of craze fibrils, the fibril diameter is strain rate dependent as shown in Figure 21. In the case of crazes I the strain rate dependence of the fibril diameter, $\bar{D}$, emerges from the strain rate dependence of the lower yield stress, σ_D^I, since $\sigma_D^I \bar{D}$ remains constant as shown by Paredes and Fischer ([110, 112], see also Sect. 3.2). In adopting a similar approach to crazes II, it

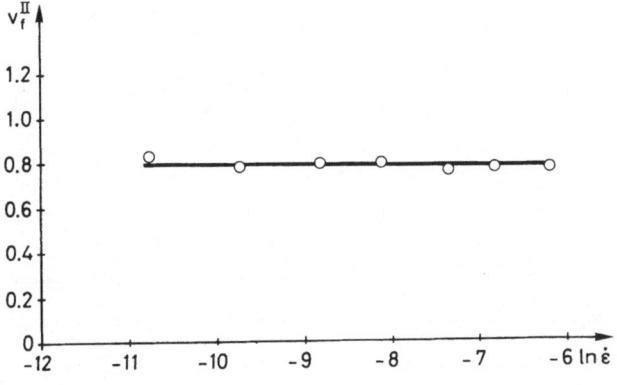

Fig. 20. Volume fraction, v_f^{II}, of fibrils within crazes II as a function of the logarithm of the true strain rate, $\dot{\varepsilon}$

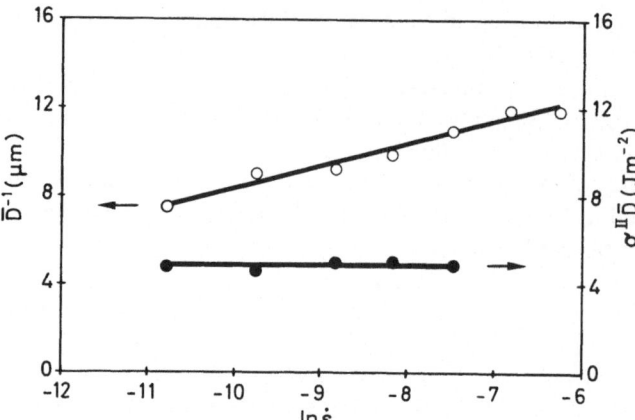

Fig. 21. Inverse of the average fibril diameter, $\bar{D}$, and product of $\bar{D}$ with the true stress, σ^{II}, at craze II initiation as a function of the logarithm of the true strain rate, $\dot{\varepsilon}$

should be noted that the local stress, $\sigma_D{}^{II}$, is essentially unknown and may be considerably lower than the tensile stress, σ^{II}, measured at intrinsic craze initiation. Thus, it is not surprising that the fibril diameters of intrinsic and extrinsic crazes are largely different. However, it is interesting to note from Figure 21 that $\sigma^{II}\bar{D}$ is independent of the strain rate $\dot{\varepsilon}$, as would be expected if σ^{II} and $\sigma_D{}^{II}$ had approximately

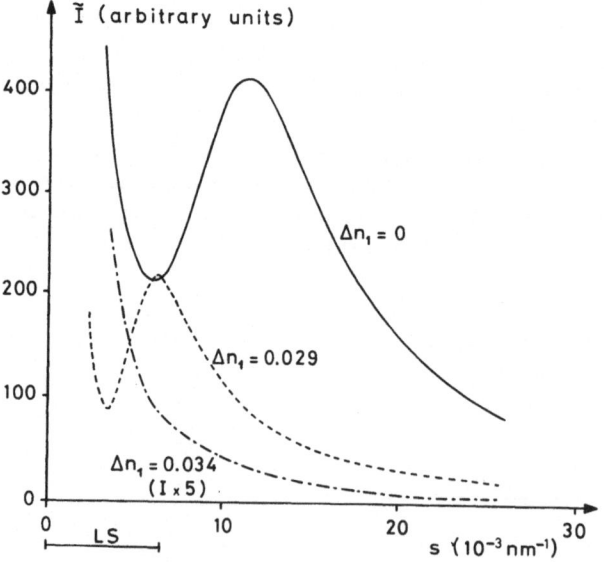

Fig. 22. Equatorial SAXS curves of pre-oriented polycarbonate. Parameter is the birefringence, Δn_1, after pre-orientation. For comparison, the range of the reciprocal vector, $s = \dfrac{2}{\lambda} n \sin \Theta$, covered by light-scattering is indicated on the abscissa ($n = 1.6, \lambda = 488$ nm, $0° \leqq 2\Theta \leqq 180°$)

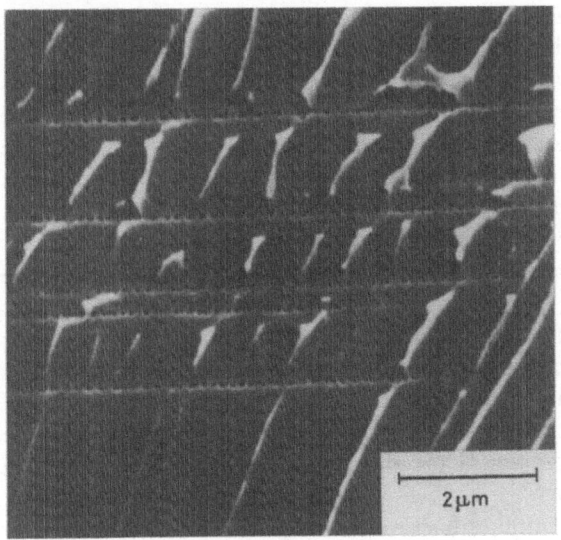

Fig. 23. Scanning electron micrograph of intrinsic cavities in pre-oriented polycarbonate ($\Delta n_I = 0.034$)

the same stain rate dependence. From the Equation $\sigma^{II}D = $ const., and from Figure 9 it follows in accordance with Figure 21 that $\bar{D}^{-1}$ is a linear function of $\ln \dot{\varepsilon}$.

Equatorial SAXS curves of pre-oriented PC are shown in Figure 22. With increasing pre-orientation the scattering maximum is first shifted to smaller reciprocal vectors and then disappears indicating that intrinsic crazes lose their regular fibrillar microstructure. This result is confirmed by the scanning electron micrograph shown in Figure 23. Even though well defined fibrils are not observed in the highly pre-oriented sample a planar organization of voids is clearly visible.

The SAXS curves which exhibit a scattering maximum have been analyzed with respect to the volume fraction of fibrils, v_f^{II}, and the average fibril diameter, $\bar{D}$.

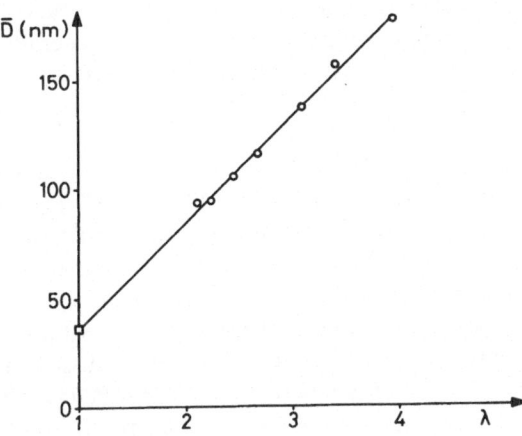

Fig. 24. Average fibril diameter, $\bar{D}$, as a function of the total extension ratio λ at craze initiation in unoriented and pre-oriented polycarbonate: crazes I ($\square$), crazes II ($\bigcirc$)

v_f^{II} remains virtually constant as a function of the pre-orientation. However, as shown in Figure 24, $\overset{\circ}{D}$ increases linearly with the total extension ration, λ, at craze initiation. It is interesting to note that the fibril diameter of $\overset{\circ}{D} = 36$ nm which is obtained by extrapolating the straight line to $\lambda = 1$ is in excellent agreement with the value of $\overset{\circ}{D} = 37$ nm measured for extrinsic crazes. This result seems to indicate that the different fibril diameter of extrinsic and intrinsic crazes primarily reflects the distinct extension ratios λ^I and λ^{II} of the matrix at craze initiation. However, further investigations are necessary to substantiate the above result.

It has previously been mentioned in Section 4.1 that intrinsic crazes which are produced in pre-oriented samples may exhibit color phenomena when irradiated by white light. The SAXS data of Figure 22 suggest the following interpretation of these phenomena. Samples with sufficiently large pre-orientation exhibit an interfibrillar interference maximum that overlaps with the light-scattering region as indicated on the abscissa of Figure 22. In the light-scattering region the overlap occurs at large scattering angles and is most pronounced for short wavelengths, i.e. for the blue spectrum of the incident light. Since these waves are also scattered most intensively, the sample appears blue in the direction of backward scattering. The complementary yellow color is observed in the forward direction.

The above results suggest the usefulness of a systematic analysis of the craze microstructure by means of light-scattering. In fact, this method could provide valuable information which is not available by SAXS. For instance, the optical anisotropy of the craze fibrils may be derived from the depolarized light-scattering component.

6 Intrinsic Cavitation in other Polymers

In this section evidence will be presented that the intrinsic crazing of PC is related to the existence of a general mode of cavitational plasticity in glassy polymers. PMMA [14, 182)] and PS are further examples of polymers in which cavitation occurs at high stresses and strains. In the case of PMMA it was observed that the presence of small amounts of monomers or oligomers greatly enhances cavity formation. It is believed that the major role of these molecules is to generate defects which allow local plastic deformation to occur. The studies on PS were conducted on pre-oriented samples to suppress the formation of surface crazes which generally cause the premature rupture of the specimen. The pre-orientation process consisted in stretching the material at 130 °C with $\dot{\varepsilon} = 0.01$ s^{-1} to an extension ratio of $\lambda_1 = 2.4$.

Figure 25 shows that the tensile behavior of PMMA and PS, stretched some 10–30 °C below T_g, is very similar to that of PC. Like PC both PMMA and PS exhibit extensive stress-whitening at high extension ratios, λ^{II}, where the peak or inflection of the stress-strain curve is observed. In Table 3, the value of λ^{II} measured for PMMA is compared to the maximum extension ratio, λ^e, of chains between entanglement points. The fact that $\lambda^{II} = \lambda^e$ provides further evidence that the cavitation in PMMA must be attributed to the same deformation mechanism as discussed in Section 4.2 for PC. The value of $\lambda^{II} = \lambda_1\lambda_2^{II} = 6.8$ measured for pre-oriented PS is higher than the theoretical value of $\lambda^e = 4.1$ included in Table 3. This result is not

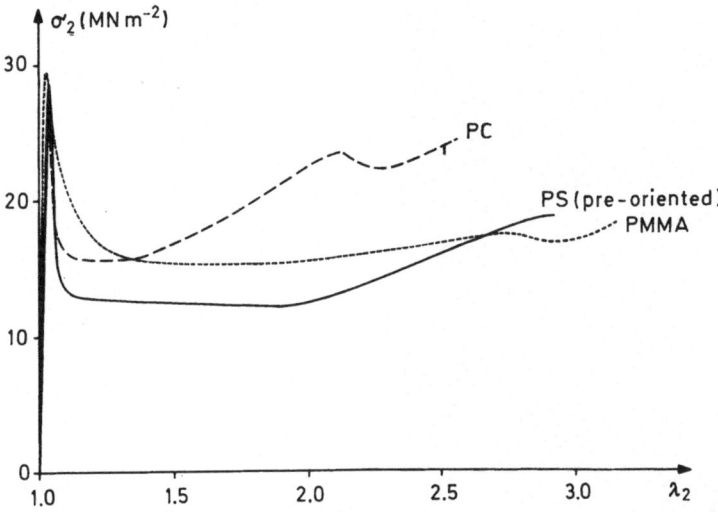

Fig. 25. Nominal stress, σ_2, versus extension ratio, λ_2, for polycarbonate, poly(methyl methacrylate) and pre-oriented polystyrene stretched with $\dot\varepsilon = 6 \times 10^{-4}$ s^{-1} at 129 °C, 80 °C and 90 °C, respectively

surprising. In fact, as previously shown for PC, the number of random links between entanglement points, and hence λ^e, may increase while the material is pre-stretched above T$_g$.

The organization of cavities in stress-whitened samples of PMMA and PS has been examined by SEM. The scanning electron micrographs presented in Figure 26 and 27 clearly reveal the high porosity of these materials. It is interesting to note that the planar organization of voids produces a structure which closely resembles that of very thin crazes. A SAXS maximum is not observed and would, in fact, not be expected in view of the large distribution in void size and in intervoid spacings seen in Figures 26 and 27.

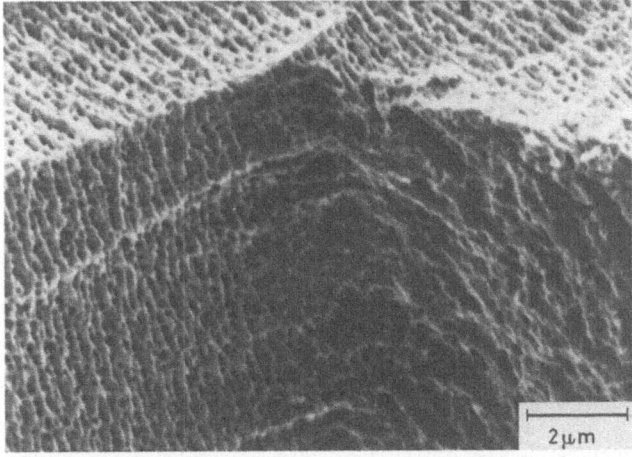

Fig. 26. Scanning electron micrograph of intrinsic cavities in poly(methyl methacrylate)

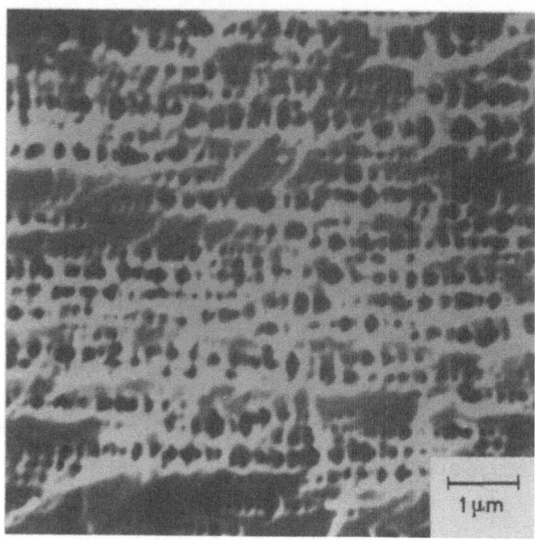

Fig. 27. Scanningelectronmicrograpph of intrinsic cavities in pre-oriented polystyrene

In discussing intrinsic cavitation in polymers stretched under high stresses another type of phenomenon seems worthy of consideration. Under certain conditions, several polymers such as poly(ethylene terephthalate) (PET), polyethylene, polypropylene and polyamide exhibit instabilities during cold drawing [134, 183-192]. In tensile tests these instabilities result in stress oscillations associated with the formation of stress-whitened zones of a fibrillar and voided microstructure as illustrated in Figures 28 and 29 for PET. The fibrils shown in Figure 29 consist of highly crystalline material. Until very recently this phenomenon was generally interpreted in terms of heat dissipation during necking. However, Pakula and Fischer [192] claim that instabilities

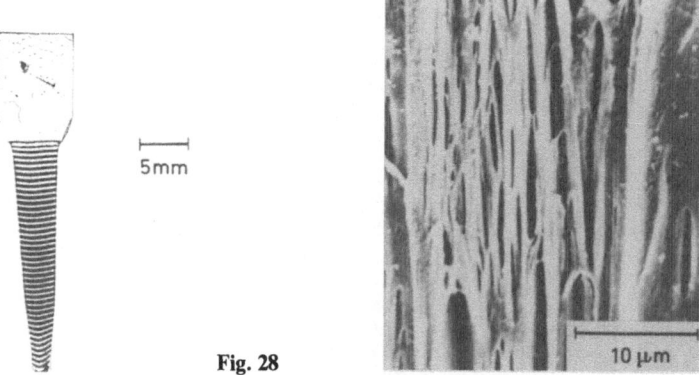

Fig. 28   Fig. 29

Fig. 28. Photograph of the alternating transparent and stress-whitened zones in poly(ethylene terephthalate) drawn under stress oscillations

Fig.29. Fibrillar microstructure of stress-whitened zones in poly(ethylene terephthalate) as revealed by scanning electron microscopy

occur even under isothermal drawing conditions. They have provided experimental evidence that stress oscillations arise from transitions between two deformation modes. If samples are drawn under constant stress the transition from the first to the second mode increases the strain rate by several orders of magnitude. The extension ratio of $\lambda = 4.4$, reported by Pakula and Fischer for PET stretched in the first mode, compares well with the value of $\lambda^{sh} = 4.3$ measured by Ward et al. [147]. The transition from the first to the second mode results in cavity formation. During the cavitation process the material is stretched to a much higher extension ratio, λ, than in the first mode, i.e. $\lambda > \lambda^{sh}$. The existence of a mode of cavitational plasticity in which material is stretched to $\lambda > \lambda^{sh}$ is a common feature between the phenomena such as observed in PET and those observed in PC, PMMA and PS. However, further investigations are necessary to clarify the precise relation between both types of phenomena.

It should be mentioned that Pakula and Fischer failed to observe the instability in PC since their experiments have been performed in a temperature range in which intrinsic crazes do not appear. It is worth noting that other deformation parameters such as the stress field which is affected for example by the sample geometry, strain rate, pre-orientation and environment, may also be of some importance. In fact, under the drawing conditions used in [11] intrinsic crazes were only initiated in slightly pre-oriented samples.

It has been shown that several polymers exhibit instabilities in their plastic deformation process. It should finally be mentioned that instabilities may also occur during the plastic deformation of metals [193]. This phenomenon which is called the Portevin-Le Chatelier effect, is generally interpreted in terms of different modes of dislocation movement depending on whether or not dislocations move by dragging along their atmosphere of impurities behind them.

7 Concluding Remarks

It has been demonstrated that under appropriate deformation conditions intrinsic crazes are initiated in PC. The intrinsic crazing of PC is a post-yield phenomenon which occurs when the material is stretched to high stresses and strains in a temperature region close to T_g. There is strong evidence that the initiation of intrinsic crazes is governed by a temperature and stress activated instability of the entanglement network frozen-in during the glass transition. The effect of pre-orientation on intrinsic craze initiation gave considerable support to this model. In fact, intrinsic craze initiation intimately reflects the modification of the entanglement network induced by pre-stretching above T_g.

The structural analysis of intrinsic crazes in PC has been carried out by SAXS. The fibrillar microstructure of these crazes gives rise to pronounced scattering effects which enable a detailed analysis of the craze structure in terms of both the volume fraction of craze fibrils and of the fibril diameter. This analysis showed that the microstructure of intrinsic and extrinsic crazes is largely different. There exists some evidence that the distinct microstructure primarily reflects the different stress-strain states of the matrix at craze initiation. Further investigations are necessary to answer the important question as to whether the two types of crazes may be interpreted on a common basis.

There is much evidence that the intrinsic crazing of PC is related to the existence of a general mode of cavitational plasticity in glassy polymers. Examples have been given for polymers which behave very similar to PC. In particular PMMA and pre-oriented PS exhibit an almost identical behavior when stretched at high stresses and strains in a temperature region close to T_g.

The author is well aware of the fact that many aspects which have been treated in the extensive literature on extrinsic crazing have not been considered in this article and that more information is needed for a comprehensive account of the observed craze phenomenon. For instance the recent work on the intrinsic crazing of PC and on related phenomena which has been reviewed here has primarily been based on structural considerations. It is believed that future work on the kinetics of craze formation and on the underlying molecular dynamics of the system may contribute considerably to a more detailed account of this phenomenon. Nevertheless, it is hoped that this work has opened up some new paths which may lead to a better understanding of the phenomenon of cavitational plasticity in polymers.

Acknowledgements: The financial support of the Swiss National Science Foundation is gratefully acknowledged. This work benefitted greatly from discussion with Prof. H. H. Kausch, and I am most grateful to his comments. I also wish to thank Profs. E. W. Fischer and E. J. Kramer for many helpful discussions and Mr. W. Dufour for his assistance in performing the experiments.

8 References

1. Rabinowitz, S., Beardmore, P.: Crit. Rev. Macromol. Sci. *1*, 1 (1972)
2. Kambour, R. P.: J. Polym. Sci., Macromol. Rev. *7*, 1 (1973)
3. Kausch, H. H.: Polymer Fracture, Berlin—Heidelberg—New York, Springer 1978
4. Kramer, E. J.: Environmental Cracking of Polymers, in: Developments in Polymer Fracture-1 (ed. Andrews, E. H.), London, Applied Science Publ. 1979, p. 55
5. Argon, A. S.: Pure and Appl. Chem. *43*, 247 (1975)
6. Argon, A. S., Hannoosh, J. G.: Phil. Mag. *36*, 1195 (1977)
7. Rehage, G., Goldbach, G.: Angew. Makromol. Chem. *1*, 125 (1967)
8. Murray, J., Hull, D.: J. Polym. Sci., Part A-2, *8*, 1521 (1970)
9. Hoare, J., Hull, D.: Phil. Mag. *26*, 443(1972)
10. Lainchbury, D. L. G., Bevis, M.: J. Mat. Sci. *11*, 2235 (1976)
11. Dettenmaier, M., Kausch, H. H.: Polymer *21*, 1232 (1980)
12. Kausch, H. H., Dettenmaier, M.: Polymer Bulletin *3*, 565 (1980)
13. Dettenmaier, M., Kausch, H. H.: ibid *3*, 571 (1980)
14. Dettenmaier, M., Kausch, H. H.: Colloid and Polym. Sci. *259*, 937 (1981)
15. Brown, H. R., Kramer, E. J.: J. Macromol. Sci. B *19*, 487 (1981)
16. J. Macromol. Sci. B *12* (1976); Faraday Disc. Chem. Soc. *68* (1979)
17. Wendorff, J. H.: Polymer *23*, 543 (1982)
18. Ballard, D. G. H., Burgess, A. N., Cheshire, P., Janke, E. W., Nevin, A., Schelten, J.: ibid *22*, 1353 (1981)

19. Gawrisch, W., Brereton, M. G., Fischer, E. W.: Polymer Bulletin 4, 687 (1981)
20. Flory, P. J.: Statistical Mechanics of Chain Molecules, New York, Wiley 1969
21. Yoon, D. Y., Flory, P. J.: Polymer Bulletin 4, 693 (1981)
22. Cotton, J. P., Decker, D., Benoit, H., Farnoux, B., Higgins, J., Jannink, G., Ober, R., Picot, C., des Cloizeaux, J.: Macromolecules 7, 863 (1974)
23. Kirste, R. G., Kruse, W. A., Ibel, K.: Polymer 16, 121 (1975)
24. Lieser, G., Fischer, E. W., Ibel, K.: J. Polym. Sci. B 13, 39 (1975)
25. Kirste, R. G., Lehnen, B. R.: Makromol. Chem. 177, 1137 (1976)
26. Dettenmaier, M.: J. Chem. Phys. 68, 2319 (1978)
27. Fischer, E. W., Stamm, M., Dettenmaier, M., Herchenröder, P.: ACS-Prepr. 20, 219 (1979)
28. Wignall, G. D., Longmann, G. W.: J. Mat. Sci. 8, 1439 (1973)
29. Gupta, M. R., Yeh, G. S. Y.: J. Macromol. Sci. B 15, 119 (1978)
30. Schubach, H. R., Nagy, E., Heise, B.: Colloid and Polym. Sci. 259, 789 (1981)
31. Wendorff, J. H., Fischer, E. W.: Kolloid Z. 251, 884 (1973)
32. Lin, W., Kramer, E. J.: J. Appl. Phys. 44, 4288 (1973)
33. Dettenmaier, M., Fischer, E. W.: Makromol. Chem. 177, 1185 (1975)
34. Renninger, A. L., Wicks, G. G., Uhlmann, D. R.: J. Polym. Sci., Polym. Phys. Ed., 13, 1247 (1975)
35. Dettenmaier, M.: Progr. Coll. Polym. Sci. 66, 169 (1979)
36. Fischer, E. W., Dettenmaier, M.: J. Non-crystall. Solids 31, 181 (1978)
37. Klement, J. J., Geil, P. H.: J. Macromol. Sci. B 6, 31 (1972)
38. Yeh, G. S. Y.: ibid B 6, 451 (1972)
39. Geil, P. H.: ibid B 12, 173 (1976)
40. Yeh, G. S. Y.: ibid B 7, 729 (1973)
41. Meyer, M., Van der Sande, J., Uhlmann, D. R.: J. Polym. Sci., Polym. Phys. Ed., 16, 2005 (1978)
42. Thomas, E. L., Roche, E. J.: Polymer 20, 1413 (1979)
43. Kaempf, G., Orth, H.: J. Macromol. Sci. B 11, 151 (1975)
44. Grosskurth, K. P.: Gummi, Asbest, Kunststoffe 25, 1159 (1972)
45. Grosskurth, K. P.: private communication
46. Grosskurth, K. P.: Colloid and Polym. Sci. 259, 163 (1981)
47. Dettenmaier, M., Kausch, H. H.: ibid 259, 209 (1981)
48. Yannas, I. V., Lunn, A. C.: J. Macromol. Sci. B 4, 603 (1970)
49. Yannas, I. V., Sung, N.-H., Lunn, A. C.: ibid B 5, 487 (1971)
50. Yannas, I. V., Doyle, M. J.: J. Polym. Sci. A-2 10, 159 (1972)
51. Yannas, I. V.: J. Macromol. Sci. B 6, 91 (1972)
52. Jansson, J.-F., Yannas, I. V.: J. Polym. Sci., Polym. Phys. Ed., 15, 2103 (1977)
53. Robertson, R. E.: J. Appl. Polym. Sci. 7, 443 (1963)
54. Bauwens, J. C., Bauwens-Crowet, C., Homès, G.: J. Polym. Sci. A-2 7, 735 (1969)
55. Bauwens-Crowet, C., Bauwens, J. C., Homès, G.: ibid 7, 1745 (1969)
56. Matz, D. J., Guldemond, W. G., Cooper, S. L.: J. Polym. Sci., Polym. Phys. Ed., 10, 1917 (1972)
57. Bauwens-Crowet, C., Bauwens, J. C., Homès, G.: J. Mat. Sci. 7, 176 (1972)
58. Bauwens-Crowet, C., Ots., J.-M., Bauwens, J. C.: J. Mat. Sci. Lett. 9, 1197 (1974)
59. Brady, T. E., Yeh, G. S. Y.: J. Macromol. Sci. B 9, 659 (1974)
60. Wu, W., Turner, A. P. L.: J. Polym. Sci., Polym. Phys. Ed. 13, 19 (1975)
61. Narisawa, I., Ishikawa, M., Ogawa, H.: ibid 16, 1459 (1978)
62. Glasstone, S., Laidler, K. J., Eyring, H.: The Theory of Rate Processes, New York, Mc Graw-Hill, 1941
63. Argon, A. S.: J. Macromol. Sci. B 8, 573 (1973)
64. Argon, A. S.: Phil. Mag. 28, 839 (1975)
65. Argon, A. S., Bessonov, M. I.: ibid 35, 917 (1977)
66. Argon, A. S., Bessonov, M. I.: Polym. Eng. Sci. 17, 174 (1977)
67. Robertson, R. E.: J. Chem. Phys. 44, 3950 (1966)
68. Kambour, R. P., Robertson, R. E.: The mechanical properties of plastics, in: Polymer Science, (ed. Jenkins, A. D.), Amsterdam—London, North-Holland Publ. Co. 1972, p. 687
69. Argon, A. S.: Stability of Plastic Deformation, in: The Inhomogeneity of Plastic Deformation, Metals Park, Ohio, ASM, 1973, p. 161

70. Bowden, P. B.: Phil. Mag. *22*, 455 (1970)
71. Bowden, P. B.: The Yield Behaviour of Glassy Polymers, in: The Physics of Glassy Polymers, (ed. Haward, R. N.), London, Applied Science Publ., 1973, p. 279
72. Wu, W., Turner, A. P. L.: J. Polym. Sci., Polym. Phys. Ed., *11*, 2199 (1973)
73. Ishikawa, M., Narisawa, I., Ogawa, H.: Polymer J. *8*, 181, 391 (1976)
74. Donald, A. M., Kramer, E. J.: J. Mat. Sci. *16*, 2967 (1981)
75. Donald, A. M., Kramer, E. J.: ibid *17*, 1871 (1982)
76. Donald, A. M., Kramer, E. J.: Polymer *23*, 1183 (1982)
77. Le Grand, D. G.: J. Appl. Polym. Sci. *13*, 2129 (1969)
78. Allen, G., Morley, D. C. W., Williams, T.: J. Mat. Sci. *8*, 1449 (1973)
79. Hull, D., Owen, T. W.: J. Polym. Sci., Polym. Phys. Ed., *11*, 2039 (1973)
80. Parvin, M., Williams, J. G.: Int. J. Fract. *11*, 963 (1975)
81. Adams, G. A., Cross, A., Haward, R. N.: J. Mat. Sci. *10*, 1582 (1975)
82. Mills, N. J.: ibid *11*, 363 (1976)
83. Pitman, G. L., Ward, I. M., Duckett, R. A.: ibid *13*, 2092 (1978)
84. Wellinghoff, S. T., Baer, E.: J. Appl. Polym. Sci. *22*, 2025 (1978)
85. Petrie, S. P., Dibenedetto, A. T., Miltz: J., Polym. Eng. Sci. *70*, 385 (1980)
86. Brown, H. R.: J. Mat. Sci. *17*, 469 (1982)
87. Wyzgoski, M. G., Yeh, G. S. Y.: J. Macromol. Sci. B *10*, 441 (1974)
88. Watts, D. C., Perry, E. P.: Polymer *19*, 248 (1978)
89. Yee, A. F.: Proc. 5th Int. Conf. on Deformation, Yield and Fracture of Polymers, Cambridge 1982
90. Taylor, G. I.: Proc. Roy. Soc. A *201*, 192 (1950)
91. Argon, A. S., Salama, M.: Mat. Sci. Eng. *23*, 219 (1976)
92. Parvin, M., Williams, J. G.: J. Mat. Sci. *10*, 1883 (1975)
93. Fraser, R. A. W., Ward, I. M.: Polymer *19*, 220 (1978)
94. Kambour, R. P., Holik, A. S., Miller, S.: J. Polym. Sci., Polym. Phys. Ed., *16*, 91 (1978)
95. Narisawa, I., Ishikawa, M., Ogawa, H.: J. Mat. Sci. *15*, 2059 (1980)
96. Pitman, G., Ward, I. M.: ibid. *15*, 635 (1980)
97. Walker, N., Hay, J. N., Haward, R. N.: Polymer *20*, 1056 (1979)
98. Walker, N., Haward, R. N., Hay, J. N.: J. Mat. Sci. *16*, 817 (1981)
99. Brown, N.: J. Polym. Sci., Polym. Phys. Ed., *11*, 2099 (1973)
100. Kastelic, J. R., Baer, E.: J. Macromol. Sci. *7*, 679 (1973)
101. Imai, Y., Brown, N.: J. Polym. Sci., Polym. Phys. Ed., *14*, 723 (1976)
102. Spurr, O. K., Niegisch, W. D.: J. Appl. Polym. Sci. *6*, 585 (1962)
103. Kambour, R. P., Gruner, C. L., Romgosa, E. E.: Macromolecules *7*, 249 (1974)
104. Jacques, C. H. M., Wyzgoski, M. G.: J. Appl. Polym. Sci. *23*, 1153 (1979)
105. Morgan, J. R., O'Neal, J. E.: Polymer *20*, 376 (1979)
106. Kambour, R. P.: ibid *5*, 143 (1964)
107. Wyzgoski, M. G., Yeh, G. S. Y.: J. Macromol. Sci. B *10*, 647 (1974)
108. Durtnal, G.: Polymer *16*, 549 (1975)
109. Thomas, E. L., Israel, S. J.: J. Mat. Sci. *10*, 1603 (1975)
110. Paredes, E., Fischer, E. W.: Makromol. Chem. *180*, 2707 (1979)
111. Donald, A. M., Kramer, E. J.: J. Polym. Sci., Polym. Phys. Ed., *20*, 899 (1982)
112. Paredes, E., Fischer, E. W.: ibid *20*, 929 (1982)
113. Lauterwasser, B. D., Kramer, E. J.: Phil. Mag. *39*, 465 (1979)
114. Verheulpen-Heymans, N.: Polymer *20*, 356 (1979)
115. Harris, J. S., Ward, I. M.: J. Mat. Sci. *5*, 573 (1970)
116. Beardmore, P., Rabinowitz, S.: ibid *10*, 1763 (1975)
117. Maxwell, B., Rahm, L. F.: Ind. Engng. Chem. *41*, 1988 (1949)
118. Hull, D., Hoare, L.: Plast. Rubber, Mat. and Appl., May 1976, p. 65
119. Bucknall, C. B., Clayton, D.: J. Mat. Sci. *7*, 202 (1972)
120. Heikens, D., Sjoerdsma, S. D., Coumans, W. J.: ibid *16*, 429 (1981)
121. Kambour, R. P.: Appl. Polym. Symp. *7*, 215 (1968)
122. Kambour, R. P., Kopp, R. W.: J. Polym. Sci. A-2 *7*, 183 (1969)
123. Mackay, M. E., Teng, T. G., Schultz, J. M.: J. Mat. Sci. *14*, 221 (1979)
124. Moet, A., Palley, I., Baer, E.: J. Appl. Phys. *51*, 5175 (1980)
125. Miles, M. J., Baer, E.: J. Mat. Sci. Lett. *14*, 1254 (1979)

126. Kambour, R. P., Miller, S.: J. Mat. Sci. *11*, 1220 (1976)
127. Haussy, J., Cavrot, J. P., Escaig, B., Lefebvre, J. M.: J. Polym. Sci., Polym. Phys. Ed., *18*, 311 (1980)
128. Argon, A. S., Salama, M. M.: Phil. Mag. *36*, 1217 (1977)
129. Ward, I. M. (ed.), Structure and Properties of Oriented Polymers, London, Applied Science Publ., 1975
130. Pinnock, P. R., Ward, I. M.: Trans. Faraday Soc. *62*, 1308 (1966)
131. Allison, S. W., Ward, I. M.: Brit. J. Appl. Phys. *18*, 1151 (1967)
132. Raha, S., Bowden, P. B.: Polymer *13*, 174 (1972)
133. Gent, A. N., Thomas, A. G.: J. Polym. Sci. A-2 *10*, 571 (1972)
134. Haward, R. N.: The Post-Yield Behaviour of Amorphous Plastics, in: The Physics of Glassy Polymers, (ed. Haward, R. N.), London, Applied Science Publ., 1973, p. 340
135. Moon, P. C., Barker, R. E.: J. Polym. Sci., Polym. Phys. Ed., *11*, 909 (1973)
136. Fellers, J. F., Kee, B. F.: J. Appl. Polym. Sci. *18*, 2355 (1974)
137. Wellinghoff, S., Baer, E.: J. Macromol. Sci. B *11*, 367 (1975)
138. Kusy, R. P., Turner, D. T.: Polymer *15*, 394 (1974)
139. Robertson, R. E.: Chem. Sci. *154*, 89 (1976)
140. Kusy, R. P., Katz, M. J.: J. Mat. Sci. *11*, 1475 (1976)
141. Lainchbury, D. L. G., Bevis, M.: ibid *11*, 2222 (1976)
142. Kramer, E. J.: ibid *14*, 1381 (1978)
143. Southern, J. H., Ballman, R. L., Buroughs, J. A., Paul, D. R.: J. Polym. Sci., Polym. Lett. Ed., *16*, 157 (1978)
144. Fellers, J. F., Chapman, T. F.: J. Appl. Polym. Sci. *22*, 1029 (1978)
145. Bersted, B. H.: ibid *24*, 37 (1979)
146. Fellers, J. F., Huang, D. C.: ibid *23*, 2315 (1979)
147. Rietsch, F., Duckett, R. A., Ward, I. M.: Polymer *20*, 1133 (1979)
148. Jud, K., Kausch, H. H.: Polymer Bulletin *1*, 697 (1979)
149. Jud, K., Kausch, H. H., Williams, J. G.: J. Mat. Sci. *16*, 204 (1981)
150. Zanker, H., Bonart, R.: Colloid and Polym. Sci. *259*, 87 (1981)
151. Turner, D. T.: Polymer *23*, 626 (1982)
152. Kausch, H. H., Dettenmaier, M.: Colloid and Polym. Sci. *260*, 120 (1982)
153. Nguyen, T. Q., Kausch, H. H., Jud, K., Dettenmaier, M.: Polymer, *23*, 1305 (1982)
154. Porter, R. S., Johnson, J. F.: Chem. Rev. *66*, 1 (1966)
155. Graessley, W. W.: Adv. Polym. Sci. *16*, 1 (1974)
156. Ferry, J. D.: Viscoelastic Properties of Polymers, New York, Wiley, 1980
157. Vinogradov, G. V., Malkin, A. Ya.: Rheology of Polymers, Berlin—Heidelberg—New York, Springer, 1980
158. De Gennes, P. G.: J. Chem. Phys. *55*, 572 (1971)
159. Doi, M., Edwards, F.: J. Chem. Soc. Faraday II *74*, 1789 (1978)
160. Seitz, J. T.: 50th Golden Jubilee Rheology Soc., Boston, 1979
161. Wagner, M. H., Meissner, J.: Makromol. Chem. *18*, 1533 (1980)
162. Kahar, N., Duckett, R. A., Ward, I. M.: Polymer *19*, 136 (1978)
163. Retting, W.: Colloid and Polymer Sci. *257*, 689 (1979)
164. Treloar, L. R. G.: The Physics of Rubber Elasticity, London—New York, Clarendon Press, 1958
165. Jernigan, R. L., Flory, P. J.: J. Chem. Phys. *50*, 4185 (1969)
166. Engelaere, J. C., Cavrot, J. P., Rietsch, F.: Polymer *23*, 766 (1982)
167. Gent, A. N.: J. Mat. Sci. *5*, 925 (1970)
168. Porod, G.: Kolloid Z. *124*, 83 (1951)
169. Guinier, A., Fournet, G.: Small-Angle Scattering of X-Rays, New York, John Wiley, 1955
170. Porod, G.: Fortschr. Hochpolymer-Forsch. *2*, 363 (1961)
171. Pilz, I., Glatter, O., Kratky, O.: Methods in Enzymology *61*, 147 (1979)
172. Oster, G., Riley, D. P.: Acta Cryst. *5*, 272 (1952)
173. Heyn, A. N.: J. Appl. Phys. *26*, 519, 1113 (1955)
174. Heikens, D.: J. Polym. Sci. *35*, 139 (1959)
175. Kratky, O., Schwarzkopf-Schier, K.: Chemie *94*, 714 (1963)
176. Bonart, R.: Kolloid Z. *211*, 14 (1966)
177. Perret, R., Ruland, W.: J. Appl. Cryst. *2*, 209 (1969)

178. Harget, P. J., Krimm, S.: Acta Cryst. A *27*, 586 (1971)
179. Stillinger, F. H., Di Marzio, E. A., Kornegay, R. L.: J. Chem. Phys. *40*, 1564 (1964)
180. Beahan, P., Bevis, M., Hull, D.: Proc. R. Soc. Lond. A *343*, 525 (1975)
181. Beahan, P., Bevis, M., Hull, D.: J. Mat. Sci. *8*, 162 (1972)
182. De Brossin, A., Dettenmaier, M., Kausch, H. H.: Helv. phys. acta, *55*, 213 (1982)
183. Müller, P. H., Entgelter, A.: Kolloid Z. *150*, 156 (1957)
184. Hookway, D. C.: Proc. Text. Inst. *49*, 7 (1958)
185. Roth, W., Schrith, R.: Faserforsch. Textiltech. *11*, 365 (1960)
186. Andrianova, G. P., Kardash, G. G., Kechekyan, A. S., Kargin, V. A.: Vysokomol. soyed. A *10*, 1990 (1968)
187. Kramer, E. J.: J. Appl. Polym. Sci. *14*, 2825 (1970)
188. Andrianova, G. P., Kechekyan, A. S., Kargin, V. A.: J. Polym. Sci., Part A-2, *9*, 1919 (1971)
189. Richards, R. C., Kramer, E. J.: J. Macromol. Sci. B *6*, 243 (1972)
190. Barenblat, G. I.: Methods of the Combustion Theory in the Mechanics of Deformation, Flow and Fracture of Polymers, in: Deformation and Fracture of Polymers (ed. Kausch, H. H., Hassell, J. A., Jaffee, R. I.), New York—London, Plenum 1973, p. 91
191. Andrianova, G. P., Arutyunov, B. A., Popov, Yu. V.: J. Polym. Sci., Polym. Phys. Ed., *16*, 1139 (1978)
192. Pakula, T., Fischer, E. W.: ibid *19*, 1705 (1981)
193. Dieter, G. E.: Mechanical Metallurgy, New York, Mc Graw-Hill 1976

Received December 1, 1982
H. Kausch (editor)

Optical Interference Measurements and Fracture Mechanics Analysis of Crack Tip Craze Zones

W. Döll
Fraunhofer-Institut für Werkstoffmechanik, Wöhlerstr. 11, D-7800 Freiburg/Breisgau, FRG

Advances in Polymer Science 52/53
© Springer-Verlag Berlin Heidelberg 1983

List of Symbols

a	Crack length
$\dot{a}$	Crack speed
da/dN	Crack propagation rate in fatigue
E	Young's modulus
G_I	Strain energy release rate in mode I
K_I	Stress intensity factor in mode I
K_{Ic}	K_I at onset of critical fracture
K_{I0}	K_I at onset of crack growth
ΔK_I	Relative stress intensity factor in fatigue
M_w	Weight average molecular weight
M_n	Number average molecular weight
N	Cycle number in fatigue
n	Fringe number
n_1	Fringe number of loaded craze
n_0	Fringe number of unloaded craze
r	Coordinate
r_p	Plastic zone size
s	Craze length
T	Temperature
t	Time
U_0	Viscoelastic strain energy
$2v(x)$	Displacement of elastic-plastic boundary
$2v$	Maximum displacement of the craze zone at the crack tip
$2v_c$	2v at fracture
$2v_0$	2v in the unloaded state
x, y, z	Coordinates
ε	Strain
λ	Wave length of light
μ	Refractive index of loaded craze
μ_0	Refractive index of unloaded craze
ν	Poisson's ratio
σ	Stress
σ_c	Craze stress
σ_y	Yield stress
φ	Polar coordinate
ω	Frequency in fatigue

1 Introduction

In glassy thermoplastics the process of fracture is preceded by the formation of a characteristic zone of heavily deformed material, the craze zone. Just ahead of the crack tip, where the stresses are particularly concentrated, molecular chains of the polymer are drawn out of their amorphous arrangement in the bulk material into bundles under the action of the principal tensile stress component acting normal to the crack plane. These bundles are interspersed by voids, leading to a material with rubber-like properties. Crazing as a normal yielding phenomenon thus differs significantly from conventional plasticity in other materials where it is essentially a volume-conserving, shear phenomenon.

Owing to their structure, crazes have a lower density and a lower refractive index than the bulk polymer. Thus it is possible to measure their sizes and shapes using optical interference, provided that the separations involved are comparable in order of magnitude to the wavelength of light, that the refractive index of the craze is significantly different from that of the bulk material, that the boundary between the two regions is sharp and last but not least that the material is transparent.

Since the pioneer work of Kambour [1, 14] and Bessenov and Kuvshinskii [2] the optical interference method has been widely used in different glassy thermoplastics to determine characteristic craze dimensions and critical displacements at the crack tip before and during fracture. Amongst the critical displacements, the maximum length to which the bundles of molecular chains spanning the tip of the crack in the craze zone can be stretched is of special interest, since the maximum length of stretched molecular bundles is one of the essential parameters governing the resistance against crack propagation in glassy thermoplastics. The other essential parameter is a characteristic stress which the bundles can sustain. It is the merit of fracture mechanics, especially of the plastic zone model due to Dugdale [3] and to Barenblatt [4], that quantitative information on the craze stress can be derived from interference optical measurements. Hence, the connection with fracture mechanics has made the optical interference method a powerful tool in the determination of those properties of the material in the micro region around the crack tip which directly relate to the fracture process (e.g. [5−7]).

The aim of this review is to concentrate mainly on these fundamental aspects of the fracture behavior of glassy thermoplastics. In the first Section, following an outline of the relevant fracture mechanics theory, the optical interference method is described and the nature of the results obtainable from it is discussed. The next Section then considers the behavior of cracks and crazes in specimens subjected to quasistatic loading, whilst the final Section examines the role of crazing associated with fatigue crack growth.

2 Analytical Description and Experimental Determination of Craze Zones at Crack Tips

2.1 Fracture Mechanics Approach

Fracture mechanics describes the behavior of sharp cracks in loaded materials and thus enables one to characterize the strength of materials and components. Since Griffith's [8] early considerations and Irwin's [9] basic work, fracture mechanics has been systematically developed and has now reached a high level of sophistication.

There is a correspondingly extensive literature on the subject and the interested reader should consult, for example, References [10-12] for general aspects of the subject or References [13-17] for specific considerations of polymer fracture. The approach here will be limited to introducing those concepts and relationships which are germane to this review.

2.1.1 Linear Elastic Material Behavior

For a crack in an infinite, linear elastic, isotropic body loaded by forces the stresses σ_{ij} at a point P (polar coordinates r, φ) near the crack tip may be described in the following form:

$$\sigma_{ij} = \frac{K}{\sqrt{2\pi r}} f_{ij}(\varphi) \qquad i, j = x, y, z . \tag{1}$$

r is the distance of point P from the crack tip, f_{ij} is an angular function and K is known as the stress intensity factor; x, y, z are the cartesian coordinates.

In this chapter only crack openings in response to tensile stresses normal to the crack plane will be considered in detail. This is known as mode-I loading, and under these conditions the normal stresses σ_{yy} in the crack plane (φ = 0) are given

$$\sigma_{yy} = \frac{K_I}{\sqrt{2\pi r}}. \tag{2}$$

The dependence of σ_{yy} on the distance r from the crack tip is qualitatively shown in Fig. 1, which also indicates how K_I determines the stress intensity at the crack tip. The stress intensity factor K_I describes the combined effect of the external stress σ and the crack length a:

$$K_I = \sigma \cdot \sqrt{a} \cdot Y \tag{3}$$

where Y is a correction factor which takes into account the particular configuration of specimen and crack geometry (e.g. [18]).

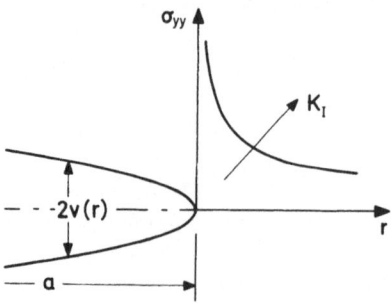

Fig. 1. Crack opening and tensile stress near the crack tip in an ideal elastic material

The displacements at $\varphi = \pi$ and hence the opening $2v(r)$ of a crack in an linear elastic material are given by

$$2v(r) = \frac{8 \cdot K_I}{E^*} \cdot \sqrt{\frac{r}{2\pi}} \tag{4}$$

where E^* is the reduced modulus ($E^* = E$ in plane stress and $E^* = E/(1 - v^2)$ in plane strain, $v =$ Poisson's ratio).

From Eq. (4) and also from Fig. 1 it can be seen that near its tip the crack opens parabolically; the magnitude of the opening $2v(r)$ at a fixed distance is influenced by the elastic properties of the material and governed by the stress intensity factor K_I.

The behavior of propagating cracks is often described in terms of the elastic strain energy required to create unit area of fracture surface: the strain energy release rate G_I. Between stress intensity factor K_I and strain energy release rate G_I the following relationship exists:

$$K_I^2 = G_I \cdot E^* \tag{5}$$

This relationship was first derived by Irwin[9] in calculating the work needed for the closure of the crack along an inifinitesimal crack length by the aid of equations (2) and (4).

2.1.2 Plastic Zone Size — Dugdale Model

Equation (3) indicates that near the crack tip the stresses can exceed the yield stress σ_y of real materials. Therefore, in a certain zone around the crack tip the material undergoes plastic deformations. To account for this two basically different models of plastic zones have been developed.

In the model derived by McClintock and Irwin[19] the shape and size of the plastic zone were calculated by a combination of the stress field at the crack tip (e.g. Eq. (2)) with a yield criterion (e.g. von Mises, Tresca). This leads to the well known "dog-bone" type of plastic zone showing the influence of stress state. Its form is often approximated to by a circle of radius r_p, where

$$r_p = \frac{1}{2\pi} \cdot \frac{K_I^2}{\sigma_y^2} \tag{6}$$

giving a basis for rough estimations of the extent of crack tip yielding.

Two comments should be made: first, the yield criteria involved e.g. von Mises, Tresca are based on yielding in response to shear stresses; secondly the above solution — although established on physical grounds — is mathematically not exact.

On the other hand, in glassy polymers the zone of crack tip yielding is often found to be a thin wedge rather than a circle. It is now well documented that a good description of the shape and size of this yielded zone at the crack tip can be provided by the plastic zone size model proposed by Dugdale[3] and by the cohesive force model of Barenblatt[4]. Similar solutions and further developments have been contributed by other authors[20-22].

Here the solution of Dugdale will be followed since it clearly shows a significant feature of the model. A flat ellipse of length 2c is considered in an infinite plate loaded in tension by a stress σ remote from and normal to the ellipse. The ends of the ellipse terminate in small plastic zones whose boundaries are under uniform internal pressure stresses (see Fig. 2). For these internal stresses static equilibrium is achieved by imposing equal and compressive opposite stresses σ_c. This elastic problem of an ellipse with its boundaries partly under pressure stress had previously been solved by Muskhelishvili [23]. Dugdale applied the solution to the limit of a flat ellipse, a crack, and determined the length s of the plastic zones by the condition that there should be no stress singularity at the tip of the crack:

$$s = a \cdot \left\{ \frac{1}{\cos \dfrac{\pi}{2} \dfrac{\sigma}{\sigma_c}} - 1 \right\} \tag{7a}$$

a = crack length ($= c{-}s$)

For small scale yielding (i.e. $\sigma \ll \sigma_c$) this can be expressed in fracture mechanics terms as:

$$s = \frac{\pi}{8} \cdot \frac{K_I^2}{\sigma_c^2} \tag{7b}$$

An analytical expression for the displacements 2v of the elastic-plastic boundaries in front of and behind the crack tip was derived by Goodier and Field [21]:

$$2v(x) = \frac{2c\,\sigma_c}{\pi \cdot E^*} \cdot \left\{ \cos \theta \cdot \ln \frac{\sin^2 (\theta_2 - \theta)}{\sin^2 (\theta_2 + \theta)} + \right.$$

$$\left. + \cos \theta_2 \cdot \ln \frac{(\sin \theta_2 + \sin \theta)^2}{(\sin \theta_2 - \sin \theta)^2} \right\} \tag{8a}$$

x = coordinate parameter ($x < |c|$), $\theta = \arccos \dfrac{x}{c}$, $\theta_2 = \arccos \dfrac{a}{c} = \dfrac{\pi}{2} \dfrac{\sigma}{\sigma_c}$; the position of the crack tip is at $x = a$.

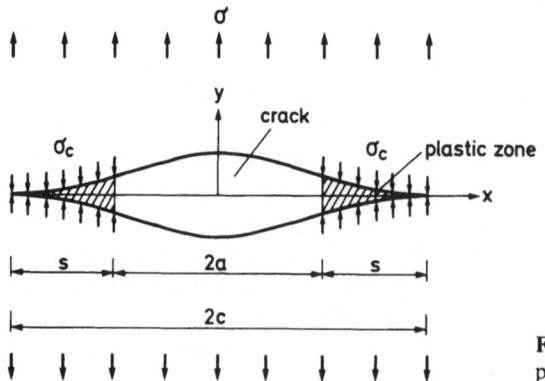

Fig. 2. Dugdale-Muskhelishvili model of plastic zones at the ends of a loaded ellipse

A simpler expression for the displacements of the plastic zone boundaries alone has been derived by Rice [22]. It has the form

$$2v(x_1) = \frac{8\sigma_c \cdot s}{\pi \cdot E^*} \left\{ \xi - \frac{x_1}{2s} \ln \left(\frac{1 + \xi}{1 - \xi} \right) \right\} \tag{8b}$$

where $\xi = \left(1 - \frac{x_1}{s} \right)^{1/2}$ with x_1 = coordinate parameter, crack tip at $x_1 = 0$.

In Fig. 3 the dependence of the displacements on the distance from the crack tip is shown using a normalized scale and assuming $\sigma = 0.1\sigma_c$. At a distance behind the crack tip the elastic-plastic boundary has an almost parabolic form, whilst directly at the crack tip — in contrast to the ideal elastic solution (Eq. (4)) — there is a certain displacement which is also the maximum width of the plastic zone. Here, and in the following text, the width of the plastic zone at the crack tip will be denoted by 2v without any coordinate parameter. In fracture mechanics terms it can be expressed as

$$2v = K_I^2/(\sigma_c \cdot E^*) \tag{8c}$$

Hence the displacement 2v at the crack tip is governed not only by the materials' properties (E^*, σ_c) but also, and more severely, by the stress intensity factor K_I.

Using the relationship between K_I and G_I Equation (8c) may be rearranged in the form:

$$G_I = 2v \cdot \sigma_c \tag{8d}$$

which is of special importance for the interpretation of fracture in polymers on molecular basis.

A note should be added concerning the two parameters which chiefly characterize the shape of the plastic zone: length s and maximum width 2v.

Whilst both depend strongly on K_I (compare Eq. (7b) and (8c)) the relationship of maximum width to length does not:

$$\frac{2v}{s} = \frac{8}{\pi} \cdot \frac{\sigma_c}{E^*} \tag{9}$$

It is only dependent on the material's properties.

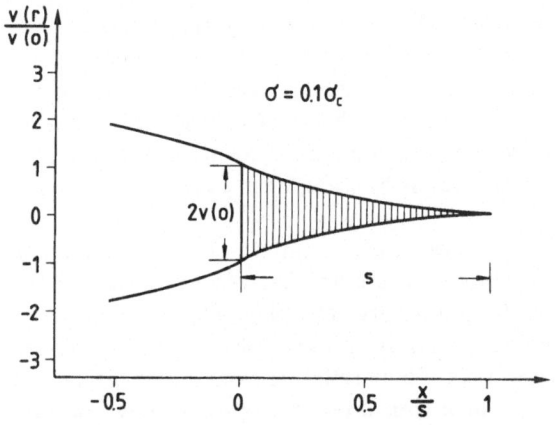

Fig. 3. Elastic-plastic boundary as calculated from the Dugdale model

The outstanding importance of the Dugdale-Barenblatt-model lies in the fact that yielding occurs by tensile stresses (normal yielding). This becomes obvious if, in the limit of the ellipse becoming a crack, the stresses at the boundary of the plastic zone are considered: they are oriented normal to the crack plane. Now, it was shown by Sternstein et coworkers [75,142] that crazing is a phenomen which essentially occurs under the action of normal stresses. The fibrils in the craze are oriented under the action of the maximum principal tensile stress component and the propagation direction of the craze zone is perpendicular to it and parallel to the axis of the minor principal tensile stress.

In this context is has also to be pointed out that in the original Dugdale model the material behavior is assumed to be linearly elastic and perfectly plastic; the latter assumption leads to a uniform stress distribution in the plastic zone. This may be a simplified situation for many materials; to model, however, the material behavior in the crack tip region where high inhomogeneous stresses and strains are acting is a rather complex task if nonlinear, rate-dependent effects in the continuum outside and strain hardening or softening in the plastic zone are to be taken into account. The situation gets even more complex if in addition, for fast moving cracks, dynamic and thermal effects have to be included. On the other hand solutions are available which permit certain important aspects to be taken into account, e.g. effects of viscoelasticity of the bulk material on crack growth [13,16,24,25] as well as of the craze material on craze growth [26], and effects of work hardening and rate sensitive plastic deformation [27], and of dynamics due to running cracks [28-30].

2.2 Interference Optics

The classical methods of light optical interference are well known in the determination of small dimensions which are of the order of the wave length of light. A new scientific field has been opened up by their application to investigations of the craze behavior at crack tips in transparent glassy thermoplastics [1,2].

A typical arrangement for interference investigations is schematically depicted in Fig. 4. The fracture mechanics specimen is illuminated in reflection with monochromatic light under normal incidence. Typical fringe systems are shown in Fig. 5 for an unloaded and a loaded crack in a PMMA-specimen of high molecular weight. The crack propagation direction is from left to right; two fringe systems are apparent, both with decreasing spacing towards the crack tip: those to the left arise from interference between reflections from the two crack faces, whilst those to the right arise from reflections at the two boundaries between crazed and uncrazed material. In order to calculate the crack openings and craze thicknesses the positions of the individual fringes are determined accurately by scanning in a microdensitometer (Fig. 5) [5,7,31].

The microdensitometer trace of the interference fringe pattern of a craze at a crack tip in polystyrene (PS) shows an interesting feature as observed by Doyle [31]: the bright fringes show a pattern of alternating intensity. The phenomenon seems to be characteristic of crazes in PS although not generally in other polymers. It it explained in the way that the optical interference of rays of light reflected from the wedge-shaped craze layer are superimposed on those of a thin layer of approximately constant

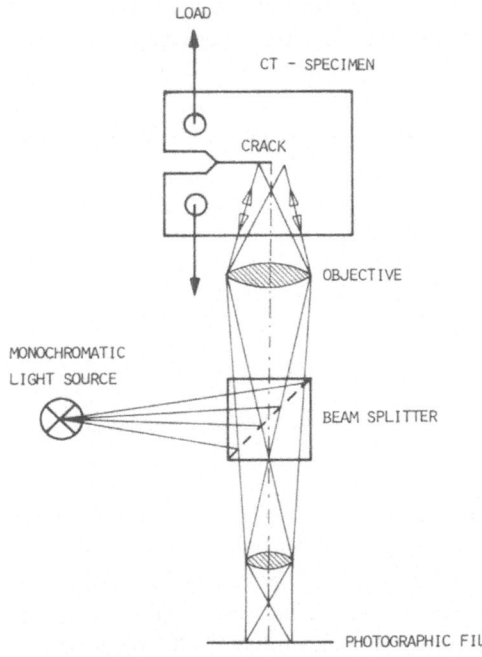

Fig. 4. Sketch of optical interference arrangement for displacement measurements at the crack tip

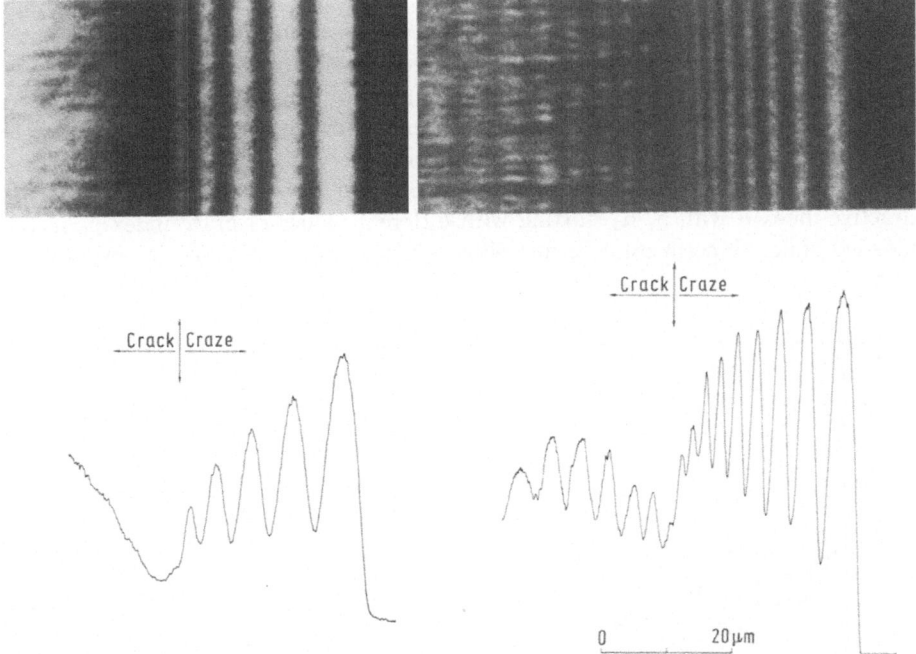

Fig. 5. Interference fringe patterns ($\lambda = 546$ nm) and microdensitometer traces of an unloaded and loaded ($K_I = 20{,}8$ N/mm$^{3/2}$) crack and craze in PMMA

thickness at the median plane of the wedge. A basis for this model was the direct observations by Beahan et al. [32] using transmission electron microscopy of a layer of 50 to 100 nm thickness in the structure of a polystyrene craze. This midrib layer was subsequently investigated in more detail [33].

The thicknesses $2v(x)$ of craze and crack opening are given by basic interference theory

at a bright fringe by $\qquad 2v(x) = \dfrac{\lambda}{2\mu}\,(n - 1/2)$ $\hfill$ (10a)

and $\hfill n = 1, 2, 3 \ldots$

at a dark fringe by $\qquad 2v(x) = \dfrac{\lambda}{2\mu}\,n$ $\hfill$ (10b)

where n is the order of fringe at position x, λ is the wavelength of monochromatic light and μ is the refractive index.

The crack opening can be determined ($\mu = 1$), if the fringe order n is known. This method has been used in investigations of stationary cracks to determine static [34] and dynamic [35] stress intensity factors, the latter being induced by a shock wave.

For the calculation of the craze thickness knowledge of the refractive index of the craze and its variation with strain is required. Refractive indices of unloaded crazes have been determined in different glassy polymers by Kambour [36-38] by measuring the critical angle of total reflection of light at the craze/bulk polymer interface; thus e.g. in PMMA the measured craze refractive index was $\mu_0 = 1.32$. Under the assumption that the craze extends with complete lateral constraint and using the Lorentz-Lorenz equation, which relates polarizability to the refractive index, the density of the craze material [37] and the craze refractive index μ [38] have been determined as a function of strain. For the evaluation of interference fringe patterns it is more useful to use the numbers of fringes n_1 and n_0 of the loaded and unloaded craze respectively instead of strain or extension ratio. Figure 6 shows the decrease of craze refractive index μ with n_1/n_0 starting with different values of craze index μ_0 in the unloaded state. Experimental results show typical values of n_1/n_0 in PMMA [5,7]

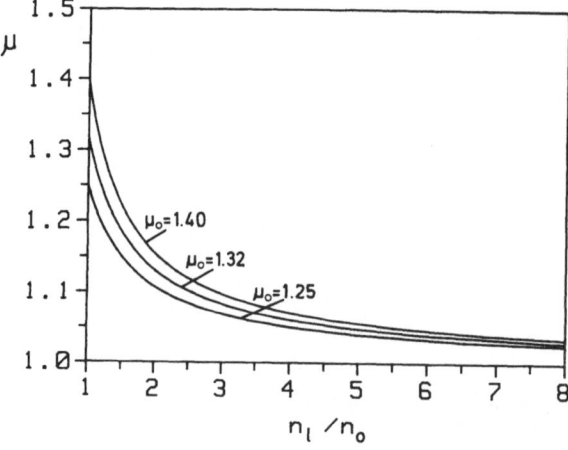

Fig. 6. Refractive index μ of the craze zone as a function of relative fringe number n

at break in the range of 2–3 and in PC [39] of 1.4–1.5 leading to craze refractive indices μ of 1.15–1.09 and 1.19–1.12 respectively. A similar restriction on the range of values of n_1/n_0 is found in the experimental results for quasistatic and cyclically loaded cracks in the subsequent Sections of this review. Hence, with reference to Fig. 6, the variation in craze refractive index is not nearly as great as might initially have been expected.

There is no direct information available on any possible temperature dependence of the craze refractive index. However, it might be expected that the temperature dependence is similar to that of the bulk material, which e.g. in PMMA increases by less than 1 % in the temperature range of 60 °C to −30 °C [40]. Also, measurements of the refractive index of the broken craze layer in PMMA at 25° and 60 °C showed a constant value of 1.32 ± 0.01 [41] which is just the same as for the unloaded craze at room temperature.

2.3 Some Results and Comments

Figure 7 shows an example of the experimentally determined shapes (points) of the crack opening and the craze zone in PMMA of high molecular weight. Taking the different scales of the vertical and horizontal axes into account it becomes evident that the craze is a long thin wedge. In addition the arrangements of molecules and of stretched and broken fibrils are schematically indicated. In Fig. 7 the measured points indicate that the crack tip is blunted and that in this position the craze width 2v is larger than the crack opening.

The similarity between the measured craze zone and the calculated Dugdale plastic zone is apparent when Figs. 3 and 7 are compared. Before applying the Dugdale formula to the measured craze zone some points of detail should be noted. Whilst the position of the crack tip is known fairly precisely, the location of the craze tip and hence the craze length s cannot be measured directly, but can — in a similar way to

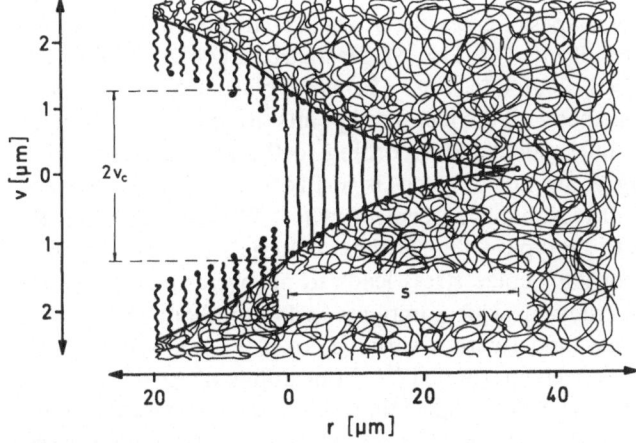

Fig. 7. Measured craze and crack opening (points) in PMMA (Arrangement of molecules and deformation behavior of fibrils are schematically indicated)

the maximum craze width 2v and the crack opening COS — only be obtained by extrapolation. Such an extrapolation can be made by "hand", or by a polynominal fit or more substantiated using the Dugdale model. With the latter extrapolation method the information of all measured points along the craze contour can be fully utilized. In one such procedure all the experimental points for a particular craze are used together with Eqs. (7b) and (8b) to calculate by iteration that value of s (and hence the corresponding values of E and σ_c) which minimizes the variation in E along the length of the craze [42]. The thus extrapolated position of the craze tip is shown in Fig. 7 as the open circle. Also shown are the lines corresponding to the displacements 2v(x) calculated from Eq. (8a) using the fitted values of s, E and σ_c. Comparing the calculated with the measured values it can be seen that there is good agreement in the craze zone (as might have been expected) but the calculated curve for the crack opening is displaced from the experimental points. This apparent discrepancy between the Dugdale model and the experimentally determined crack opening is removed, however, if it is taken into account that the model provides the locus of the displacements of the elastic plastic boundary not only ahead of the crack tip but also behind it. This means that, in thermoplastics, to the measured crack opening there must be added the thickness of the layers of craze material which remains on the fracture surfaces. This thickness is of the same order as the apparent discrepancy; e.g. in PMMA the thickness of the surface layer was determined to be 0.58 μm [41] (although this does vary with molecular weight [7]).

In a thermoplastic material it is, therefore, important to distinguish between the crack opening stretch (= COS) and the maximum craze width 2v. To characterize plastic deformation and fracture behavior of a thermoplastic material the maximum length of stretched fibrils and hence the maximum craze width 2v is a more fundamental parameter than the crack opening stretch. The latter, in addition, depends on the relaxation behavior of the broken remnants on the fracture surface.

It has been reported [5,42] that in PMMA the Dugdale model provides a good quantitative description of the craze zone at the crack tip. (The derived material parameters (E, σ_c) will be examined in some detail in Sect. 3). On the other hand it has also been reported that the description by the Dugdale model is not so good in some other materials. In Fig. 8 some shapes of crack tip crazes available from the literature are compiled: PES (Polyethersulfone) ([43], Fig. 8), plasticized PMMA ([5], Fig. 12), PC (polycarbonate of M_w = 17900) ([39], Fig. 15a) and plasticized PVC (polyvinylchloride) ([44], Fig. 5). Together with the measured points the individual fits of the Dugdale model are given as lines calculated according to the procedure described above. Clearly the fits in Fig. 8 are not as good as in PMMA in Fig. 7. PES and plasticized PMMA show a still reasonable agreement, whilst PVC indicates and PC exhibits deviations. It seems to be typical that in these materials the application of the Dugdale model with a constant craze stress leads to an overestimation of the maximum craze width at the crack tip which amounts in PC to about 15%.

Using a modified [27] Dugdale model with a variable craze stress along the craze zone this effect has qualitatively been interpreted [45]. At positions where the constant stress Dugdale model gives displacements higher than the measured ones the actual craze stress must be higher. In the case of PC a closer inspection reveals a stress peak at the crack and craze tip. Kambour [46] predicted just such a stress distribution in a craze from the analysis of the stress distribution around a craze (without a crack)

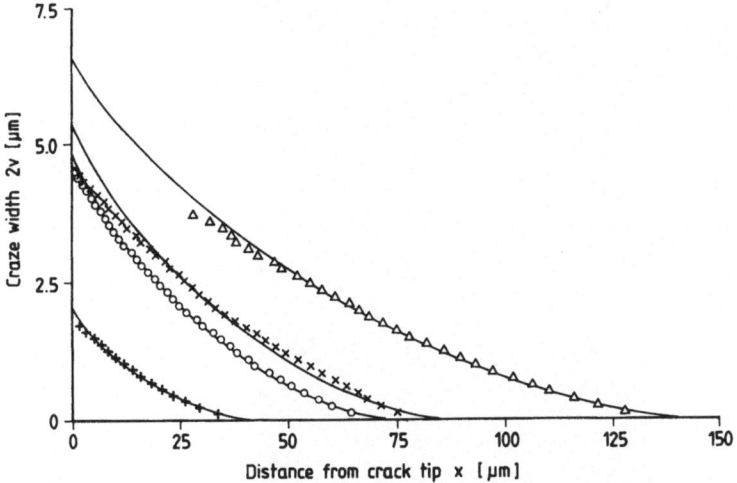

Fig. 8. Measured craze zone at the crack tip and fit by the Dugdale model ($-$) for different polymers: + PES [43], o plasticized PMMA [5], x PC [39], Δ plasticized PVC [44]

given by Knight [47]. In recent comprehensive studies Kramer and coworkers (e.g. [33,48-51] see also Sect. 1), using quantitative transmission electron microscopy in thin films, determined displacement profiles of crazes in different polymers for both liquid environments and air. The stress distribution along the craze was calculated using either the Fourier transform or the distributed dislocation method. They thus determined stress profiles along the craze zone (in PS) and also along the plastic deformation zone (in PC) at tiny cracks in polymer films. These showed a moderate increase at the crack tip and a pronounced peak at the tip of the craze and of the plastic zone, respectively, which in PC amounted to more than double the average stress along the zone [50]. The zone in front of the crack tip in PC has been clearly identified to be a plastic zone rather than a craze [51,52]. Hence, it may be that the especially high stresses are due to this other type of deformation. With PS is has to be noted, however, that the craze zone observed by Kramer et al. had extremely low craze stresses and was a factor of 5 smaller in length and almost a factor of 20 smaller in maximum width than those observed by Kambour [2] and Doyle [31].

However, the behavior of the material under high stresses and strains in the micro region at the crack tip should also reflect specific features determined in macroscopic experiments. In thermoplastic materials the dependence on strain and time is of prime importance. These questions will be addressed in the following Section.

3 Quasistatic Loading Conditions

In many different materials under mode-I loading conditions three regions of fracture behavior can be distinguished as a function of the stress intensity factor K_I:
— no crack growth for $K_I < K_{Io}$
— slow crack growth for $K_{Io} < K_I < K_{Ic}$
— rapid crack propagation for $K_I > K_{Ic}$

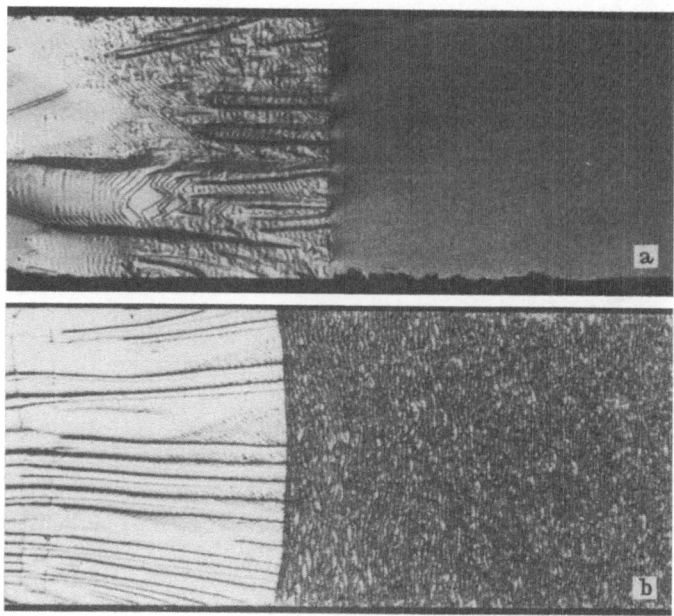

Fig. 9a and b. Fracture surfaces of **a** high ($M_w = 8 \cdot 10^6$) and **b** low ($M_w = 1.15 \cdot 10^5$) molecular weight PMMA showing transition from slow to fast crack propagation (left to right) [54]

For many polymers the transition from slow to rapid crack growth is relatively sharp. This is exemplified by PMMA. At the critical value of stress intensity factor, K_{Ic}, a jump in crack speed is observed which ranges from about 0.1 ms^{-1} at the end of the slow growth region to about 10 to 100 ms^{-1}, depending on testing conditions, at the beginning of the rapid crack propagation region (e.g. [53,54]). Associated with this jump is a significant change in the fracture surface morphology which can be seen in Fig. 9.

At present there is only very little and indirect information available on the crazing behavior at the tips of fast running cracks in glassy thermoplastics (e.g. [55,56,141]). Hence this Section will confine itself to a discussion of the behavior of single crazes at the tips of stationary or slowly moving cracks.

3.1 Moving Cracks

The propagation behavior of moving cracks is usually described by measuring crack resistance (characterized by K_I) as function of crack speed $\dot{a}$ of the crack tip. Figure 10 shows an example of the results of different authors [57-59] who studied the crack propagation in a commerical grade PMMA using different test methods (notched tension, parallel cleavage, tapered cleavage, compact tension, double torsion). Although the stress intensity factor K_I depends upon specimen geometry and loading conditions there is a good agreement between the results of all the different test methods. The crack speed $\dot{a}$ ranges from nearly 10^{-7} mm/s to 10^2 mm/s, thus covering

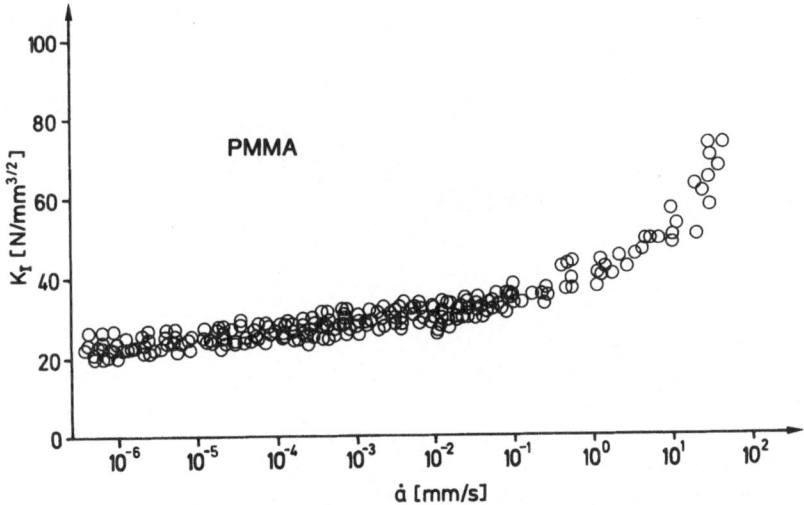

Fig. 10. Fracture toughness K_I as function of crack speed $\dot{a}$ in PMMA at room temperature as reported by different authors [57-59]

8 decades in crack speed; in contrast to this, K_I increases relatively moderately by a factor of less than 4. This relatively small variation in K_I compared to that in $\dot{a}$, together with the spread of data shown in Fig. 10, make it very important experimentally to measure K_I and the corresponding value of $\dot{a}$ simultaneously. If these are to be correlated with measurements on craze zones during crack propagation, then all three sets of data need to be recorded together for accurate work.

All thermoplastics are known as viscoelastic materials and hence exhibit time dependent properties. It is known that during crack propagation the influence of time is also involved such that slow crack speeds correspond to long times and high crack speeds to short times. In the following we will be concerned with this problem in more detail.

Another aspect of viscoelastic behavior is the influence of temperature and in the slow crack propagation region is has also been well documented. Marshall et al. [60] observed that in PMMA the crack speed curves are shifted to lower K_I-values with increasing temperature and that also K_{Ic} decreases in the temperature range from -60 °C to 80 °C.

Such a curve of K_I vs. $\dot{a}$ is naturally dependent on the material investigated and also on material specific parameters as molecular weight; in PMMA this leads to a shift to lower K_I-values or equivalently G_I-values with decreasing molecular weight [61, 62].

It should be noted that slow crack propagation curves of quasi-brittle materials have been used by different authors [e.g. [17,58,63,64] to predict creep life curves: a very interesting and promising application.

3.1.1 Craze Dimensions

In PMMA of high molecular weight the craze dimensions have been measured at moving crack tips in the speed range of 10^{-8} mm/s $< \dot{a} < 10^2$ mm/s (see Fig. 10) at room temperature using an experimental set-up especially developed to apply the

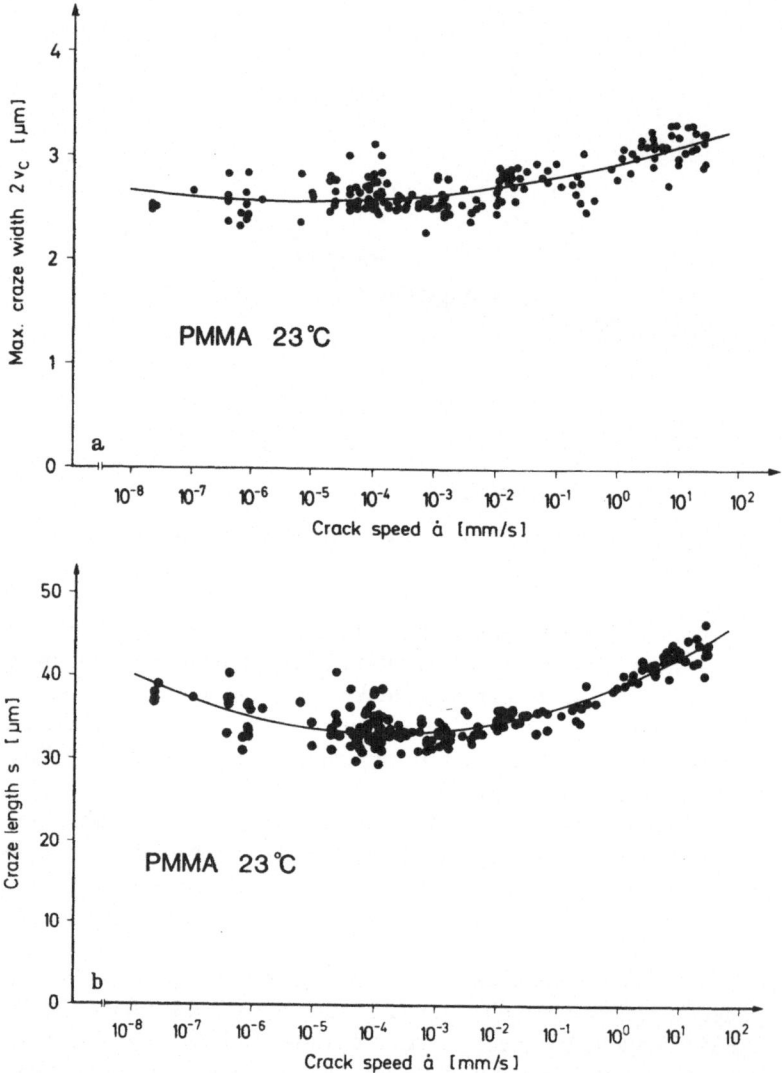

Fig. 11a and b. Variation with crack speed of the dimensions of the craze zone at the tip of a propagating crack in PMMA [5]. **a** maximum craze width $2v_c$; **b** craze length s

microscopic interference technique under short time conditions [59]. It should be emphasized that in these experiments the following relevant parameters were measured simultaneously: interference fringe pattern, load, crack length and crack speed. Thus a complete characterization of any instant during the fracture process could be obtained.

Figure 11a shows the maximum craze width $2v_c{}^1$ as a function of crack speed $\dot{a}$ and Fig. 11b the corresponding craze length s. Initially the results were interpreted as

---

[1] The subscript "c" in 2v will be used to indicate a critical magnitude.

suggesting that the craze dimensions remained practically constant, independent of crack speed [59]. The mean values obtained were s = (34.9 ± 3.1) μm and $2v_c$ = (2.7 ± 0.2) μm. More recent data, however, has prompted a re-examination of these results which reveals that both these craze dimensions go through a minimum at speeds of about 10^{-4}—10^{-3} mm/s. The effect is more pronounced for the craze length (Fig. 11 b) than for the craze width (Fig. 11 a). It will be shown later that an increase in craze dimensions during crack propagation is to be expected at lower speeds due to a long-time-effect (see Sect. 3.2.2) and at higher speeds due to a temperature effect (see Sect. 3.1.3, 3.3).

A much more pronounced decrease in craze length with crack speed has been reported by Vavakin et al. [6] who investigated the craze zones at moving crack tips in PMMA (characterized as coloured plastic) using the interference method in the crack speed range of 10^{-5} to 10^{-1} mm/s. Their results show a decrease in craze length from 70 to 40 μm and also a decrease in maximum craze width from 3.4 to 2.4 μm. Although a rough agreement may be seen between the results there is disagreement in one essential point. Based on the data in Fig. 10 one would expect K_I to increase from 24 to 34 N/mm$^{3/2}$ in this range of crack speeds. However, Vavakin et al. report a practically constant K_I-value of 35 N/mm$^{3/2}$. This is in a way unexpected and possibly due to differences in the investigated materials. Support is given by the qualitative observations of Brown and Ward [5] that the moving craze in PMMA shows only a slight decrease in maximum craze thickness whilst in plasticized PMMA (with 7% dibutyl phthalate) both the length and thickness of the craze were seen to be very speed dependent, the moving craze always possessing smaller dimensions.

3.1.2 Derived Material Parameters as Functions of Crack Speed and Time

In Section 2.3 it was pointed out, that in PMMA the Dugdale model gives a good description of the craze shape at the crack tip. For the calculation of the Dugdale plastic zone size, in addition to an accurate K_I-value, Young's modulus (E) and craze stress (σ_c) must be known. These mechanical parameters are time-dependent in a viscoelastic material like PMMA and hence for propagating cracks a time-dependence, or equivalently, a crack speed-dependence of these "moduli" will be involved. By fitting the Dugdale model to the measured craze-zones, as shown in Fig. 11 a, b σ_c and E are found to be dependent on crack speed $\dot{a}$ as shown in Fig. 12a, b [65]. In the investigated crack speed range σ_c and E increase from about 60 to 120 N/mm^2 and from 2000 to 3400 N/mm^2, respectively. This increase in σ_c is the reason, that at nearly constant maximum craze width the strain energy release rate G_I increases with crack speed [66].

Attempts have been made to correlate crack speed $\dot{a}$ with time t. An analysis of the fracture behavior of thermoplastics shows that it is essentially determined by craze formation and stretching the fibrils up to fracture. Therefore, the time involved in this process is considered to be the relevant time t which may be calculated by [16]:

$$t = \frac{s}{\dot{a}} \tag{11}$$

Thus in Fig. 12a, b the crack speed axis can be converted into a time scale ranging from about 10^4 to 10^{-4} seconds. On this basis a comparison has been made [67]

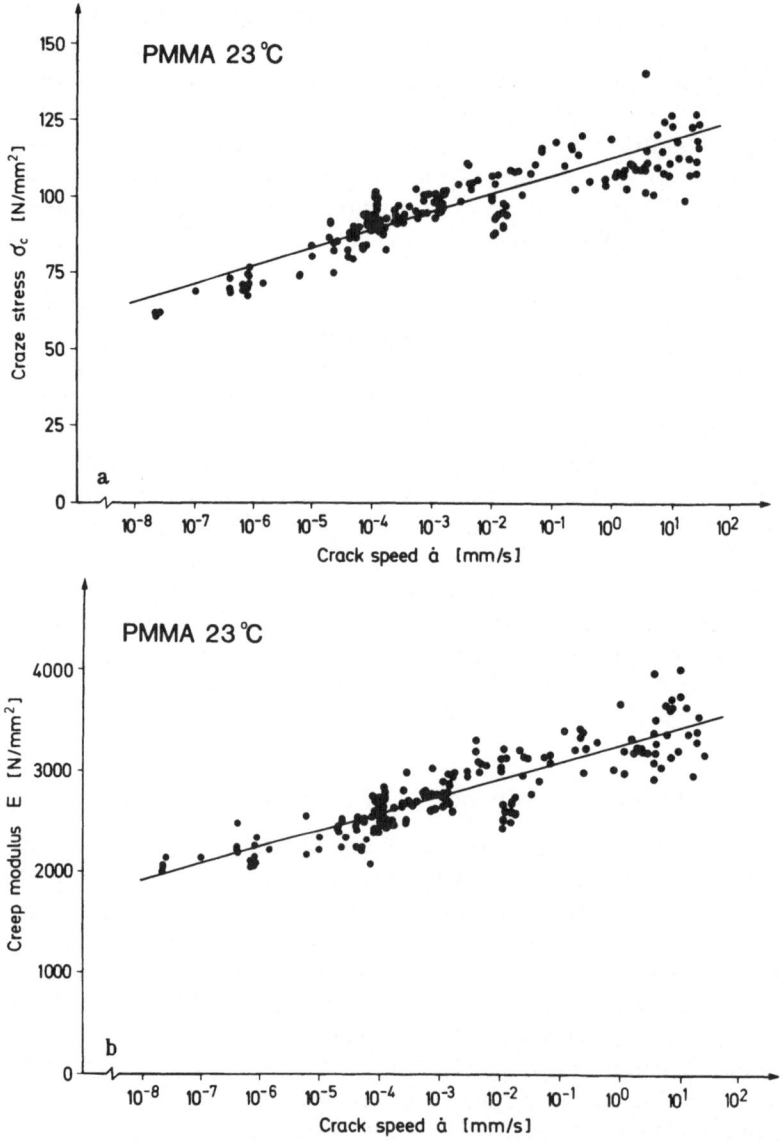

Fig. 12a and b. Material data of the micro region at the crack tip as derived by the application of the Dugdale model to measured craze sizes (Fig. 11) [65]. **a** craze stress σ_c; **b** creep modulus E

with data determined in macroscopic experiments. Figure 13 represents the time dependent creep moduli E as measured by different authors [67–69]. It can be seen that the creep modulus curve obtained from the material behavior in the micro region near the crack tip corresponds to a curve of about 1 % strain. Using the fracture mechanics approach, the strain ε_y near the crack tip can be calculated from [67]

$$\varepsilon_y = \frac{1}{E(t)} \cdot \frac{K_I}{\sqrt{2\pi r}} \cdot (1 - \nu - \nu^2) \tag{12}$$

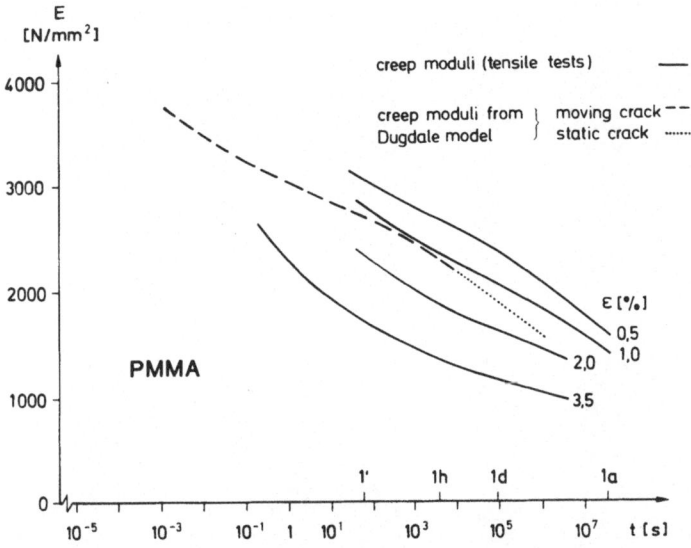

Fig. 13. Creep moduli E at different strains ε as a function of the loading time t determined in macroscopic tensile tests [68,69] and in the micro region at the crack tip [67]

in plane strain. Thus, for a moving crack the strain at the craze tip (r = s) may be estimated to be 0.8–0.9 %. Hence, in the field of macroscopically determined creep moduli in Fig. 13 the position of that curve which characterizes the strain field near and around the craze zone seems to be reasonable.

3.1.3 Influence of Temperature and Molecular Weight

In Fig. 14 the effects of temperature and molecular weight on the craze size at the crack tip can be estimated from the different fringe pattern in two grades of PMMA. The size of the craze in the low molecular weight PMMA remains almost unchanged with temperature whereas the craze size in the high molecular weight material increases with temperature; at all measured temperatures the craze size in low molecular weight material has been found to be smaller than that in the high molecular weight material [66].

In the context of the results for moving cracks discussed in the previous section it should be noted that the craze sizes reported in this section have been measured from just after onset of slow crack propagation to speeds up to 10^{-3} mm/s [66,70-72]. Quantitative results of maximum craze width $2v_c$ as a function of temperature T are compiled in Fig. 15 for PMMA [66,70,71], PC [72] and PVC [66]. At a first glance there seems to be no consistent pattern of behavior. In the two grades of polycarbonate PC (1) (Makrolon, M_w = 20600) and PC (2) (Lexan, M_w = 17000) the maximum craze width $2v_c$ remains constant for each material over the temperature ranges used; the higher molecular weight material, PC (1), has the bigger craze width. In three of the four PMMA-grades the maximum craze widths are nearly constant in the measured temperature range and thus show a similar behavior to PC. Of these materials PMMA (2) is known [66] as a low molecular weight material (M_w = 120000). PMMA (1),

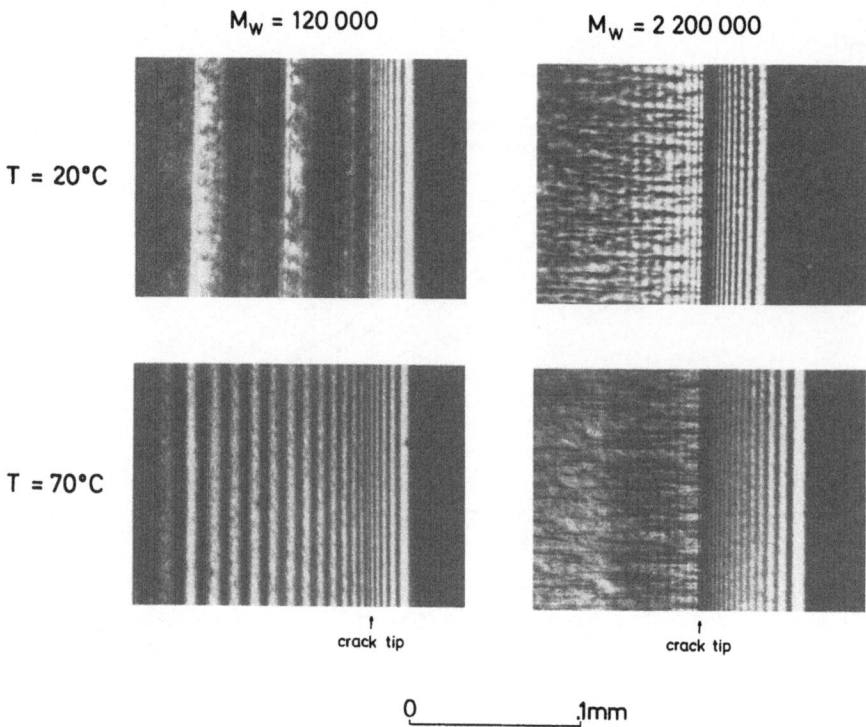

Fig. 14. Interference fringe patterns in PMMA of low and high molecular weight at different temperatures

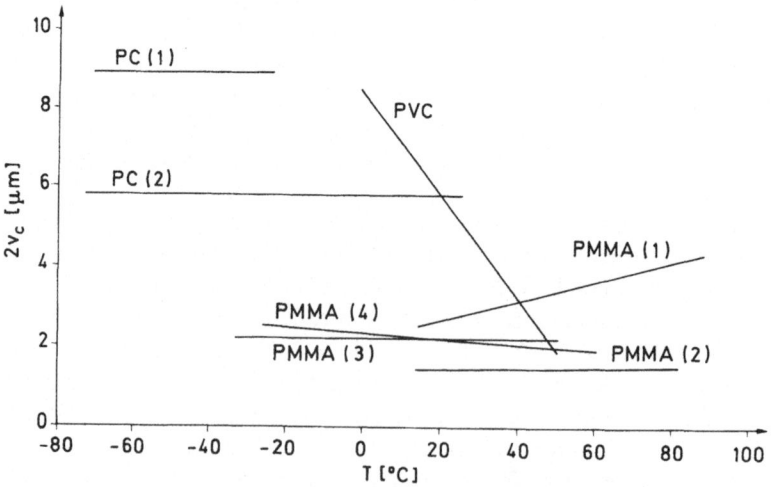

Fig. 15. Maximum craze width $2v_c$ measured at different temperatures T in various polymers: PC (1) $\triangleq$ ($M_w = 20600$), PC (2) $\triangleq$ ($M_w = 17000$) [72]; PVC, PMMA (1) $\triangleq$ ($M_w = 2.2 \cdot 10^6$), PMMA (2) $\triangleq$ ($M_w = 1.2 \cdot 10^5$) [66]; PMMA (3) [70]; PMMA (4) [71]

which had a high molecular weight ($M_w = 2200000$), shows a distinct increase in craze width. The molecular weights of PMMA (3) and (4) are not specified. However, from the data in Fig. 17 it may be inferred that these materials were also of low molecular weight.

Two parameters seem to govern the temperature behavior of the maximum craze width: the relationship between the length of stretched molecular chain and the maximum craze width [7] and the mechanical behavior of the molecular chain. In the five polymer grades which exhibit a constant $2v_c$ with temperature the length of stretched molecular chain is much smaller than the maximum craze width (= length of stretched fibrils). In the case of high molecular weight PMMA the possible length of a stretched chain is larger than the measured length of stretched fibrils. Thus the high molecular weight PMMA is endowed with a reserve in extensibility of the chains, leading to an increase in stretched fibrils' length with temperature. The temperature behavior of the stretched fibrils in PVC can be related to the structure of the molecular chain. The chloride side group in PVC is much smaller than that of the ester in PMMA or the aromatic rings in PC and hence as the temperature increases, the entanglements are more easily loosened under stress leading to the observed decrease in maximum craze width.

The effect of molecular weight on craze dimensions has been investigated in more detail in PC [39] and PMMA [7, 66, 73]. The results of Pitman and Ward [39] in PC measured at $-30\ °C$ are given in Fig. 16 and show an increase by a factor of ten in maximum craze width as the weight average molecular weight M_w doubles. The effect is less pronounced with craze length which increases by a factor of four.

In PMMA the effects of molecular weight and of molecular weight distribution on craze dimensions and on crack opening have been intensively studied [7,73,74]. In Fig. 17 the results of the maximum craze width $2v_c$ are represented as function of M_w for two different temperatures. It can be seen that materials with a broad molecular weight distribution, denoted as polydisperse PMMA and characterized by open symbols, show a marked increase of craze width at low molecular weights, a levelling off above a molecular weight of about 200000. At low molecular weights there is only a small increase in craze width as the temperature is raised from

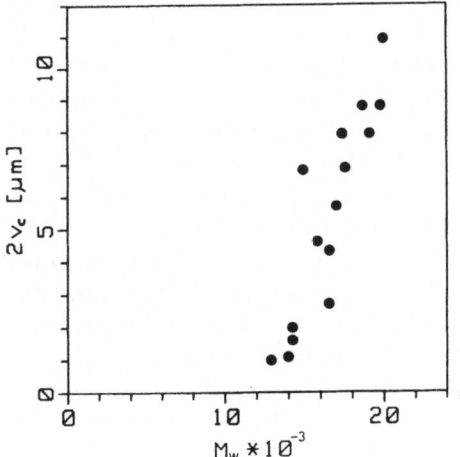

Fig. 16. Maximum craze width $2v_c$ in polycarbonate measured at $-30\ °C$ as a function of molecular weigth M_w [39]

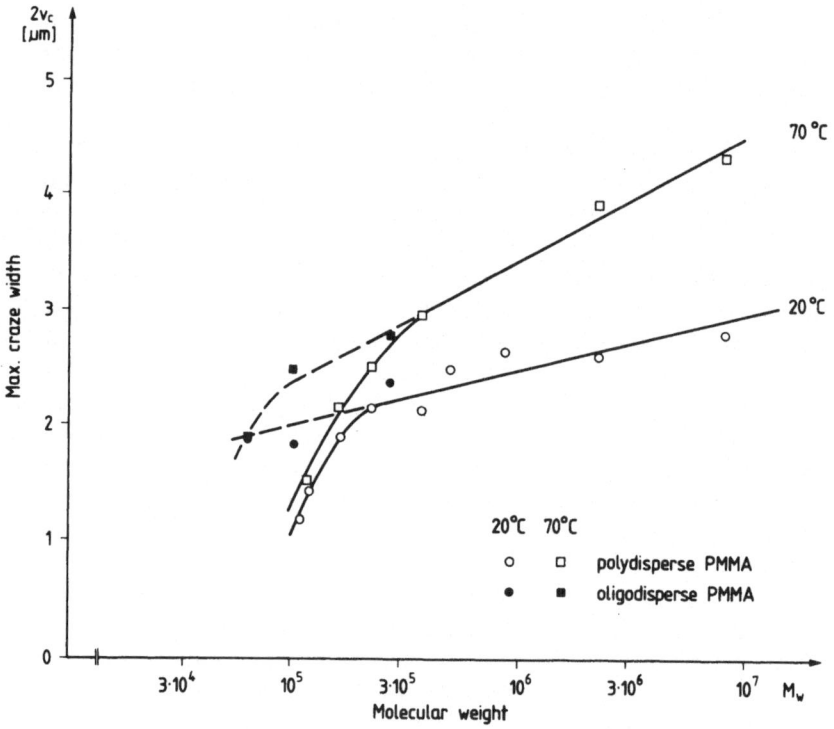

Fig. 17. Maximum craze width $2v_c$ in PMMA of different molecular weight distributions as a function of molecular weight M_w at different temperatures [74]

20° to 70 °C. In contrast, at higher molecular weights, the craze width shows a large increase at the higher temperature. This effect is seen to be due to the larger extensibility of longer chains at higher M_w; e.g. the theoretical ratios between the length of a fully stretched chain and maximum craze width are about 0.2 for $M_w = 100\,000$ and nearly 7 for $M_w = 8\,000\,000$ both at room temperature.

The influence of molecular weight distribution has been examined in that region of molecular weight showing the marked increase in craze dimensions at constant temperature. In addition to the polydisperse PMMA-materials which had broad molecular weight distributions ($M_w/M_n \approx 2$ for $M_w < 5 \cdot 10^5$ and $M_w/M_n \geqq 3.5$ for $M_w > 5 \cdot 10^5$) in Fig. 17 the results of oligodisperse PMMA-materials ($M_w/M_n \approx 1.25$) are represented by closed symbols. A linear extrapolation of the $2v_c$-M_w-curves established for $M_w > 300\,000$ in polydisperse materials shows that the values of $2v_c$ measured in obligodisperse PMMA fit these extrapolated straight lines and that the marked increase in craze width is shifted to lower M_w. Hence one can conclude that the marked increase in the $2v_c$-M_w-curves of polydisperse PMMA is due to the presence of very low molecular weight fractions in the molecular weight distribution; this may be a reflection of the increased defect density in the fibrils with very short molecular chains.

3.1.4 Derived Material Parameters as Functions of Temperature

From the reported craze dimensions, the tensile creep moduli E and craze stress σ_c have been derived by the aid of the Dugdale model. For PMMA the thus evaluated creep moduli E [42, 71] are shown as a function of temperature T in Fig. 18 together with results of compliance measurements [70]. Before a comparison is made the different test conditions have to be taken into account. The measurements on propagating cracks were performed at two different crack speeds $\dot{a} \approx 10^{-3}$ mm/s [71] and $\dot{a} \approx 10^{-5}$ mm/s [42]. From the results shown in Fig. 12b a higher value of E is expected at the higher crack speed [71]. In the compliance tests [70] the elastic behavior of the specimen was measured under small strains leading to a modulus which in a non-linear elastic material is expected to be on a higher value than the creep modulus (compare Fig. 13); however, for an exact classification the loading times in the experiments are needed which are not published.

In Fig. 19 the derived craze stresses σ_c are shown as a function of temperature T for various polymers [42, 66, 70–71]. For the PMMA results reported by Schirrer, Goett [71] and Weidmann, Döll [42] the differences in crack speed have to be taken into account (compare Fig. 12a) as mentioned above.

For temperatures below 10 °C the results of Morgan and Ward [70] on PMMA suggest a constant craze stress. However, this may reflect their use of the Young's modulus in their evaluation, which shows a steeper variation with temperature than

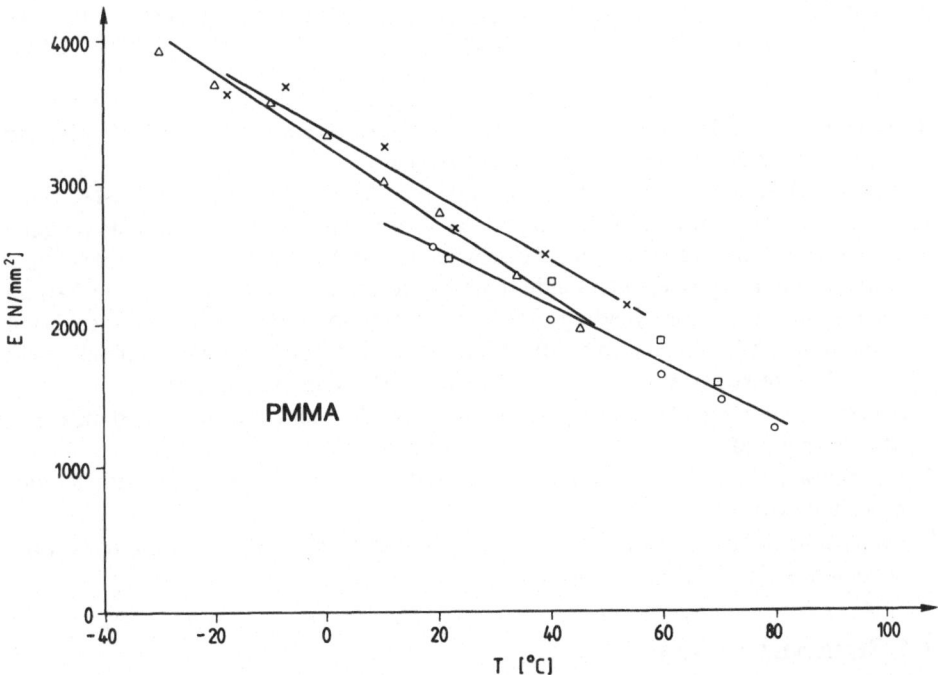

Fig. 18. Influence of temperature T on modulus E determined by interference fringe pattern and Dugdale model ($\times$ [71], low ($\square$) and high ($\bigcirc$) molecular weight material [42]) and compliance measurement ($\triangle$ [70])

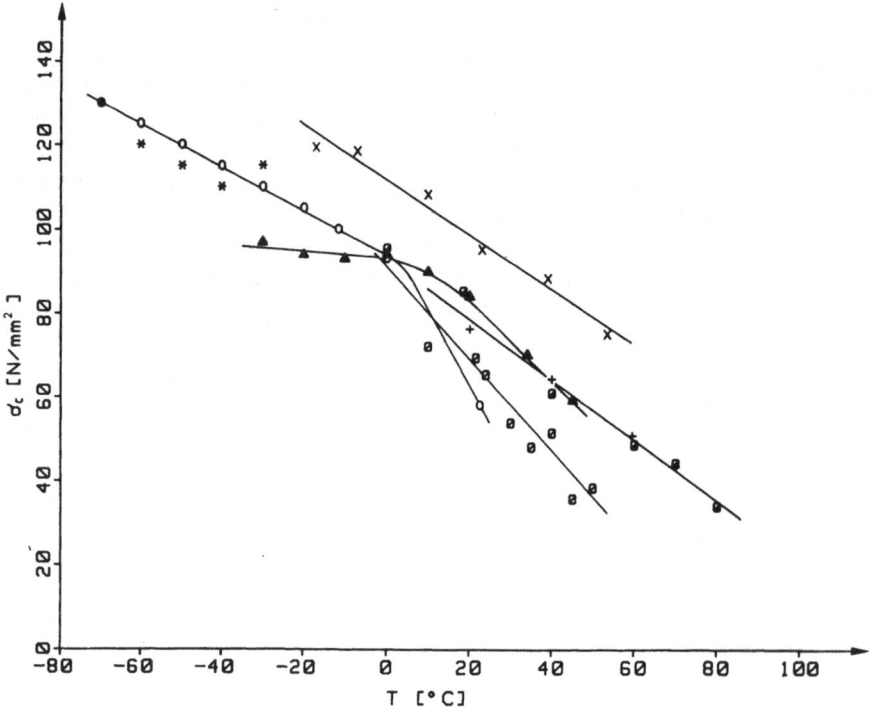

Fig. 19. Craze stress σ_c as a function of temperature T in different polymers: PMMA ($\times$ [71]; [70], (+ ($M_w = 1.2 \cdot 10^5$), ($M_w = 2.2 \cdot 10^6$) [42])), PC (* ($M_w = 17\,000$), ($M_w = 20\,600$) [72])), PVC ($\boxtimes$ [66])

the other results in this range (see Fig. 18). The two different grades of PC [72] differing slightly in molecular weight fit quite well together.

It is interesting to note that the results for the craze stress in PMMA show practically no effect of molecular weight over a very wide range of molecular weights, whilst for PC σ_c increases by a factor of 3 when the molecular weight is doubled [39].

Different attempts have been made to correlate craze stress σ_c with the yield stress or the fracture stress after yielding. For PMMA a good agreement has been reported [5,42] whilst in PVC considerable differences have been found [66]. Qualitatively yield stress and craze stress have been found to be of the same magnitude.

However, there is a lack of adequate quantitative data and new, improved measurements are required which should fulfill the following:
1. full characterisation of the fracture behavior including its variation with K_I and especially with $\dot{a}$
2. evaluation of the measured craze shape by numerical methods using a variable craze stress along the craze contour.

3.2 Stationary Cracks

The growth behavior of surface crazes has been intensively studied in various polymers (e.g. [14,76−80]). The fracture mechanics approach provides a basis for an analytical description of the growth in front of a crack tip (e.g. [16,81−83]). In this Section attention will be confined to investigations into growing craze zones in front of stationary cracks. This implies that the acting K_I-value had to be lower than that necessary for slow crack propagation.

3.2.1 Craze Growth in Liquid Environments

An example of the growth behavior of crazes in a liquid environment is shown in Fig. 20 which is taken from the results of Williams and Marshall [83]. They measured the craze length versus loading time at different K_I-values in PMMA specimens immersed in methanol. The time-dependent craze-behavior was interpreted in terms of a plasticisation mechanism incorporating the effect of the fluid [16]. Due to its porous nature the craze has a very high area to volume ratio so that penetration of the fluid by only a small distance leads to a complete plasticisation of the fibrils and a subsequent drop in the load carrying capacity σ_c of the fibrils; the material effectively behaves as one with a lower craze stress $\alpha\sigma_c$ ($\alpha < 1$).

Williams [16] modelled this behavior using the Dugdale model and derived a growth law of the form:

$$r_p \propto K_I \cdot t^{1/2} \tag{13}$$

which was confirmed by results such as those shown in Fig. 20. As has been pointed out, however, the crazes are seen to arrest after some time and show little subsequent relaxation controlled growth.

3.2.2 Craze Growth in Air

In this Section the kinetics of craze growth at crack tips in air will be considered in some detail. We shall not be concerned with the initiation phase and any micro mechanism (e.g. [84–86]) leading to craze initiation.

Differences in the growth behavior between surface crazes and the single craze at a crack tip have been reported.

The growth behavior of crazes in unnotched samples has been widely investigated. Figure 21 shows examples of the observed growth of surface crazes in terms of the

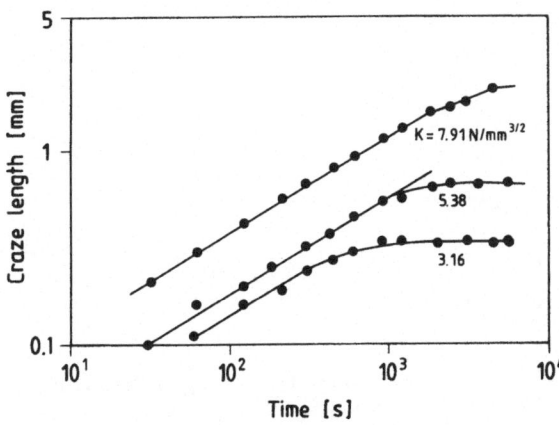

Fig. 20. Craze growth at a stationary crack tip in PMMA in methanol [83]

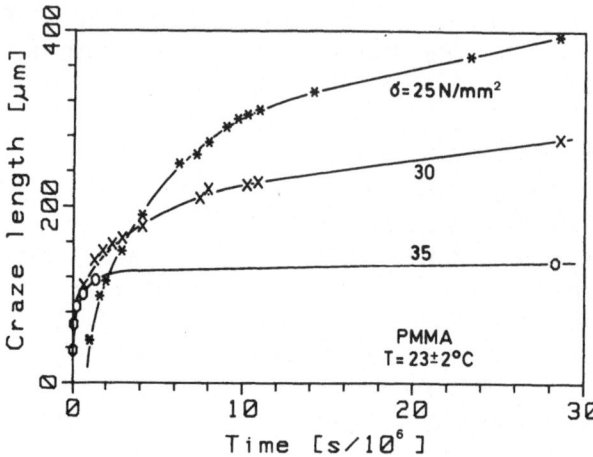

Fig. 21. Growth of surface crazes in unnotched specimens in air [80)]

average craze lengths as a function of time in PMMA (high molecular weight) loaded in a creep test in air [80)]. Two stages of growth can be discerned: initially that of free and subsequently that of impeded propagation due to the mutual influences of the crazes which lead to diminished growth rates. It can be seen that even in PMMA the crazes may become very long. This raises the interesting question of under what circumstances the craze breaks leading to a system crack/craze which is usually hard to resolve from a crack; however, this problem has yet not been solved satisfactorily.

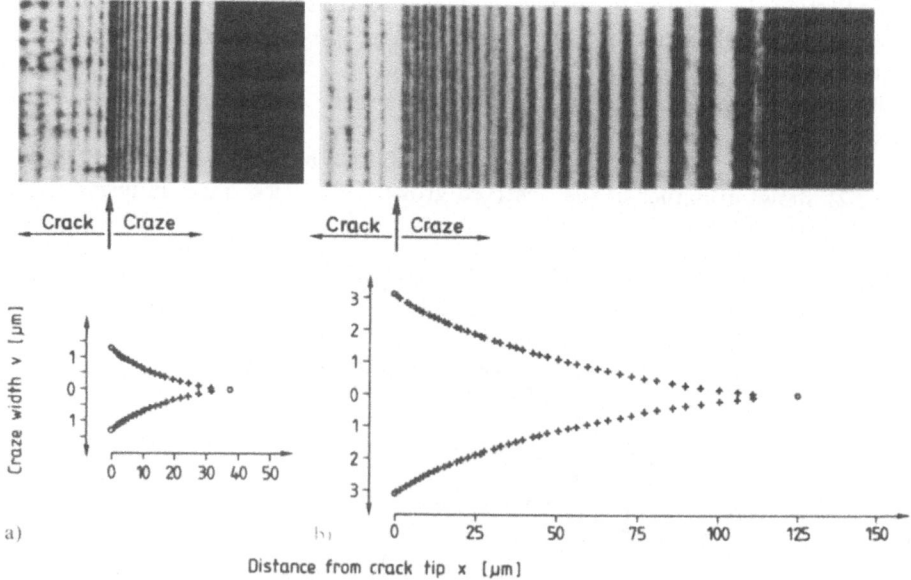

Fig. 22a and b. Two stages of primary craze growth at a stationary crack tip in PMMA in air, $K_I = 20$ N/mm$^{3/2}$ [87)]. **a** loading time $1.5 \cdot 10^2$ seconds; **b** loading time $2 \cdot 10^6$ seconds

In all papers mentioned above the characterization of the craze growth has been performed by measuring the craze length. By application of the optical interference method the essential information on the growth of craze width is available. Figure 22 shows the interference fringe patterns and measured craze zone sizes at a stationary crack tip in PMMA loaded in a creep test at a constant K_I-value at loading times of 150 s and $2 \cdot 10^6$ s [87].

From Fig. 22 it can be seen that the crack tip craze grows significantly in length as well as in width with loading time. To observe such craze growth behavior it is necessary to perform the experiments at K_I-levels below K_{Io} at which very slow crack propagation starts; e.g. in high molecular weight PMMA K_{Io} is about 21 N/mm$^{3/2}$ at room temperature. The quantitative growth behavior is shown in Fig. 23a, b. Starting with a craze length s of about 25 µm in the time region of seconds the crack tip craze grows to about six times of its original length with increasing loading time. The growth rate $\dot{s}$, however, decreases from more than 10^{-8} mm/s to less than 10^{-9} mm/s at craze lengths of 100 µm and more. It should be noted that this propagation behavior differs from that reported previously [83] which showed a linear increase in a log-log-plot.

In that and a subsequent paper [16] the increase in craze length was assumed to occur without a significant increase in maximum craze width and the latter remaining well below the critical displacement at fracture. In Fig. 23b the measured time-dependent behavior of maximum craze width 2v is shown. Contrary to the above assumption the maximum craze width exhibits an almost dramatic growth, in reaching three times its critical value at fracture after times under load approaching 10^7 s (c.f. 3.1). Initiation of fracture, however, was not observed in these experiments. Moreover, to initiate fracture in specimens with a wide craze at the crack tip K_I-values were needed which were distinctly higher than K_{Io}. This may be due to the crack tip blunting which is associated with a wider craze width.

In the light of these results previously formulated fracture criteria [16] need to be revised. A critical displacement at the crack tip seems not to be a unique criterion, and a fracture criterion formulated in K_I-terms alone e.g.

$$K_I > K_{Io}$$

is insufficient for all conditions as the above example of the blunted crack tip shows. Therefore, instead of a macroscopic continuum approach an attempt is being made to use a molecular approach based on the life-time of fibrils at the crack tip [88].

Craze growth at the crack tip has been qualitatively interpreted [87] as a cooperative effect between the inhomogeneous stress field at the crack tip and the viscoelastic material behavior of PMMA, the latter leading to a decrease of creep modulus and yield stress with loading time. If a constant stress on the whole craze is assumed then time dependent material parameters can be derived by the aid of the Dugdale model. An averaged curve of the creep modulus E(t) is shown in Fig. 13 as a function of time, whilst the craze stress σ_c is shown in Fig. 24.

There are some further aspects to the craze growth behavior at stationary crack tips in PMMA [67,87]. There is a temperature dependence, the rate of craze growth increasing as the temperature is raised. The growth rate also depends on K_I when

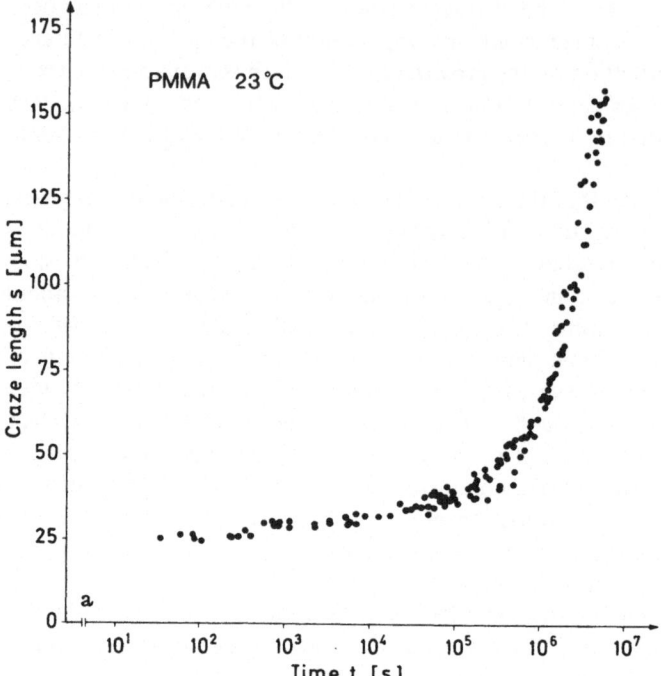

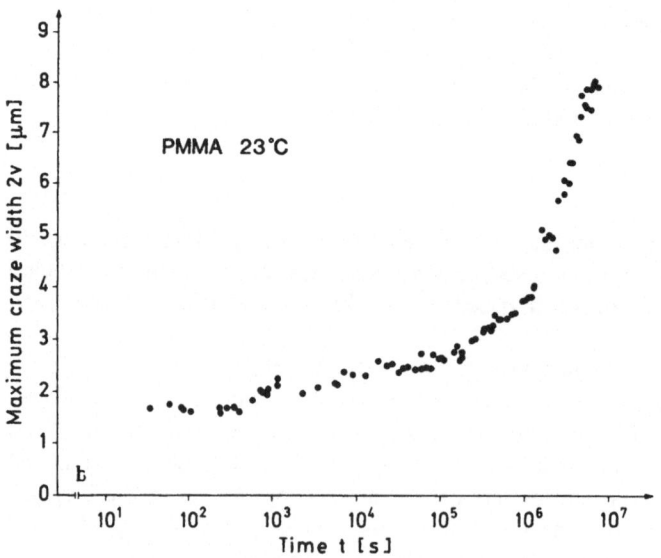

Fig. 23a and b. Growth of a primary craze at a stationary crack tip in PMMA in air, $K_I = 19\,N/mm^{3/2}$ [87]. **a** craze length s as a function of loading time t; **b** maximum craze width 2v as a function of loading time t

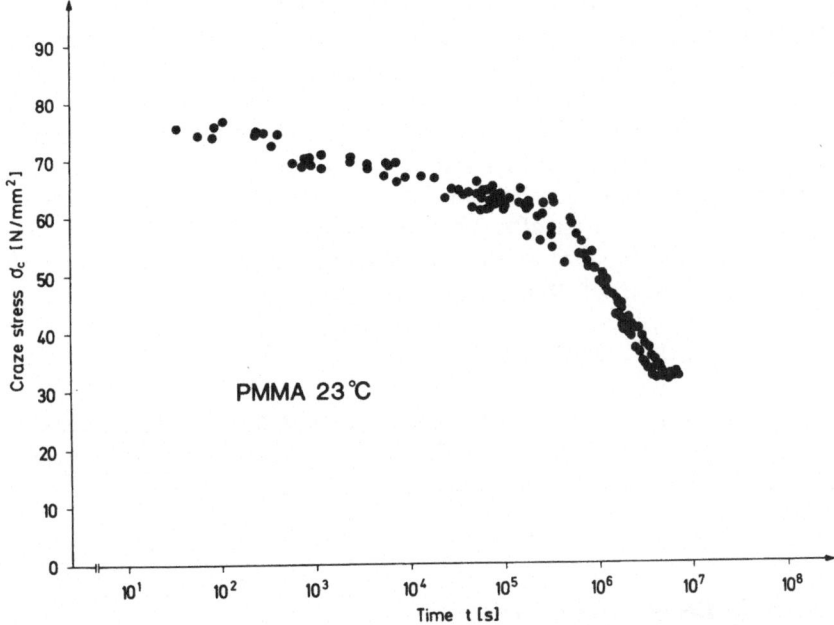

Fig. 24. Craze stress σ_c as a function of loading time t [87]

$K_I < K_{Io}$. This latter dependence, however, only becomes pronounced at longer loading times (e.g. at room temperature at times of the order of 10^5–10^6 s). At shorter times it is observed that the growth behavior depends on the initial craze size which in turn is strongly related to the method by which the crack and associated craze is introduced into the specimen (i.e. fatigue, cleavage or slow crack propagation). Thus, starting with a large initial craze, further growth will only occur once sufficient time has elapsed so that the time-dependent craze stress characteristic of that craze size is equal to or smaller than the imposed one [67]. This can be seen from the data in Fig. 23 and 24.

Observations on a low molecular weight PMMA have yielded some intriguing results [87]. It was established earlier in this work (Sect. 3.1.3) that the craze dimensions for crack propagation in low molecular weight PMMA are significantly smaller than those in the high molecular weight material (see Fig. 17). In contrast to this, crazes at stationary crack tips under steady, subcritical loads, behave in a remarkably similar way in the two materials, showing comparable increases in craze dimensions with time. This has important consequences for any molecular interpretation of the process since, for example, after 10^7 s at 20 °C and $K_I = 20$ N/mm$^{3/2}$ the fibrils spanning the crack tip in low molecular weight PMMA have reached a length corresponding to 25 fully stretched molecular chain lengths.

In order to gain an insight into the deformation behavior of the fibrils in the craze and the time-dependence of the fibrillation process it is useful to examine the growth in the width and the length of the craze.

An analysis of the ratio of 2v to s as a function of time (Fig. 23) shows that on a logarithmic time scale it decreases almost linearly from about 0.077 at short times to

0.045 at 10^7 seconds [89]. The short time ratio of 0.077 is the same as that already found over a wide range of crack speeds (Fig. 11). The decrease in 2v/s for stationary cracks indicates that the growth in the length s is larger than in width 2v. It is evident that the increase in length with time has to occur by the fibrillation of fresh bulk material. In order to get some idea of the contribution of fibrillation to the craze width, measurements were performed on unloaded crazes after they had been loaded for different times. The results showed that the unloaded craze width $2v_o$ increased with loading time from about 1 μm after 100 s to 3.5 μm after 10^7 s [89]. This indicates that the growth in maximum craze width 2v with time occurs by fibrillation of fresh bulk material. The strain ε(t) of fibrils, defined by:

$$\varepsilon(t) = [2v(t) - 2v_o(t)]/2v_o(t) \tag{14}$$

decreases, however, by a factor of 2 in the time range investigated. This decrease in fibrilar strain provided an explanation for the growth in craze width being smaller than that in craze length and also suggested that the fibrils neither break nor are the molecules pulled out [89].

3.3 Stress-strain Behavior of a Craze

The mechanical properties of a craze were first investigated by Kambour [46] who measured the stress-strain curves of crazes in polycarbonate (Lexan, $M_w = 35000$) which had first been grown across the whole cross-section of the specimen in a liquid environment and subsequently dried. Figure 25 gives examples of the stress-strain curves of the craze determined after the 1st and 5th tensile loading cycle and in comparison the tensile behavior of the normal polymer. The craze becomes more and more elastic in character with increasing load cycles and its behavior has been characterized as similar to that of an opencell polymer foam. When completely elastic behavior is observed the apparent craze modulus is 25% that of the normal polymer [46].

The mechanical behavior of a single craze at a stationary crack tip has been studied in PMMA [71,90] using the optical interference method for measuring the shapes of the craze under different external load levels up to the onset of slow crack propagation. The stresses on the craze have been calculated with the aid of the Dugdale model. Since the behavior of the fibrils spanning the crack tip is of special interest their displacements were considered. Figure 26 shows the thus determined stress-displacement curves in the temperature range from 20° to 80 °C [90]. All the curves are practically linear and they terminate at the onset of crack propagation. The maximum length of the stretched fibrils increases with temperature in this material ($M_w = 2200000$) whilst the stress which causes rupture of the fibrils at the crack tip decreases markedly with temperature (see Fig. 15 and Fig. 19). It has been pointed out [42] that these stresses correspond to the yield stresses of PMMA (see Sect. 3.1.4). It should also be noted that in PMMA it is often difficult to distinguish the yield stress from the fracture stress after yielding.

The influence of molecular weight M_w on the deformation behavior of the fibrils can be seen in Fig. 27. This shows that the stresses at rupture are approximately the

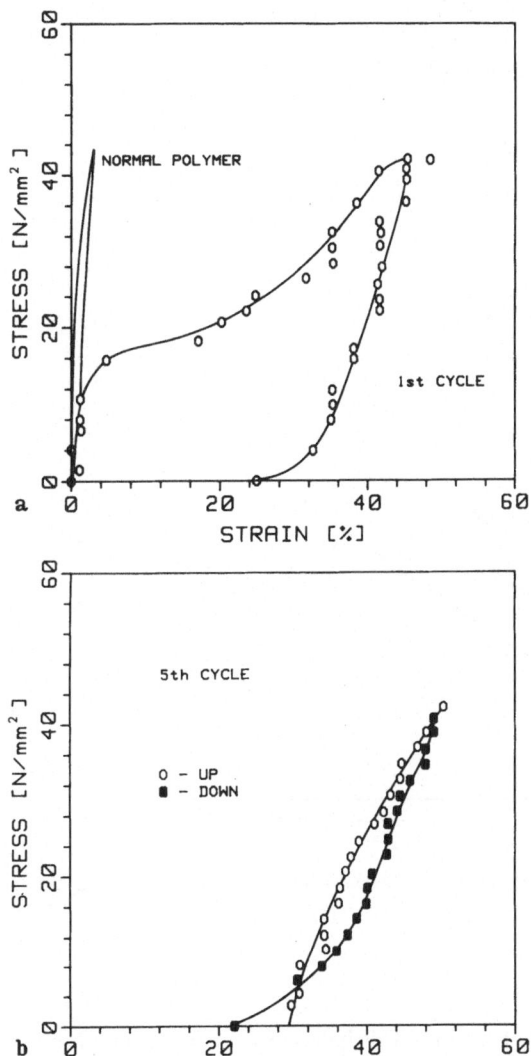

Fig. 25a and b. Stress-strain curves of a craze in polycarbonate grown across the total cross-section of the specimen [46]

same for both materials (see Fig. 19) and that the extensibility of fibrils as calculated from Eq. (14) is by more than a factor of 2 higher in the high molecular weight material. This implies that the fibrils in the lower molecular weight material are stiffer. Apparent Young's moduli of the craze material may be derived from Fig. 27. These are found to be about 1 % and 2 % respectively of the values of the bulk high and low molecular weight PMMA. The Young's moduli of the bulk, in contrast, show practically no dependence on molecular weight (see Fig. 18). The temperature depencence of the modulus is quite similar in bulk and craze material, both showing a decrease with temperature [90] above a critical temperature T_c [71]. It has been observed in a variety of different thermoplastics that, above this material specific temperature T_c a single craze will exist at the crack tip whilst below T_c multiple crazing in

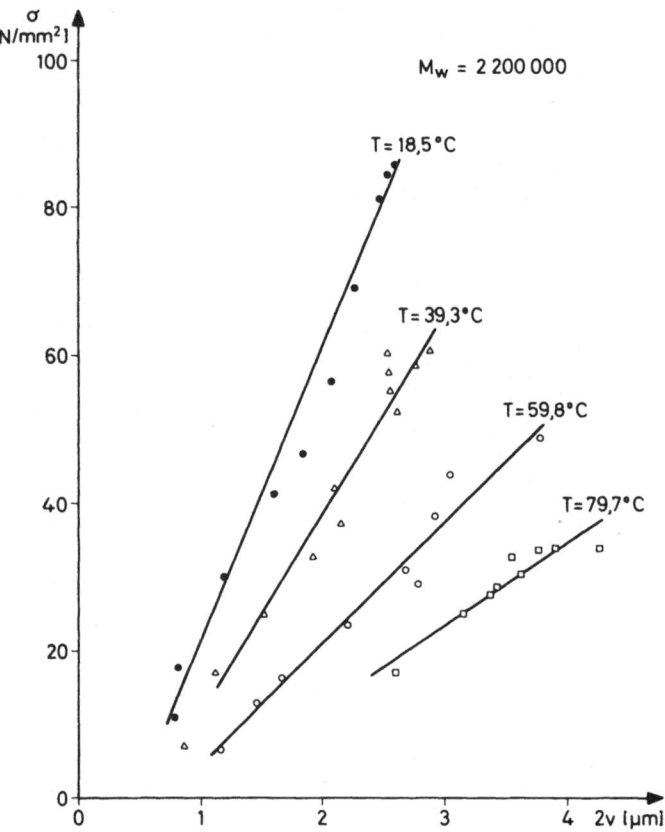

Fig. 26. Influence of temperature on the deformation behavior of a craze at the crack tip in PMMA [92]

form of sheaves of crazes will occur [91]. Recently, this behavior has been related to the different stiffness of the craze fibrils [71].

Although the modulus of the PMMA craze is very low in comparison to that of the bulk a considerable amount of energy can be stored in the viscoelastic deformation of the craze due to its large extensibility. This energy U_0 has been determined [90] and is shown as a function of temperature for two different molecular weight PMMA materials in Fig. 28 together with the corresponding strain energy release rate G_I as calculated from Eq. 8d using the experimental data for the onset of crack propagation ($\dot{a} \approx 10^{-5}$ mm/s). It can be seen that both these energies exhibit a similar temperature dependence and that U_0 amounts to 30%–40% of the total fracture energy G_I. The question of which form of energy the major part of G_I is converted into was considered in an earlier publication [92]. Based on the determination of the heat outputs from fast fractures in PMMA [93–95] it was shown that at the transition to fast fracture in PMMA nearly 60% of the strain energy release rate is directly converted into heat. Thus one can establish an energy balance for slow crack

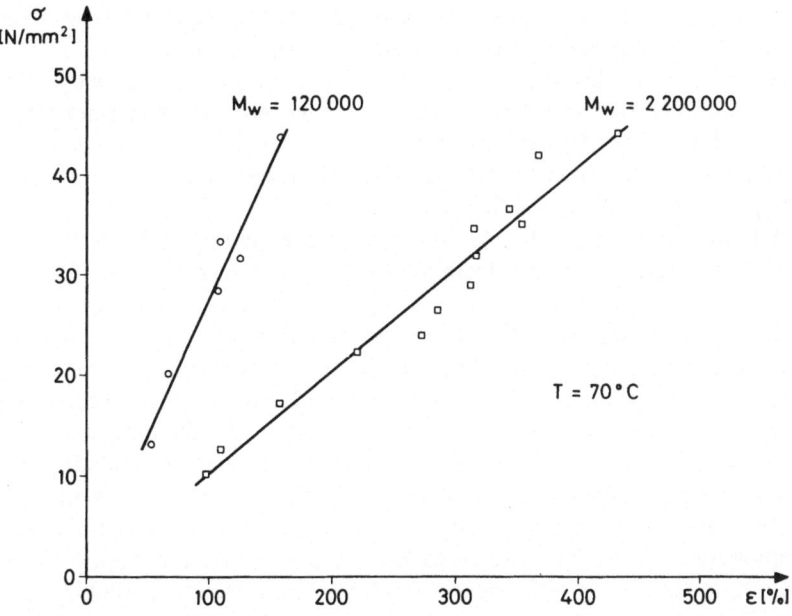

Fig. 27. Influence of molecular weight on the deformation behavior of crazes at the crack tip in PMMA [92]

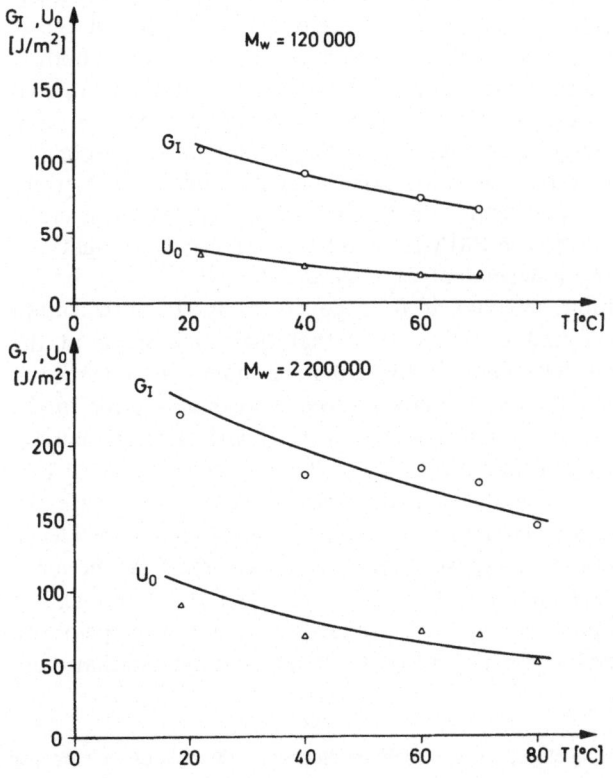

Fig. 28. Strain energy release rate G_I and viscoelastic energy U_0 of the craze as functions of temperature in PMMA [92]

propagation in PMMA which includes the principal energy components and the deformation behavior of the craze material at the crack tip as follows:

— during the formation of the crazes (phase I) fibrils are formed and drawn out of the bulk; during this frictional deformation heat is generated which amounts to about 60% of the strain energy release rate.

— in phase II the fibrils are heavily stretched (by about 200 to 400%) requiring a further 30% to 40% of the strain energy release rate.

— in phase III the stretched fibrils are ruptured by the advancing crack tip and snap back like relaxing springs[2] forming the oriented layer on the fracture surface.

The local heat output due to the fibrillation process in the craze zone induces a rise in temperature which depends on time, on the magnitude and size of the heat source, on the thermal properties of the material and also on the speed of the moving source. For a given heat source the temperature rise in the craze zone will reach a maximum when the conditions change from isothermal to adiabatic, that is to fast crack speeds. Williams [96,47] first used the isothermal-adiabatic effect to explain the observed jump in crack speed which is associated with a change in fracture surface morphology (Fig. 9). The rise in temperature leads to a softening of the material and a reduction in its resistance to fracture. The excess in available strain energy thus arising produces a sharp increase in the acceleration of the crack. It should also be mentioned that a second mechanism, the β-relaxation in PMMA, has been invoked [98] to explain the transition in crack speed behavior based on a correlation between the temperature dependence of a "time to failure" inferred from fracture experiments and the temperature variation of the reciprocal frequency of the β-relaxation peak. Support for both these mechanisms or even a combination of both [16] is to be found in the literature. By interpreting the crack jump as an isothermal-adiabatic transition, however, bounds of the crack speed immediately prior to the transition could be calculated [53] using a simplified model of a moving cylindrical heat source [99]. These data are found in good agreement with experimental values [53,54]. Further support for the isothermal-adiabatic transition mechanism for the jump in crack speed is to be found in the results for PMMA of different molecular weights [54]. The β-relaxation is substantially independent of molecular weight [139] and, hence, so also should be the transition behavior if it is governed by the β-relaxation mechanism. However, it was found experimentally that the crack speed at the transition was molecular weight dependent [54], the lower the molecular weight (and smaller craze zone size) the smaller was the crack speed immediately prior to the transition. In the context of the above paragraph it is worthwile to reexamine the data in Fig. 11 showing the measured craze dimensions with crack speed in PMMA.

At the lower end of the crack speed range the craze dimensions increase the slowlier the crack moves. This is to be seen as a reflection of the time dependence of craze growth discussed in Sect. 3.2.2 (slower speeds = longer times). At the other end of the speed range, at speeds rising to the transition speed ($\sim 10^2$ mm/s), the increasing build-up of heat in and around the craze leads to a rise in temperature. For example, using the data for PMMA (1) in Fig. 15, it can be estimated that there

---

[2] This leads to small pulses generating elastic waves which are damped in the bulk of the specimen.

is a temperature rise of 24 K at a crack speed of 10^2 mm/s. This is turn produces an increase in the craze dimensions.

Temperature rises far higher than this have been measured for fast running cracks when the conditions are fully adiabatic. Thus, in experiments using infrared spectroscopy on cracks in PMMA running at about 500 m/s temperature rises of some 450 K were observed [100].

4 Fatigue Loading

The crack growth behavior of polymers under cyclic loading has been intensively studied and the state of knowledge is also well documented by excellent, recently published review articles [101–103] among which the comprehensive book on "Fatigue of Engineering Plastics" by R. W. Hertzberg and J. A. Manson [101] should be explicitly mentioned.

In this contribution attention will be concentrated on the relationship between crack and craze growth in fatigue, and there will only be some brief introductory comments on macroscopic aspects of fatigue crack propagation in polymers. The main points will be the microscopic aspects of the so-called "continuous" and "discontinuous" crack propagation in fatigue. These two types of fatigue crack growth differ in their formation and growth mechanisms. In "continuous" crack propagation each loading cycle causes an increment in crack length, whilst in "discontinuous" crack propagation the crack tip remains stationary for many loading cycles and then at some particular cycle it advances by an increment². In the following review emphasis will be placed on new or improved results most of which have been obtained by the application of interference optics to these problems.

4.1 Macroscopic Aspects of Crack Propagation

It is now well established that fracture mechanics is an excellent tool to characterize the growth of a fatigue crack from an existing crack under model-I loading conditions. It allows the determination of the increase in crack length da per number of cycles dN as a function of relative stress intensity factor ΔK_I (see Fig. 31). Examples of the relationship da/dN versus ΔK_I are given in Fig. 29 for different polymers as measured by Hertzberg et al. [101,104]. In addition to the chemical composition of the polymer the configuration of the macromolecule plays an important role as can be seen from the effect of molecular weight on the fatigue crack resistance. In Fig. 30 the results on PVC [105,106], PMMA [107] and PC [108] are given using an evaluation by Hahn et al. [109]. Although the constant ΔK_I-values applied had different amplitudes the fatigue crack propagation behavior is quite similar for the three polymers. They all show

---

² Basically both types of fatigue crack propagation are discontinuous; hence the so-called "discontinuous" type would be better characterized as "retarded" crack growth [114].

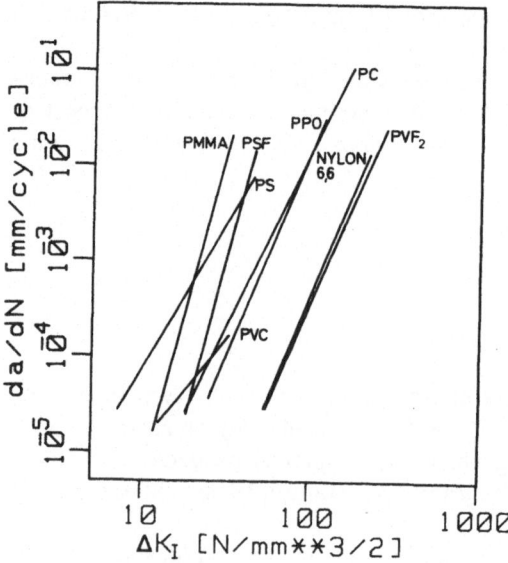

Fig. 29. Fatigue crack propagation rate da/dN as a function of relative stress intensity factor ΔK_I in different polymers (Hertzberg et al. [101,104])

a sharp decrease in the crack propagation rate with increasing molecular weight and — if an appropriate linear scale is taken — a levelling off at higher molecular weights [109]. Copolymerisation and plasticizing can also effect fatigue crack propagation as can be seen from the data on PMMA [107] and PVC [105]. Additives, e.g. stabilizers, may also have an influence. It is often due to differences in these intrinsic parameters that test results in the literature on commercially available polymers from different manufacturers are not fully comparable.

The influence of mechanical parameters such as mean stress or ratio between minimum and maximum load on fatigue crack propagation has been investigated for

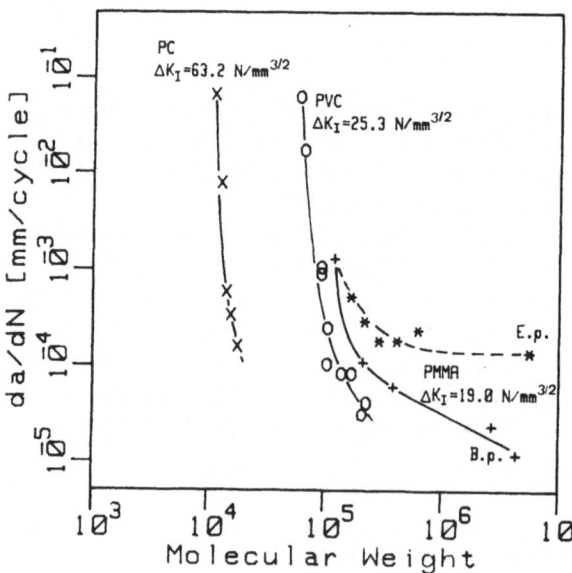

Fig. 30. Effect of molecular weight on crack propagation rate da/dN in PC [108], PVC [105,106] and PMMA [107]. E.p. = Emulsion polymerized, B.p. = Bulk polymerized

various polymers. These effects are usually incorporated into the well known Paris law using empirical parameters. It is also interesting that an influence of test frequency has been observed for different polymers in which the fatigue crack propagation rate decreases with increasing frequency. No details will be discussed here since most of the available information is given in the already mentioned reviews [101–103]. However, a point worthy of mention in the context of the following Sections is the way in which the frequency sensitivity effect has been interpreted physically [101]. Heat is produced in the plastic zone at the crack tip due to the fatigue loading. At higher frequencies the increase in temperature becomes sufficient to enhance the yielding process leading to an increase in the crack tip radius. This larger radius at the crack tip reduces the effective ΔK_I which in turn, produces a decrease in the fatigue crack propagation rate. No direct evidence of the enhanced radius at the crack tip — the crack opening displacement — has ever been cited. However, in principle it should be possible to investigate this problem using a more elaborate version of the optical interference technique.

4.2 Interference Optics in Fatigue

The optical interference method used in fatigue is essentially the same as that already applied to static and continuously moving cracks under static load and discussed in earlier sections of this review. Two procedures have been applied which differ in technical complexity and in the information which can be derived from them. The more straightforward procedure consists of applying the optical interference technique to specimens after they have been loaded in fatigue. In this way the lengths of craze zones have been determined in vinylurethane [110], PVC [111] and PC [108]. Using

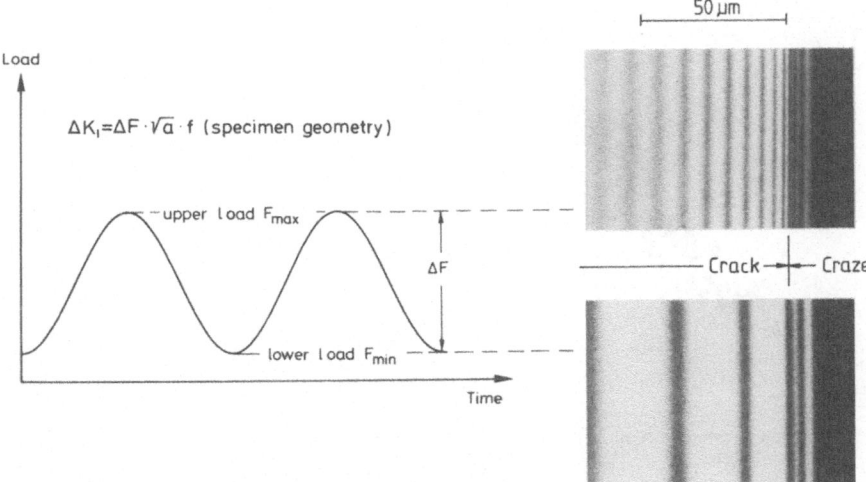

Fig. 31. Example of interference fringe patterns measured during fatigue crack propagation ($\omega = 50\,\mathrm{Hz}$; $da/dN = 1.2 \cdot 10^{-5}$ mm/cycle) in PMMA at upper and lower loads [114]

the other, more elaborate experimental arrangement [112, 113] the following relevant parameters can be simultaneously recorded during fatigue crack propagation: the interference fringe pattern, the load acting, the crack length a, and the number of load cycles N. From this the craze zone dimensions (length, width and contour), the stress intensity factor K_I, the crack propagation rate da/dN and crack speed $\dot{a}$ can be derived. Thus a full characterization of crack and craze growth under fatigue loading can be obtained. Figure 31 shows an example of the interference fringe systems generated during fatigue crack propagation at a loading frequency of 50 Hz in PMMA photographed at the upper and the lower load. The evaluation of the interference fringe systems is performed as described in Sect. 2.2.

4.3 Microscopic Aspects of Continuous Crack Propagation

4.3.1 Craze Zone Size

Whilst Fig. 29 shows the macroscopic fatigue crack growth behavior, Fig. 32 sheds some light on the material response in the micro region at the crack tip by representing the maximum craze width $2v_c$ as function of crack propagation rate da/dN. It can be seen that during fatigue crack propagation in high molecular weight PMMA the maximum craze width $2v_c$ (and hence also the COS) is not constant but increases

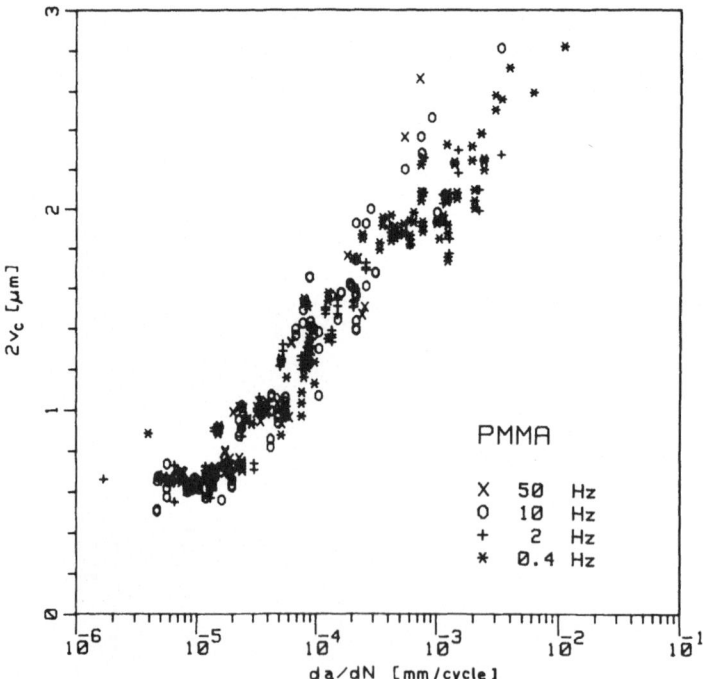

Fig. 32. Maximum craze width $2v_c$ at the crack tip as a function of crack propagation rate da/dN [114]

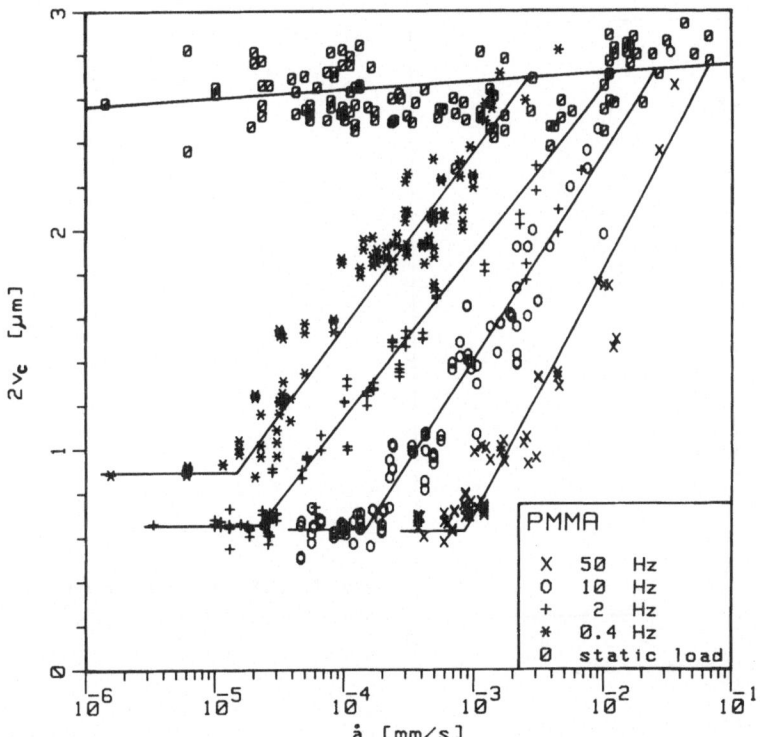

Fig. 33. Maximum craze width $2v_c$ as a function of crack speed $\dot{a}$; reference curve (at the top) is for continuously moving cracks under quasistatic load (comp. Fig. 11a) [114]

by a factor of nearly 5 (from 0.6 μm to about 2.8 μm) in the predominantly investigated crack propagation range between 10^{-5} and 10^{-2} mm/cycle [114]. It should be mentioned that the plotted points were obtained at four frequencies in a range from 0.4 to 50 Hz. In all about 350 individual points were measured. Figure 32 suggests a lower limit of $2v_c$ at crack propagation rates smaller than 10^{-5} mm/cycle. As has been shown [113,114], an upper limit can be derived by comparing the curve with the results for continuously moving cracks under constant load using the crack speed $\dot{a}$ as a common measure. Since $\dot{a}$ is equal to da/dN · ω (ω = frequency) the data shown in Fig. 32 will be separated into different curves if they are replotted in Fig. 33 according to their different frequencies. At lower frequencies the increase in $2v_c$ occurs at lower crack speeds $\dot{a}$. At the top of Fig. 33 the maximum craze width $2v_c$ for continuously moving cracks is nearly constant at 2.7 ± 0.2 μm [59]. This "quasistatic curve" obviously forms the upper limit of the maximum craze width $2v_c$ which can be attained under fatigue loading.

The craze length s in high molecular weight PMMA is shown in Fig. 34 as a function of ΔK_I. This curve is also based on about 350 individual measurements. The craze length s increases linearly with ΔK_I, from about 10 μm up to about that value measured for a continuously propagating crack, i.e. $s = 35 \pm 3$ μm [59]. An increase of s with ΔK_I can also be deduced from the data on craze lengths in vinylurethane [110] and PC [108] measured in the relaxed craze state as is shown in Fig. 35. It is interesting

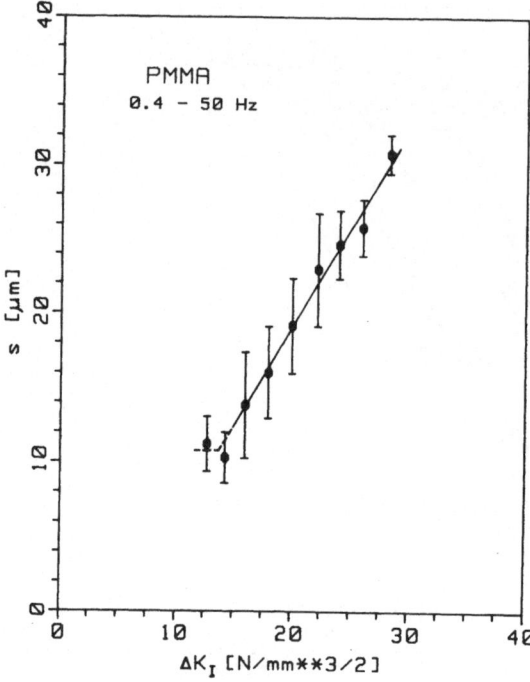

Fig. 34. Maximum craze length s at the crack tip as a function of ΔK_I [115]

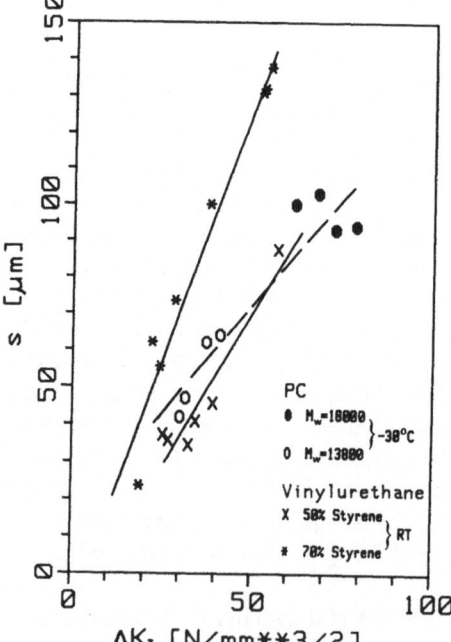

Fig. 35. Increase in craze length s with ΔK_I in PC [108] and Vinylurethane [110] measured in the unloaded state

to note that the behavior of the two PC-samples is complementary to each other, with the higher molecular weight material exhibiting the larger craze length.

In order to obtain information on the craze growth and on the fibrillation process the maximum craze widths $2v_c$ are plotted against craze lengths s for high molecular weight PMMA in Fig. 36 [114]. In addition to the individual data points from fatigue crack measurements the data from continuously moving cracks are uncluded as a large circle, which represents 113 measurements.

It can be seen from Fig. 36 that the two characteristic craze dimensions (i.e. maximum width and length) remain practically proportional to one another up to the values determined for continuous crack propagation under constant load. It is evident that the increase in craze length must occur by fibrillation of the amorphous bulk polymer. However, the thickening of the craze may either be due to additional fibrillation or to enhanced stretching of already formed fibrils or even a combination of both processes. As was shown in Chapter 1 for craze growth in thin polymer films of PS [33] and PC [116], under quasistatic tensile load the increase in craze width occurs by fibrillation of fresh bulk material. It should be emphasized, however, that there is a difference between the two types of experiment. In the fatigue crack growth experiments crazes of different sizes are compared which were generated by different load histories, i.e. different values of ΔK_I and of da/dN. In the quasistatic tensile experiments on thin films referred to above [33,116] on the other hand, the observations of the increase in craze size were performed on the same single craze[3].

Further information on craze thickening in fatigue can be obtained by comparing the width $2v_c$ of the strined craze with the craze width $2v_0$ at the lower fatigue load.

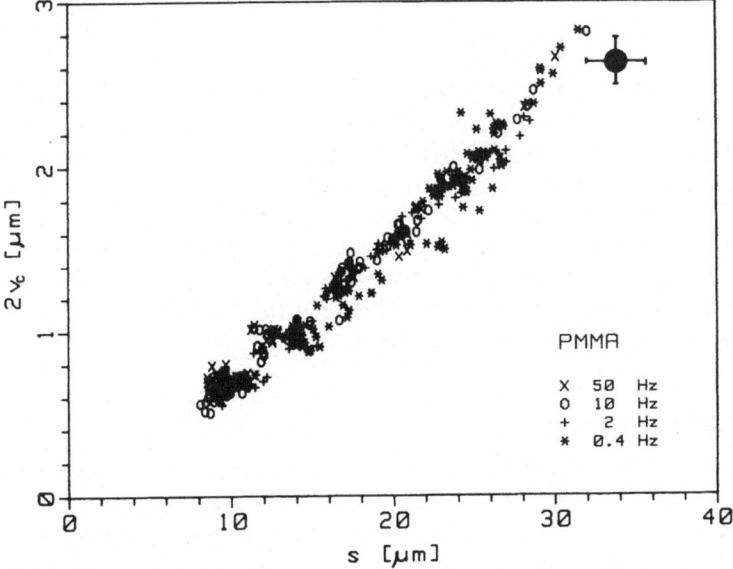

Fig. 36. Relationship between craze width $2v_c$ and length s during crack propagation under fatigue and quasistatic (circular area) loading [115]

---

[3] Note added in proof:
It was shown recently by Brown [143] that in PS the craze structures are similiar in form for thin and thick films and that, however, the fibrils are about three times larger in thin films.

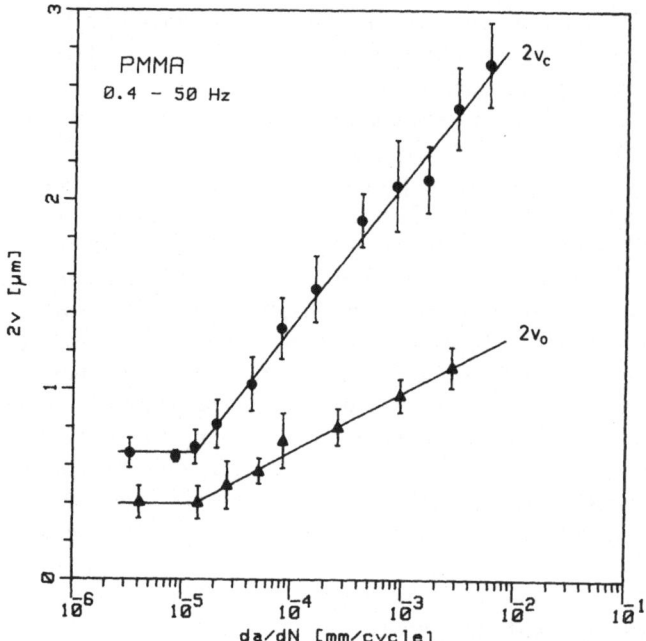

Fig. 37. Craze width at the crack tip ($2v_c$ under upper and $2v_0$ under lower load) as a function of crack propagation rate da/dN [114]

The ratio of these is assumed to be a measure of the amount of fibrillation. From such results [114] shown in Fig. 37 it may be deduced that both processes (i.e. new craze material being formed and also an increased stretching of the fibrils) occur with increasing crack propagation rate. If the strain in the craze zone is defined by $\varepsilon = (2v_c - 2v_0)/(2v_0)$ then the strain on the fibrils is nearly doubled from about 70% at low crack propagation rates of 10^{-5} mm/cycle up to 120% at 10^{-2} mm/cycle, thus approaching the magnitude measured for static cracks [90]. This clearly demonstrates, that the assumption of a constant ratio between the thicknesses of relaxed and extended crazes is not generally valid in fatigue.

4.3.2 Derived Material Parameters

For craze zones at the tips of static and of moving cracks under quasistatic loading conditions it has been shown in Sect. 3 that the normal stress σ_c acting on the craze zone and the modulus E can be derived from the measured craze dimensions using the Dugdale model.

In the case of fatigue loading much less information is available as yet. This is probably due to the difficulty of obtaining the basic data (craze width and length) in simultaneous measurements. Such measurements and evaluations have been reported for high molecular weight PMMA [114] and in Fig. 38 the thus derived craze stresses σ_c are plotted as a function of crack speed á. In this way the fatigue crack growth data can be compared with those for continuously moving cracks [65] shown by the straight line. In the crack speed range investigated σ_c increases

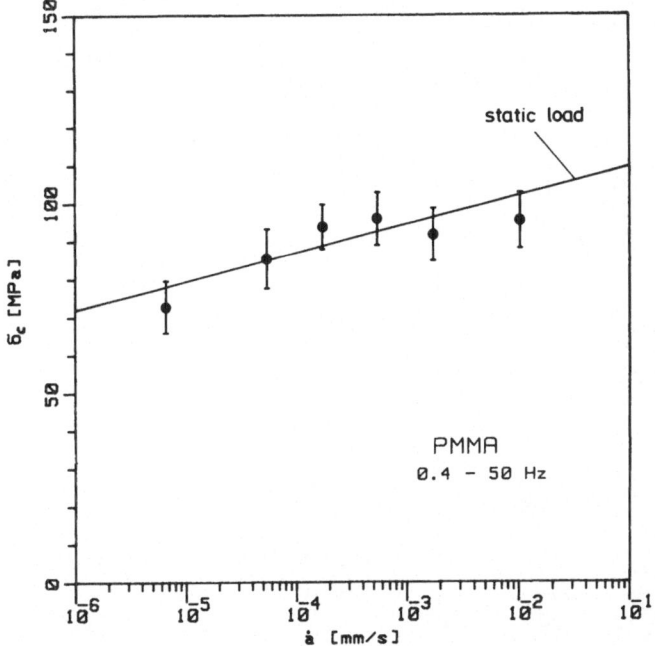

Fig. 38. Craze stress σ_c during fatigue and static loading as a function of crack speed á [114]

from about 70 to 110 N/mm² and the general tendency of the data determined under fatigue and static loading seems to be quite similiar.

The magnitude of Young's modulus as determined from the craze dimensions lies in the same range as that measured for crack propagation under constant load; the fatigue results do not show any significant change of E with crack speed. There is, however, a slight increase of E with frequency ω (Fig. 39), which has also been reported [117] for moduli measured under small strains. Since the values of the latter lie about 50% higher than of those determined here, one has to conclude that the moduli here must be characteristic for large strains. This conclusion becomes even more evident if it is recognized that the material around the craze zone is highly strained due to the high stresses around the crack tip and that the Dugdale model strictly applies to the contour of the elastic/plastic boundary in that region [115]. Thus, a similar result is derived as that already obtained for continuously moving cracks (see Sect. 3.1.2).

In this context it is relevant to consider a controversy [117-119] concerning how fatigue crack growth in polymers may be modelled. In his model Williams [118] assumes a constant crack opening displacement during fatigue and proposes a two-stage craze zone in which the newly formed craze material in a small region at the craze tip experiences very high stresses while the main part of the craze sustains a much lower stress. Hertzberg et al. [117] refer to experimental values of crazing stresses of a number of polymers and suggest that in these cases the craze experiences a uniform stress similar to that postulated in the Dugdale model.

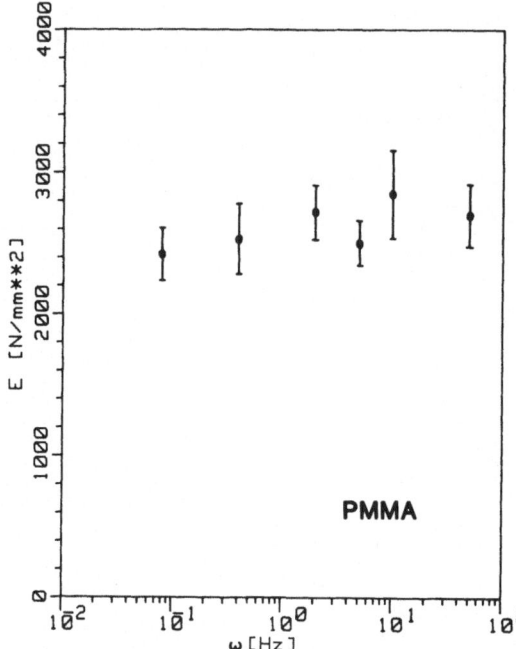

This is confirmed by the experimental data and especially the extended analysis of the measured craze contours in high molecular weight PMMA. In this material an excellent fit is achieved by using a uniform craze stress in the Dugdale model [115].

4.3.3 Fatigue Striations

In the literature on fatigue crack propagation in metals and, later, in polymers it has been well documented that fracture surfaces can exhibit ripple markings which are due to the cyclic loading process.

In polymers two sets of markings on fatigue fracture surfaces have been observed which differ distinctly in their loading genesis. One set is called striations, where each striation is found to correspond to the incremental advance of the crack front caused by just one loading cycle (this has also, somewhat confusingly, been called continuous crack growth). Figure 40 shows an example of striation in PC [122].

In metals striations were already fairly well documented when the investigation of the morphology and formation mechanism of polymer fatigue striations began. Mc Evily et al. [120] observed such striations on the fatigue fracture surfaces of PC and PE and described their formation in terms of a stress relaxation mechanism originally proposed by Laird and Smith [121] for metals. The important aspects of this model are as follows. Creation of new fracture surface occurs only during the tensile period of the loading cycle. At the same time blunting of the crack tip occurs which tends to limit the extent of crack growth. During unloading, the material surrounding the crack contracts elastically and imposes a residual compressive stress onto the severely plastically deformed material at the tip. As a consequence, a resharpening of the crack tip takes place, and the residual ductility of the material

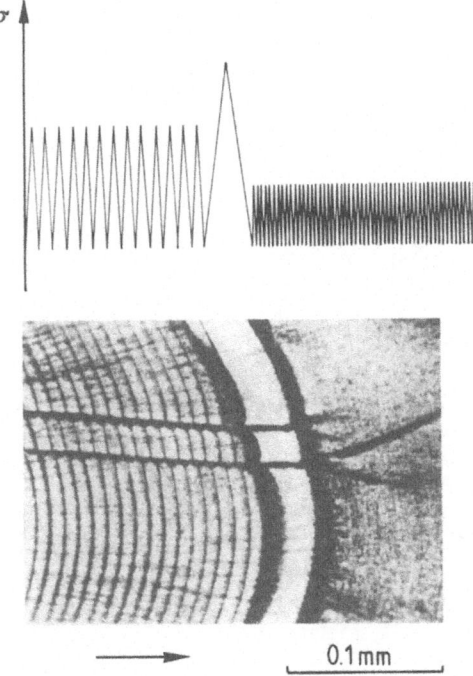

Fig. 40. Fatigue striation on the fracture of PC (lower part) due to different cyclic loading (upper part) [Jacoby, G., Cramer, Ch., Rheologica Acta 7, 23 (1968), Dr. Dietrich Steinkopff Verlag]

0.1 mm

just ahead of the crack tip is reduced because of the severe irreversible plastic deformation it has undergone. These last two factors serve to promote crack growth in the next loading cycle.

Jacoby et al. [122, 123] who intensively investigated different types of markings on the fracture surfaces of PC gave a refined version of this model. As one sees from Fig. 41 the advance of the fatigue crack front was also associated with that portion of the cycle where the load begins to rise.

A similar result for an aluminum alloy was reported by McMillan and Pelloux [124] who like Hertzberg [125] found sawtooth type striations in non-matching fracture surfaces. Other types of striations on ductile fracture surfaces in metals have been discussed by Laird [126] who pointed out the influence of different metallurgical structures and of high and low stresses.

For high-amplitude loaded PE undergoing large deformations Mc Evily et al. [120] observed troughs in the fracture surface, whilst in PMMA Feltner [127] observed a rough sawtooth crack profile with the teeth on one side of the crack fitting into notches on the other side.

To sum up, the morphology of the striations seems to be influenced by both material and mechanical loading conditions. Although the different authors generally use sophisticated arguments, they base their conclusions in most cases on subsequent inspection of the fracture surface by light or electron microscopy and on correlation with certain load programs which allow an association with crack advance (see Fig. 40). It seems, however, as if features have been ascribed to the fatigue process in polymers which may only correctly apply to metals.

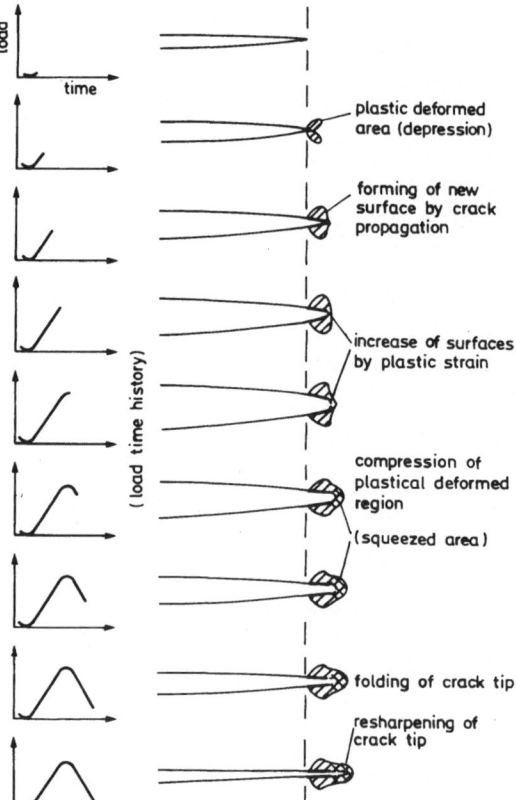

plastic deformed
area (depression)

forming of new
surface by crack
propagation

increase of surfaces
by plastic strain

compression of
plastical deformed
region

(squeezed area)

folding of crack tip

resharpening of
crack tip

Fig. 41. Formerly proposed mechanism for the forming of fatigue striations in polymers [122]

In order to develop a modified model of fatigue crack growth in glassy thermo-plastics the recently available detailed information on crazing (normal yielding), fibrillation and fibril behavior has to be taken into account [see Chapter 1 of this book]. In addition the results of recent interference optical investigations [115] may be helpful. Figure 37 shows the craze widths at the crack tip in high molecular weight PMMA under upper ($2v_c$) and lower load ($2v_0$) and hence the distances which are spanned by the fibrils during a loading cycle. On the other hand no relevant change in craze length has been measured at upper and lower loads, thus the smaller craze width at lower load must be due to a folding of the fibrils as already pointed out by Schirrer and Goett [71] and indicated in Fig. 7. To get direct information on the onset of crack propagation in PMMA during fatigue a video-film was taken using the optical interference microscopy [140]. At slow loading frequencies of a few Hz the frame rate of 25 pictures per second was high enough to get sufficient resolution. The results are shown schematically in Fig. 42. It should be emphasized that crack propagation occurs only in a small time interval before and after the load maximum. In the film it could clearly be seen that at first the crack propagated into the craze and a little later both crack and craze grew at the same rate. Thus, in complete contrast to previous assumptions the results here clearly demonstrate that a resharpening of the fatigue crack tip does not occur and that crack propagation is

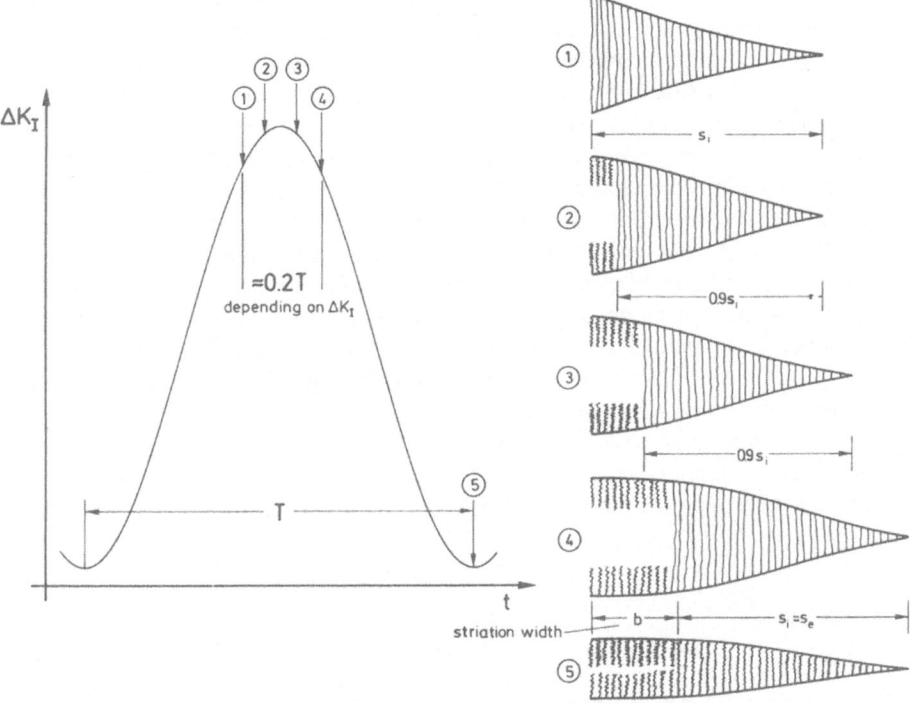

Fig. 42. Schematic representation of correlation between loading phase with crack propagation, craze growth and striation formation during continuous crack growth (in PMMA) [140]

not limited by crack tip blunting. In fact the crack starts to propagate when the crack tip has become blunted. The reason for this behavior is that is just when the crack tip is blunted that the stress on the craze zone is highest and the fibrils are stretched most severely and will fail.

The spacings between fatigue striations have often been correlated with the plastic zone size and especially its length. A direct comparison between striation spacing and length of the craze zone in PMMA can be made through the interference optical photograph [112] shown in Fig. 43. The crack was driven at a loading

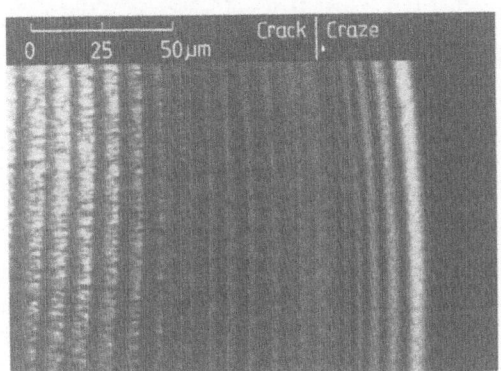

Fig. 43. Optical interference photograph showing fatigue striations on the fracture surface and craze zone in a partially broken PMMA specimen [112]

frequently of 10 Hz and moved by a speed $\dot{a}$ of nearly 10^{-1} mm/s and was then
arrested by unloading (this was necessary in order to get a sharp picture of both the
striations on the fracture surface and the interference fringes). It can be seen from
Fig. 43 that in this case the craze length is significantly larger than the striation
spacing. An indirect comparison between craze dimensions and line spacing is obtained
from Fig. 44 [114]. Here craze length s (upper curve) and maximum craze width $2v_c$
(lower curve) are plotted against crack propagation rate da/dN. In high molecular
weight PMMA the latter corresponds to the striation spacing since the crack grows
by one striation during one cycle. The additional dotted lines indicate in which
regions the dimensions of the craze would directly correspond to the striation
spacing. Thus the line at 45 degrees indicates where the increase in crack length
during one cycle is equal to the craze dimensions. It can be seen from Fig. 44 that up
to about $2 \cdot 10^{-3}$ mm even the craze width is always larger than the increase in crack
length during one cycle. Only at high crack propagation rates ($> 3 \cdot 10^{-2}$ mm/cycle)
will the striation spacing be larger than the craze length.

It can thus be concluded that during the so-called continuous mode of fatigue crack
propagation there is no simple correlation between fatigue striation spacing and craze
dimensions.

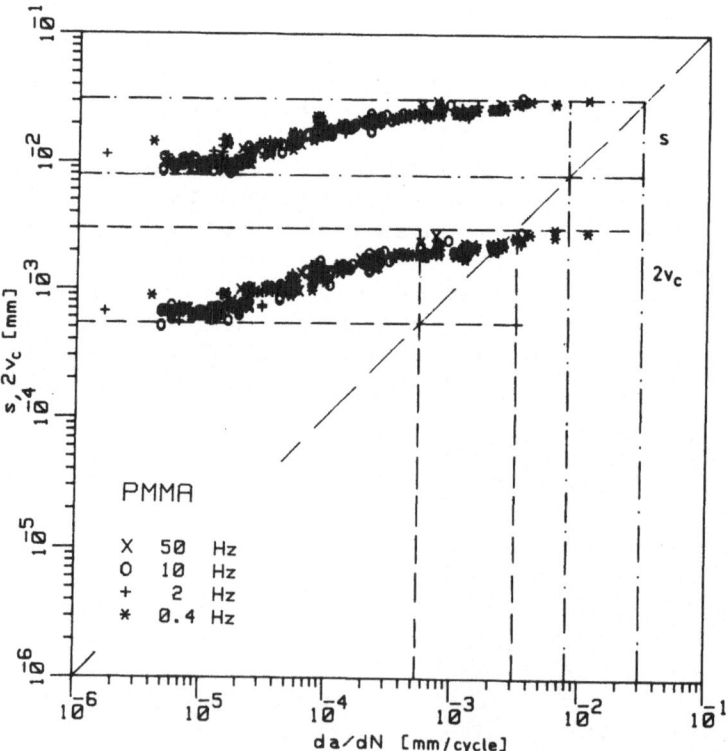

Fig. 44. Characteristic craze dimentions s, $2v_c$ versus crack propagation rate da/dN [114]

4.4 Microscopic Aspects of Discontinuous Crack Propagation

4.4.1 General Features

Elinck et al. [128] first observed the so-called discontinuous fatigue crack growth in PVC. They measured the crack length at the specimen surface as a function of time and correlated it with the markings on the fracture surface (Fig. 45). Taking the applied load frequency of 0.6 Hz into account they noted that a great number of fatigue cycles was needed for the formation of successive markings on the fracture surface. The crack propagated stepwise about once every 370 cycles.

In addition to PVC [128, 129, 111] discontinuous fatigue crack growth was subsequently observed in many other polymers, e.g., vinylurethane [110], polystyrene (PS) [130, 131], PMMA and polysulphone (PSF) [132], polycarbonate (PC) [132-134], acrylonitrile-butadiene-styrene (ABS) and polyacetal POM [135], polyethylene (PE) [136].

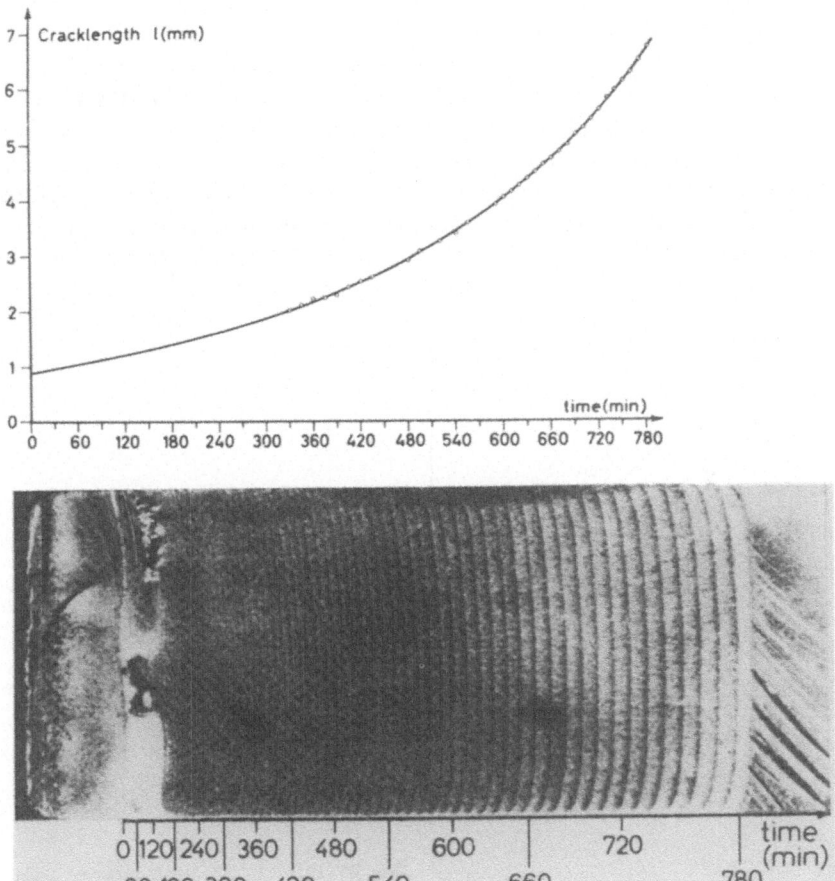

Fig. 45. Discontinuous crack growth in PVC. [Elinck, J. P., Bauwens, J. C., Homès, G., Int. J. Fracture Mech. 7, 277 (1971), Wolters-Noordhoff]

It has been pointed out by Hertzberg et al. [101, 105] that the formation of discontinuous growth bands is dependent on the molecular weight and the molecular weight distribution of the polymer and on mechanical loading conditions and that low molecular weights and low stress intensity factors ΔK_I seem to favour their formation. For example, low molecular weight PMMA may exhibit discontinuous band growth at low ΔK_I-levels, whilst high molecular weight PMMA shows striations at all ΔK_I-levels. In PVC, on the other hand, discontinuous growth bands are found at all ΔK_I-levels.

Skibo et al. [132] noted that in discontinuous crack growth the number of cycles between two successive bands depends on the polymer and strongly decreases with increasing relative stress intensity factor ΔK_I as shown in Fig. 46. An extrapolation of these data to the value of one cycle/band — that is to continuous crack growth — gives ΔK_I-levels which are too large, thus indicating that the mechanisms operating in continuous and discontinuous crack growth are fundamentally different.

From Fig. 45 it can be seen directly that the distances between discontinuous growth bands (DGB) increase with crack length and — since the load amplitude was constant — also with ΔK_I. A systematic study of the dependence of band size on ΔK_I was performed by Skibo et al. [132] whose results on different glassy polymers are shown in Fig. 47. From the slope of the curves for PS, PVC and PMMA it can be seen that the length of the band size is proportional to the second power of ΔK_I, a relationship also known to apply to plastic zone sizes. The deviation in the curves for PC and PSF is believed to result from an associated transition in fracture roughness [132]. The interpretation of the band spacing as the plastic zone length was first given by Elinck et al. [128] and yield stresses of reasonable magnitude were derived for different polymers by correlation with the Dugdale model [128, 130, 132, 135].

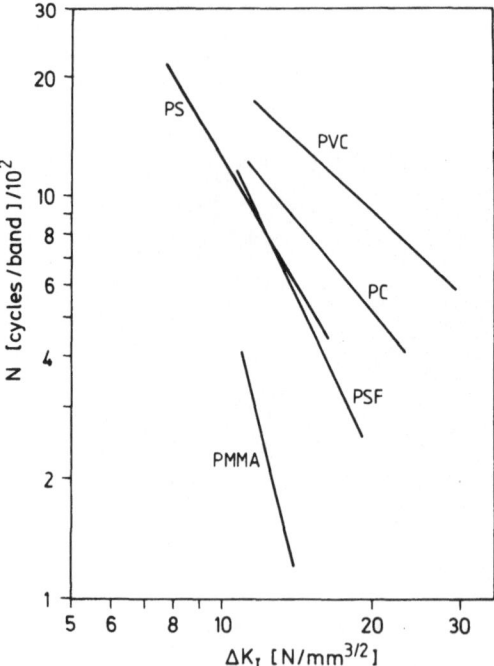

Fig. 46. Effect of ΔK_I on number of cycles N per growth band for different polymers (Skibo et al. [132])

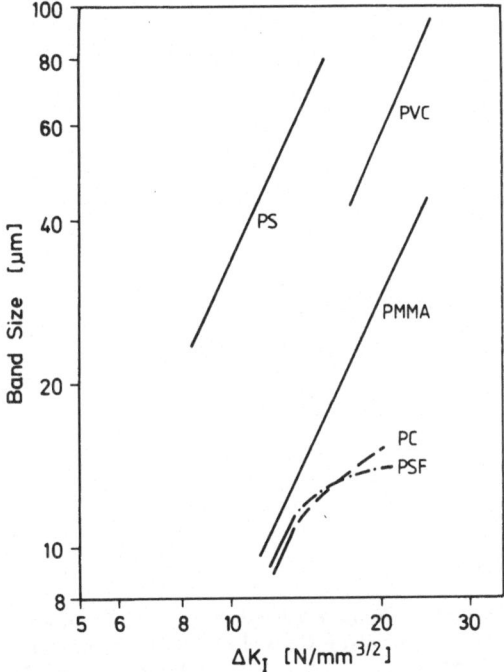

Fig. 47. Dependence of band size on ΔK_I for different polymers. (Skibo et al. [132])

4.4.2 Craze Development between Crack Jumps

Hertzberg and Manson [129] observed fatigue crack growth microscopically at the surface of a PVC specimen. They showed that the craze growth at the crack tip was continuous during fatigue and that the crack tip advanced discontinuously over a distance comparable to the extent of previous craze development (Fig. 48). From electron fractography of the fatigue growth bands on the fracture surface they concluded that the crack is arrested at the end of the craze with the crack tip being blunted by a stretching process with tear dimple formation. Their schematic representation [130] of discontinuous crack growth is also shown in Fig. 48.

Applying the optical interference method to an unloaded crack in PVC which had previously been propagated under fatigue load, Mills and Walker [111] found qualitatively that except at very low R ratios ($R = K_{min}/K_{max}$) the crack does not jump the full length of the existing craze.

Based on cinematographic studies of fatigue crack propagation in a special PVC (plasticized with 6% dioctylphthalate), which gave an increased band size and a lower yield stress [105], Hertzberg et al. [135] found that the craze grows to about 80% of its final length within the first 10% of the band's cyclic life, that then little growth occurs for a substantial portion of the total craze life, whilst the final 10% of craze length growth occurs during the final 10% of the band's cyclic life. Changes in craze thickness could not be measured.

A deeper insight into the growth behavior of the craze zone at the crack tip during retarded crack growth in PVC has been obtained by recent optical interference measurements [137] on which the following is mainly based. Figure 49 is an

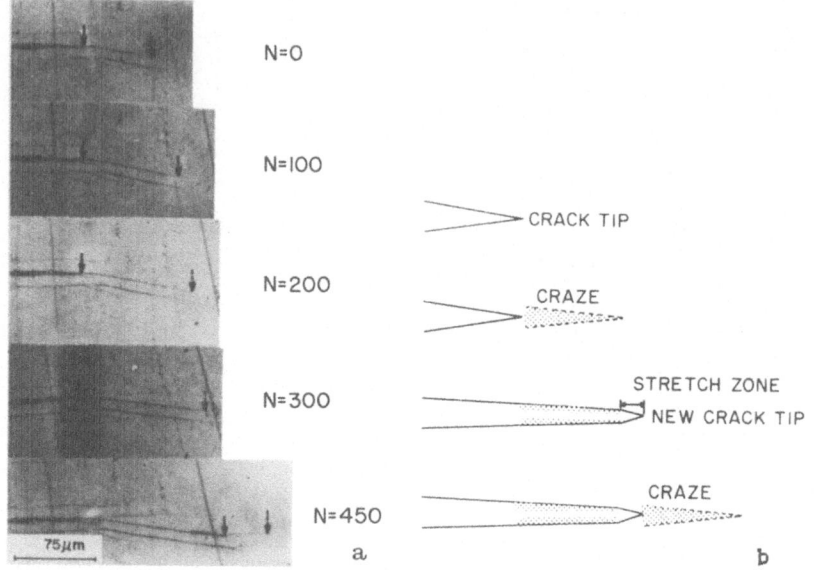

Fig. 48a and b. Discontinuous crack growth process. **a** Composite micrograph of PVC showing position of craze (↓) and crack (↓) tip at given cycles intervals; **b** Derived model of discontinuous cracking process. [Skibo, M. D., Hertzberg, R. W., Manson, J. A., Kim, S. L., J. Mater. Sci. **12**, 531 (1977), Chapman and Hall, Ltd.]

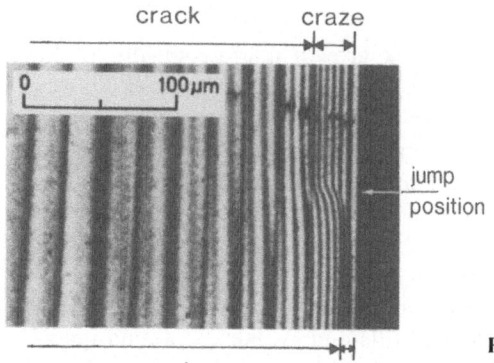

Fig. 49. Optical interference photograph taken at the moment of a crack jump in PVC [137]

optical interference photograph showing an advancing crack front in PVC. In the upper part of the photograph the crack is still stationary with the craze having its full length whilst in the lower part the crack has already moved forward. By comparing both parts it can be seen directly that in this case the crack separates the fibrillar craze matter at about two thirds of the craze length without penetrating into the front part of the craze. Additionally it can be noticed that at the position of the new crack front the maximum craze width is larger than at the original craze position. An exact evaluation of the photograph shows that a small increase in craze length must also have occurred during the crack jump.

The quantitative development of craze length s and maximum craze width 2v at the crack tip are given in Fig. 50 as functions of the cycle number N for several successive crack jumps. In this experiment ($\Delta K_I = 15.2 \text{ N/mm}^{3/2}$, R < 0.1) the crack remained stationary up to a critical cycle number N_e which is connected with a critical end length s_e or end width $2v_e$ respectively. Together with the result of Fig. 49 it is obvious that the crack does not go through the craze all the way up to the craze tip as is usually assumed [130, 131]. It stops well ahead of the craze tip, even at this low R-ratio. From Fig. 50 it can more precisely be derived that just after a crack jump the craze starts to lengthen with an initial craze length s_i of nearly 50 % of the old length; thickening begins with an initial value $2v_i$ of nearly 60 %. This relative difference between the initial length and width of the craze has to be seen as a direct effect of the crack jump and it is balanced by the following growth behavior. After an initially steeper increase in craze length, further craze growth occurs relatively linearly up to the critical end dimensions.

As a consequence of these results the model of discontinuous crack growth originally derived by Hertzberg et al. [129, 130] and reproduced in Fig. 48b must be

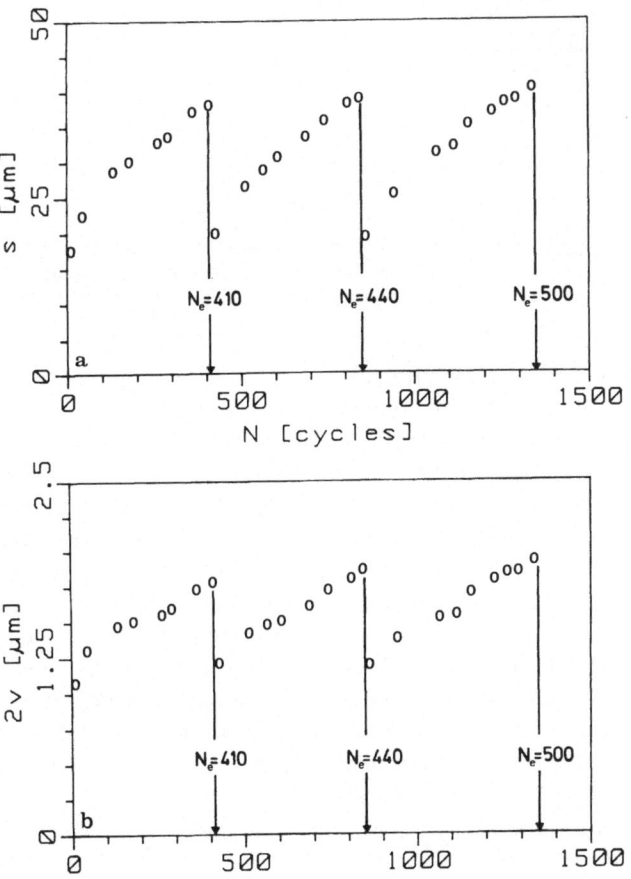

Fig. 50a and b. Craze growth between successive crack jumps in PVC at $\Delta K_I = 15.2$ N per mm$^{3/2}$ [137]. **a** Craze length s versus loading cycles N; **b** Craze width 2v direct at the crack tip versus loading cycles N

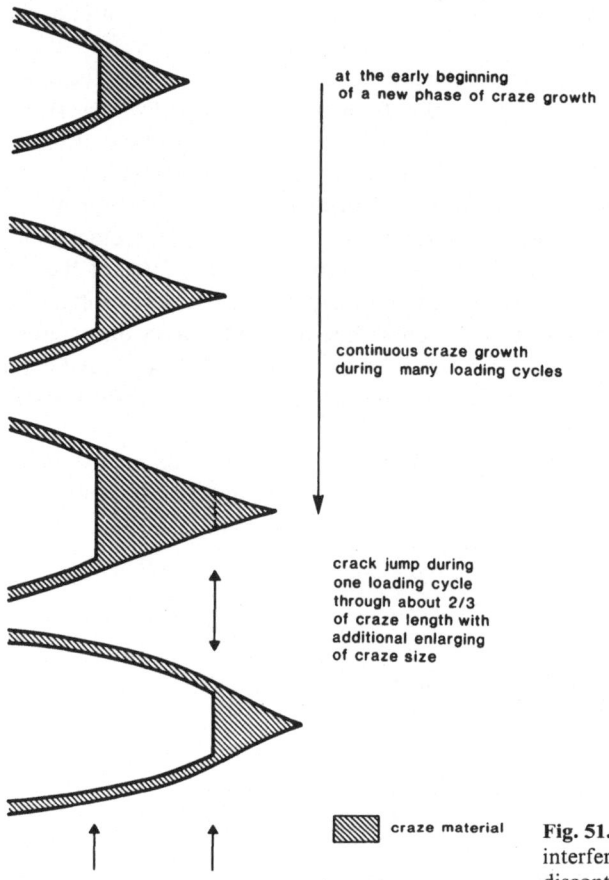

at the early beginning
of a new phase of craze growth

continuous craze growth
during many loading cycles

crack jump during
one loading cycle
through about 2/3
of craze length with
additional enlarging
of craze size

craze material

crack tip before and after jump

Fig. 51. Schematic representation of optical interference results of continuous craze and discontinuous crack growth in PVC [137]

essentially modified. A sequence of the different stages in discontinuous crack and continuous[4] craze growth is schematically represented in Fig. 51.

In order to obtain information on the sudden breakdown of the craze after a long sustaining period of several hundred loading cycles it is helpful to analyse the craze growth in detail between two successive crack jumps. Obviously the craze lengthening takes place by fibrillation of amorphous bulk material. The thickening of the craze may be due to an additional fibrillation of bulk material or due to enhanced stretching of already formed fibrils or possibly due to both processes. To investigate this question, optical interference measurements were performed on PVC and the craze width $2v_0$ at the crack tip was determined at the lower load [137]. As mentioned above $2v_0$ can be regarded as a direct measure of the amount of fibrillated material. The data strongly suggest a constant value of $2v_0 = (0.52 \pm 0.03)$ μm [137]. This result is very instructive since it implies that the

---

[4] "Continuous" craze growth has to be seen in relation to the "discontinuous" crack growth; looking at a finer scale e.g. at one loading cycle then discontinuous craze growth is to be expected.

craze thickening at the crack tip does not occur by additional fibrillation of new bulk material and hence it must take place by an enhanced stretching of the already formed fibrils. Together with the results shown in Fig. 50 the magnitude of the strain ε in the craze can be roughly estimated. The result shows that between two successive crack jumps the strain on the fibrils at the crack tip increases by the fatigue process from about 130% to about 270% at breakdown. It can be left open at this point whether the lengthening and the final breakdown of the fibrils are accompanied or even caused by a successive disentanglement (small and medium molecular weight material) or by chain scission (high molecular weight material) [101, 17].

4.4.3 Influence of Stress Intensity Factor on Craze Size

In the interval between two successive crack jumps the crack length remains unchanged and hence the stress intensity factor is constant. In order to investigate the influence of the stress intensity factor on the craze dimensions, crazes under different loads have been compared. Figure 52 shows the craze width 2v measured directly at the crack tip as a function of ΔK_I. The maximum craze widths $2v_e$ and $2v_i$ were measured under upper load just before and after a crack jump, and $2v_0$ is the craze width under lower load between two successive crack jumps. For all three characteristic craze widths a linear increase with ΔK_I can be seen. This behavior is quite similar to that already found for continuous crack growth in PMMA. However, the slopes of the curves differ to such a degree, that a quite different straining behavior of the craze fibrils must be assumed. This is brought out more clearly when the data are expressed in terms of the previously defined strain ε. Under continuous crack growth in the ΔK_I-range investigated, the strain of PMMA craze fibrils grows moderately from about 70% to about 120% with increasing ΔK_I. A completely different situation exists during discontinuous crack growth in PVC, where the strain of the craze fibrils decreases with increasing ΔK_I-values. In the situation where the individual fibrils are most severely stretched, just before a crack jump occurs, ε decreases from about 400% at $\Delta K_I = 14$ N/mm$^{3/2}$ to about 250% at $\Delta K_I = 24$ N/mm$^{3/2}$. This decrease in strain ε with increasing ΔK_I may be seen as inherently contradicting the notion that, between

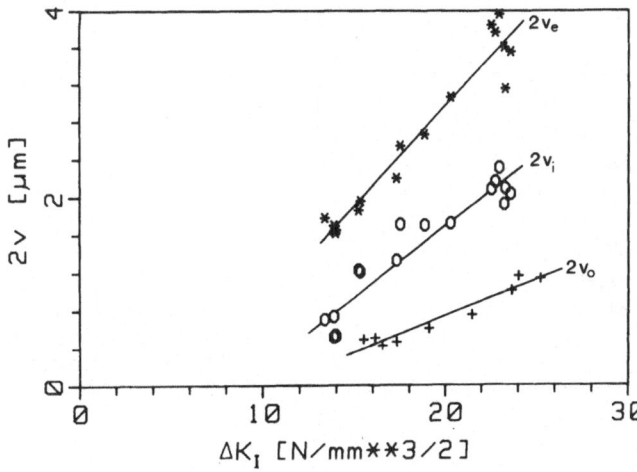

Fig. 52. Influence of ΔK_I on craze widths 2v at the crack tip in PVC during discontinuous crack growth [137]

two successive crack jumps at constant ΔK_I. the strain on the fibrils increases. The contradiction may be resolved as follows [137]. Just after a crack jump the strain on the already formed fibrils at the crack tip is governed by the fatigue effect, showing an increase in strain with the number N of cycles between two crack jumps. This cycle number N decreases with ΔK_I (see Fig. 46) leading to the related decrease of both fibril strain ε and cycle number N. An extrapolation of this relationship to just one cycle indicates a strain of about 150%, which corresponds closely to the value determined at continuous crack growth in PMMA for higher ΔK_I-levels.

The influence of ΔK_I on craze length s in PVC is shown in Fig. 53. Both characteristic craze lengths, s_e just before and s_i just after a crack jump, show a linear increase with ΔK_I, thus leading to a similar result as that observed during continuous fatigue crack growth (see Figs. 34 and 35).

In Fig. 53 the distances b of the crack jumps measured as growth bands on the fracture surface are also given. For a given ΔK_I the sum of band width b and initial craze length s_i should, in principle, be equal to the craze end length s_e. At low ΔK_I levels this is almost the case. With increasing ΔK_I, however, the sum of b and s_i increasingly exceeds s_e, so that at $\Delta K_I = 24 \text{ N/mm}^{3/2}$ the excess amounts to nearly 20 μm. This effect might be explained as follows. As already noticed in Fig. 49 and 50 an enlargement in craze length occurs in the very moment of the crack jump. The increase in craze length with ΔK_I has been quantitatively explained [137] by application of the Dugdale model. From Eq. (7b) it follows that the stress on the craze boundary and hence on the fibrils is proportional to ΔK_I and inversely proportional to the square root of craze length. From Fig. 53 it would be expected that the initial craze length as the difference between s_e and b would be nearly constant. With increasing ΔK_I a constant initial craze length requires that the fibrils would have to bear an increasing stress. The material, therefore, responds by a redistribution of stress which results in an enlarged craze length under lower stress. It is interesting to note that the measured initial craze lengths s_i lead to a nearly constant stress on the fibrils.

In this context it should be mentioned that the optical interference results on PVC indicate that the "simple" Dugdale model with a constant stress is not appropriate

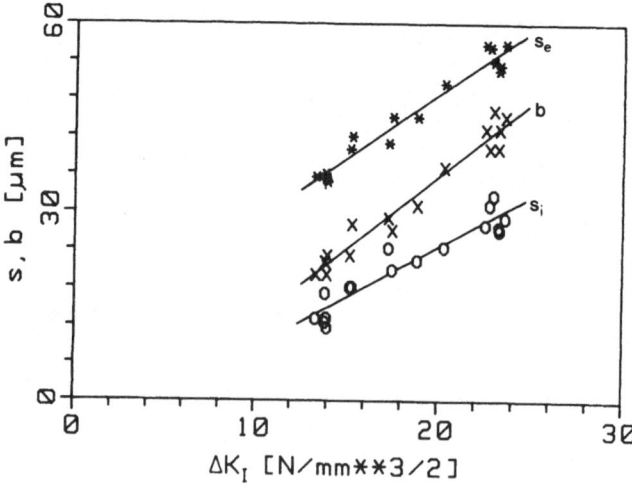

Fig. 53. Influence of ΔK_I on craze lengths s and band width b in PVC during discontinuous crack growth [137]

to describe the measured craze contour sufficiently accurately. In contrast to PMMA a variable stress distribution along the craze as determined for PS [33] seems to give a better fit to the measured craze contour in PVC. This "second order correction", however, does not change the principal conclusion of the preceding discussion.

4.4.4 Fractography of Discontinuous Growth Bands

With respect to the cracking process between two successive growth bands it has been suggested by Hertzberg and Manson [101, 104] that cracking occurs by a void coalescence mechanism with the void size distribution reflecting the internal structure of the craze just before crack propagation. The basis for this interpretation were SEM photographs (Fig. 54) of the micromorphology of the fracture surface. Many microvoids were found [101, 104] which decreased in size from 2 to 0.1 μm in the direction of crack growth. These may also be seen from the SEM photograph in Fig. 55. It has now been shown by the aid of a progressive

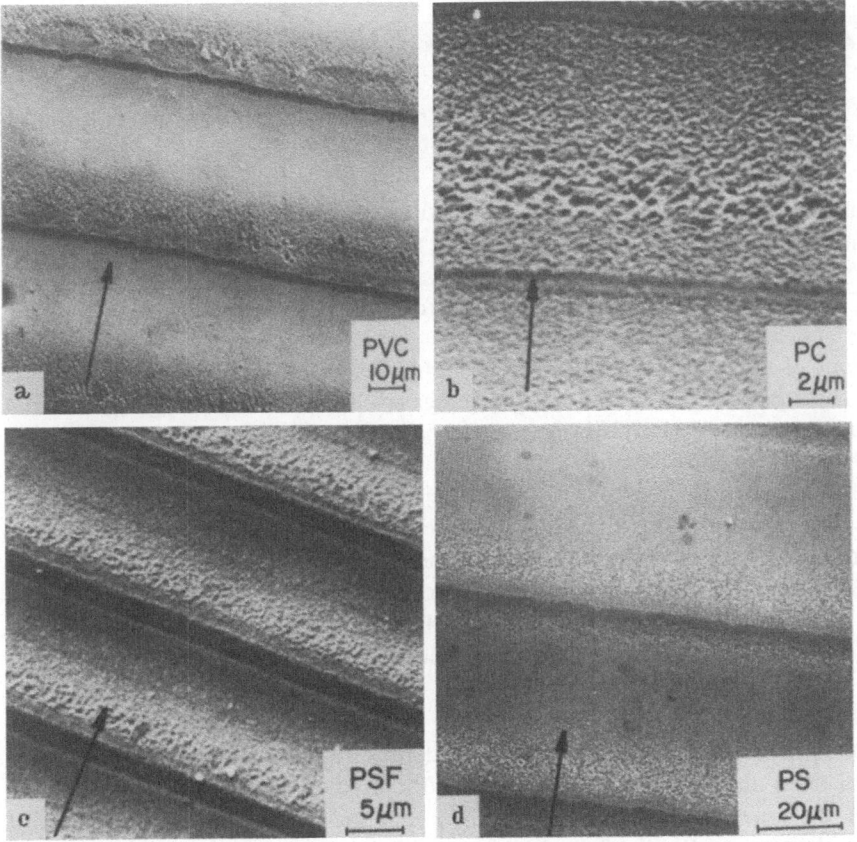

Fig. 54a–d. SEM fractographs of discontinuous growth bands in different polymers. Arrows indicate crack direction [Skibo, M. D., Hertzberg, R. W., Manson, J. A., Kim, S. L., J. Mater. Sci. 12, 531 (1977), Chapman and Hall, Ltd.]

Fig. 55. Discontinuous growth bands in PVC [138]. Crack propagation direction from left to right normal to the bands. $\Delta K_I = 17.5 \text{ N/mm}^{3/2}$

preparation technique that the typical structures in the fracture surface consist of different arrangements and differently broken types of craze fibrils [138]. This is clearly demonstrated by the SEM microphotographs in Figs. 56 and 57. In Fig. 56 details of Fig. 55 are given and it can be seen from Fig. 56a that the growth band consists of nearly flat structures.

In front of the fatigue line bundles of highly deformed and broken craze fibrils can be distinguished which are fused together whilst behind the fatigue line a flat granular structure of short fibrils can be observed. In the middle between two fatigue lines (Fig. 56b) many long curled broken single fibrils are left on the fracture surface. In this context it is instructive to refer to an earlier optical interference measurement. Figure 52 demonstrates that the maximum craze width $2v_e$ and hence the length of the stretched fibrils increases with ΔK_I. Therefore, it is to be expected that the length of the broken fibrils also increases with ΔK_I. This can be seen at least qualitatively by a comparison of the microphotographs in Fig. 56 and 57. In the latter the fracture process took place at a ΔK_I level nearly 50% higher than that in Fig. 56. The left part of Fig. 57a indicates that the growth band is followed by a mass of fused craze material together with the ends of some single fibres. Figure 57b exhibits both long and also short fibrils and thus suggests that these fibrils were not broken at their midpoints but rather fractured at or near to the interface between the craze and the bulk material. From the schematic representation of Fig. 51 a decrease in length of broken fibrils might be expected in the direction of crack growth. Figs. 56a and b provide some indications of such a process. However, the evidence is not conclusive since the fibrils are agglomerated just at the positions where their lengths should be longest.

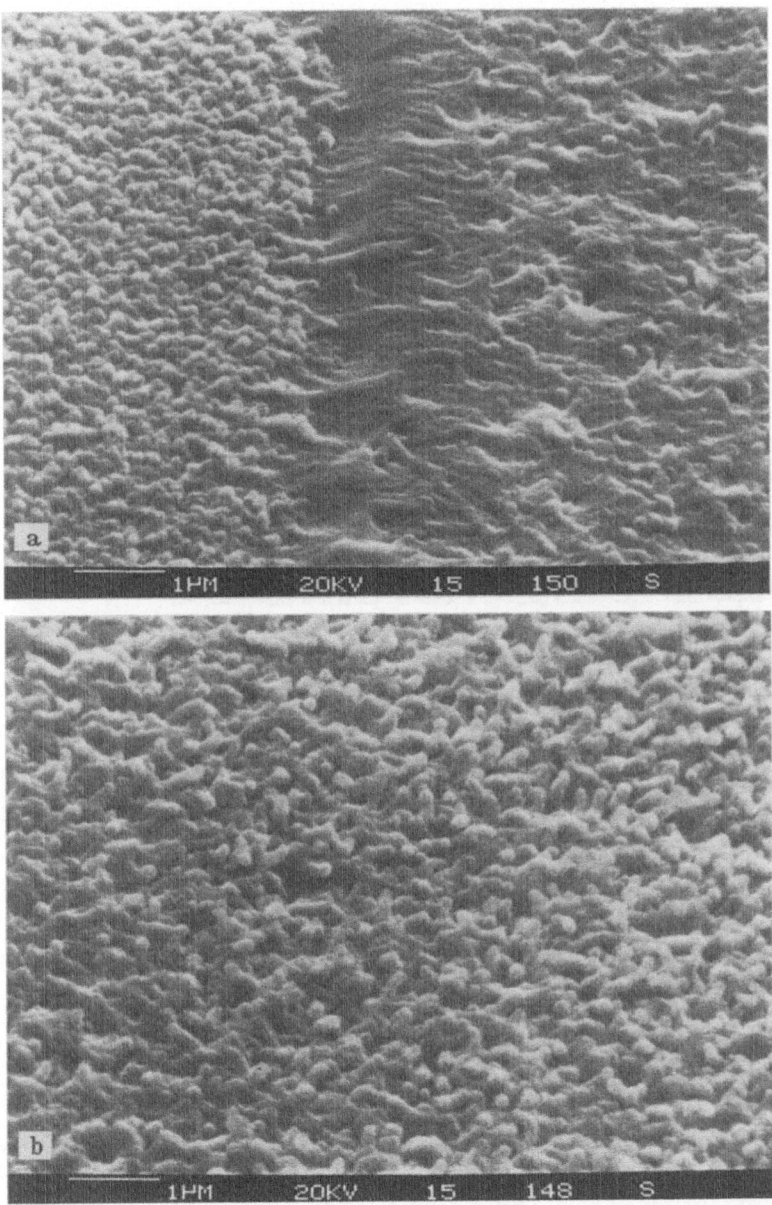

Fig. 56a and b. Micromorphology in discontinuous growth bands in PVC [138]. Details of Fig. 55. **a** Surface structure around the fatigue line; **b** Structure of fibrils at the position between two fatigue lines

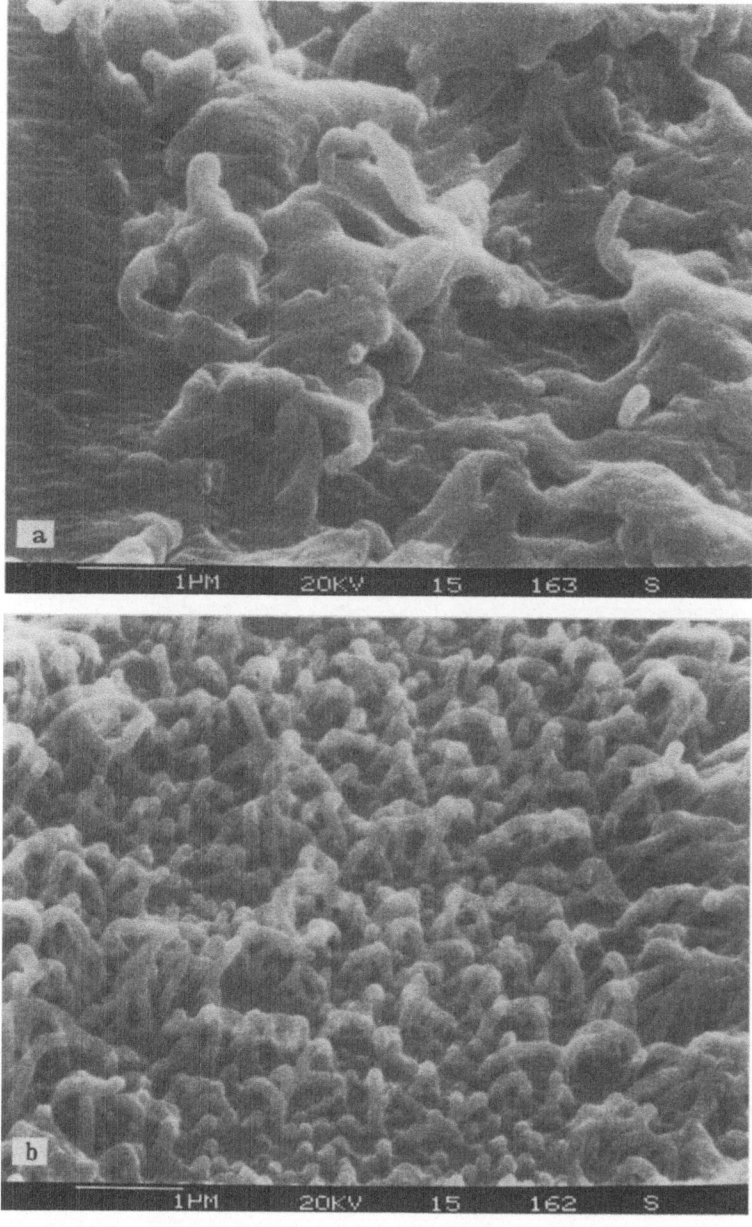

Fig. 57a and b. Micromorphology in discontinuous growth bands in PVC at $\Delta K_I = 24$ N/mm$^{3/2}$ [138].
a Position near fatigue line; **b** Position between two fatigue lines

From the different surface and fibril structures along the periodic pattern of discontinuous growth bands the following fracture mechanism for fatigue loading in PVC has been deduced [138]. At the end of a series of loading cycles, the craze fibrils directly ahead of the crack tip are most deformed and fail by thermal fatigue forming the flat structure of the narrow fatigue lines and the fused mass of fibrils close to these lines. With further crack propagation the heating effects seem to be drastically reduced and the fibrils are broken in an almost quasi-brittle manner. The stopping of the crack may then be due to an increasing resistance of the shorter and hence extensible fibrils. After crack arrest many cycles are needed to weaken these fibrils sufficiently so that the cracking process can start again.

Acknowledgements: The author wants to thank Prof. H. H. Kausch for his invitation to contribute to this volume and his continuing encouragement and support during the progress. Much of this review summarizes work performed by the polymer group in the Fraunhofer-Institut für Werkstoffmechanik and the author wishes to acknowledge his debt to all of them, especially to Dr. L. Könczöl for stimulating discussions and help in evaluation and Dr. G. W. Weidmann now at the Open University (U.K.) for critical reviewing the text and to the Deutsche Forschungsgemeinschaft (DFG) who has funded much of the work.

5 References

1. Kambour, R. P.: J. Polym. Sci. A-2 4, 349 (1966)
2. Bessenov, M. I., Kuvshinskii, E. V.: Soviet Phys. (Solid State) 3, 1957 (1961)
3. Dugdale, D. S.: J. Mech. Phys. Solids 8, 100 (1960)
4. Barenblatt, G. I.: Adv. Appl. Mechs. 7, 55 (1962)
5. Brown, H. R., Ward, I. M.: Polymer 14, 469 (1973)
6. Vavakin. A. S., Kozyrev, Yn. I., Salganik, R. L.: Mech. Solids 11, 97 (1976)
7. Weidmann, G: W., Döll, W.: Colloid and Polym. Sci. 254, 205 (1976)
8. Griffith, A. A.: Phil. Trans. Roy Soc. (London) A 221, 163 (1920)
9. Irwin, G. R., in: Handbuch der Physik 6, (ed.) Flügge, S., p. 551, Berlin-Heidelberg-Göttingen, Springer 1958
10. Liebowitz, H. (ed.): Fracture — An Advanced Treatise, Vol. I–VII New York-London, Academic Press 1968
11. Sih, G. C. (ed.): Mechanics of Fracture Vol. 1–4, Leyden, Noordhoff 1973
12. Cherepanow, G. P.: Mechanics of Brittle Fracture, New York et al., Mc Graw-Hill 1979
13. Knauss, W. G.: Appl. Mech. Reviews 26, 1 (1973)
14. Kambour, R. P.: J. Polym. Sci. Macromol. Revs. 7, 1 (1973)
15. Hertzberg, R. W.: Deformation and Fracture of Engineering Materials, New York, John Wiley 1976
16. Williams, J. G., in: Advances in Polymer Science 27, p. 67, Berlin-Heidelberg-New York, Springer 1978
17. Kausch, H. H.: Polymer Fracture, Berlin-Heidelberg-New York, Springer 1978
18. Rooke, D. P., Cartwright, D. J.: Compendium of Stress Intensity Factors, London, HMSO, Hillingdon Press 1976
19. McClintock, F. A., Irwin, G. R., in: Fracture Toughness Testing and its Applications, ASTM STP 381, p. 84, Philadelphia, ASTM 1964
20. Bilby, B. A., Cottrell, A. H., Swinden, K. H.: Proc. Roy. Soc. A 272, 304 (1963)

21. Goodier, J. N., Field, F. A., in: Fracture of Solids, (ed.) Drucker, D. C., Gilman, J. J., p. 103, New York-London, John Wiley 1963
22. Rice, J. R., in: Fracture — An Advanced Treatise, (ed.) Liebowitz, H., p. 191, New York-London, Academic Press 1968
23. Muskhelishvili, N. J.: Some Basic Problems of the Mathematical Theory of Elasticity, p. 340, Groningen, Noordhoff 1953
24. Schapery, R. A.: Int. J. Fract. 11, 141 (1975)
25. McCartney, N. L.: Int. J. Fract. 13, 641 (1977)
26. Passaglia, E.: Polymer 23, 754 (1982)
27. Rosenfield, A. R., Dai, P. K., Hahn, G. T., in: Proc. of the First International Conference on Fracture, (ed.) Yokobori, T., Kawasaki, T., Swedlow, J. L., p. 223, Sendai, The Japanese Society for Strength and Fracture of Materials, 1965
28. Broberg, B.: J. Appl. Mech. 31, 546 (1964)
29. Kanninen, M. F.: J. Mech. Phys. Solids 16, 215 (1968)
30. Lucas, R. A.: Int. J. Solids Structures 5, 175 (1969)
31. Doyle, M. J.: J. Polym. Sci., Phys. Ed. 13, 2429 (1975)
32. Beahan, P., Bevis, M., Hull, D.: J. Mater. Sci. 8, 162 (1972)
33. Lauterwasser, B. D., Kramer, E. J.: Phil. Mag. A 39, 469 (1979)
34. Sommer, E.: Eng. Fract. Mech. 1, 705 (1970)
35. Sommer, E., Soltész, U.: Eng. Fract. Mech. 2, 235 (1971)
36. Kambour, R. P.: Nature 195, 1299 (1962)
37. Kambour, R. P.: Polymer 5, 107 (1964)
38. Kambour, R. P.: J. Polym. Sci. A 2, 4159 (1964)
39. Pitman, G. L., Ward, I. M.: Polymer 20, 895 (1979)
40. Wiley, R. H., Brauer, G. M.: J. Polym. Sci. 3, 455, 647 (1948)
41. Kambour, R. P.: J. Polym. Sci. A 2, 4165 (1964)
42. Weidmann, G. W., Döll, W., in: The Physics of Non-Crystalline Solids, Fourth Intern. Conference Clausthal-Zellerfeld 1976, (ed.) Frischat, G. H., p. 606, Trans. Tech. Publications 1977
43. Hine, P. J., Duckett, R. A., Ward, I. M.: Polymer 22, 1745 (1982)
44. Brown, H. R., Stevens, G.: J. Mater. Sci. 13, 2373 (1978)
45. Seidelmann, U. et al.: Fraunhofer-Institut für Werkstoffmechanik, Z 2/79, Freiburg (1979)
46. Kambour, R. P.: Polym. Eng. Sci. 8, 281 (1968)
47. Knight, A. C.: J. Polym. Sci. A 3, 1845 (1965)
48. Kramer, E. J., in: Development in Polymer Fracture, (ed.) Andrews, E. H., Barking (UK), Appl. Sci. Publ. Ltd., 1979
49. Donald, A. M., Kramer, E. J., Bubeck, R. A.: J. Polym. Sci., Phys. ed. 20, 1129 (1982)
50. Wang, W.-C. V., Kramer, E. J.: J. Mater. Sci. 17, 2013 (1982)
51. Donald, A. M., Kramer, E. J.: J. Mater. Sci. 16, 2967 (1981)
52. Donald, A. M., Kramer, E. J.: J. Mater. Sci. 16, 2977 (1981)
53. Döll, W.: Colloid and Polym. Sci. 252, 880 (1974)
54. Döll, W., Weidmann, G. W.: J. Mater. Sci. 11, 2348 (1976)
55. Higushi, M.: Repts. Research, Inst. Appl. Mech., Kyushi Univ. 6, 173, (1958)
56. Berry, J. P.: J. Polym. Sci. 50, 107 (1961)
57. Marshall, G. P., Culver, L. E., Williams, J. G.: Plastics and Polymers, February, 75 (1969)
58. Beaumont, P. W. R., Young, R. J.: J. Mater. Sci. 10, 1334 (1975)
59. Schinker, M. G., Döll, W., in: Mechanical Properties at High Rates of Strain, 1979, Conference Series No 47, (ed.) Harding, J., p. 224, Oxford/Bristol—London, The Institute of Physics 1980
60. Marshall, G. P., Coutts, L. H., Williams, J. G.: J. Mater. Sci. 9, 1409 (1974)
61. Berry, J. P., in: Fracture Processes in Polymeric Solids, (ed.) Rosen, B., p. 195, New York-London-Sidney, Interscience Publ. 1964
62. Döll, W., in: 3rd Intern. Congress on Fracture, p. VI-515, München/Düsseldorf, Verein Deutscher Eisenhüttenleute 1973
63. Evans, A. G., Wiederhorn, S. M.: Int. J. Fract. 10, 379 (1974)
64. Döll, W., Könczöl, L.: Kunststoffe 70, 563 (1980)
65. Döll, W., Seidelmann, U., Könczöl, L.: J. Mater. Sci. 15, 2389 (1980)
66. Weidmann, G. W., Döll, W.: Int. J. Fract. 14, R 189 (1978)

67. Könczöl, L., Schinker, M. G., Döll, W., in: Vorträge der 13. Sitzung des Arbeitskreises Bruchvorgänge, p. 246, Hannover/Berlin, DVM Deutscher Verband für Materialprüfung e.V. 1981
68. Ogorkiewicz, R. M.: Eng. properties of thermoplastics, p. 225, London-New York, John Wiley-Interscience 1970
69. Menges, G., Kleinemeier, B., Döring, F.: Plastverarbeiter 30, 755 (1979)
70. Morgan, G. P., Ward, I. M.: Polymer 18, 87 (1977)
71. Schirrer, R., Goett, C.: J. Mater. Sci. Letters 1, 355 (1982)
72. Fraser, R. A., Ward, I. M.: Polymer 19, 220 (1978)
73. Döll, W., Weidmann, G. W., Kerkhof, F.: Materiaux et Techniques 67, 119 (1979)
74. Döll, W., Könczöl, L., Seidelmann, U., in: Prepr. IUPAC Makro Mainz, Vol. III, p. 1448 (ed.) Lüderwald I., Weiss, R.; 26th Int. Symposium on Macromolecules, Mainz, 1979
75. Sternstein, S. S., Ongchin, L., Silvermann, A., in: Appl. Polymer Symposia 7, p. 175, New York, Interscience 1969
76. Stuart, H. A., Markowski, G.: Werkstoffe u. Korrosion 14, 825 (1963)
77. Menges, G., Schmidt, H.: Kunststoffe 57, 885 (1967)
78. Rabinowitz, S., Beardmore, P.: CRC Crit. Rev. Macromol. Sci. 1, 1 (1972)
79. Menges, G., Rieß, R., Suchanek, H.-J.: Kunststoffe 64, 200 (1974)
80. Opfermann, J.: Investigations of Crazing and Fracture of Amorphous Polymers, Aachen, Institut für Kunststoffverarbeitung (IKV) der TH Aachen, 1978
81. Marshall, G. P., Culver, L. E., Williams, J. G.: Proc. Roy. Soc. A 319, 165 (1970)
82. Kitagawa, M., Motomura, K.: J. Polym. Sci., Phys. Ed. 12, 1979 (1974)
83. Williams, J. G., Marshall, G. P.: Proc. R. Soc. A 342, 55 (1975)
84. Gent, A. N.: J. Macromol. Soc., Phys. B 8, 597 (1973)
85. Argon, A. S., Hannoosh, J. G.: Phil. Mag. 36, 1196 (1976)
86. Kausch, H. H.: Die Angewandte Makromol. Chemie 60/61, 139 (1977)
87. Döll, W., Könczöl, L., Schinker, M. G.: Colloid and Polym. Sci. 259, 171 (1981)
88. Trassaert, P., Schirrer, R.: J. Mater. Sci. (in press)
89. Könczöl, L., Döll, W.: Fraunhofer-Institut für Werkstoffmechanik, Z 4/81, Freiburg 1981
90. Döll, W., Weidmann, G. W.: Progr. Colloid and Polymer Sci. 66, 291 (1979)
91. Schirrer, R., Goett, C.: J. Mater. Sci. 16, 2563 (1981)
92. Döll, W.: Colloid and Polymer Sci. 256, 904 (1978)
93. Döll, W.: Eng. Fract. Mech. 5, 259 (1973)
94. Fuller, K. N. G., Fox, P. G., Field, J. E.: Proc. Roy. Soc. London A 341, 537 (1974)
95. Döll, W.: Int. J. Fract. 12, 595 (1976)
96. Williams, J. G.: Appl. Mater. Res. 4, 104 (1965)
97. Williams, J. G.: Inter. J. Fract. Mech. 8, 393 (1972)
98. Johnson, F. A., Radon, J. C.: Eng. Fract. Mech. 4, 555 (1972)
99. Weichert, R., Schönert, K.: J. Phys. Mech. Solids 22, 127 (1974)
100. Fox, P. G., Fuller, K. N. G.: Nature Phys. Sci. 234, 13 (1971)
101. Hertzberg, R. W., Manson, J. A.: Fatigue of Engineering Plastics, New York-London-Toronto-Sydney-San Francisco, Academic Press 1980
102. Sauer, J. A., Richardson, G. C.: Int. J. Fracture 16, 499 (1980)
103. Radon, J. C.: Int. J. Fracture 16, 533 (1980)
104. Hertzberg, R. W., Manson, J. A., Skibo, M. D.: Polym. Eng. Sci. 15, 252 (1975)
105. Skibo, M. D. et al.: J. Macromol. Sci. Phys. B 14, 525 (1977)
106. Rimnac. C. M. et al.: J. Macromol. Sci. Phys. B 18, 351 (1981)
107. Kim, S. L. et al.: Polymer Eng. Sci. 17, 194 (1977)
108. Pitman, G., Ward, I. M.: J. Mater. Sci. 15, 635 (1980)
109. Hahn, M. T. et al.: J. Mater. Sci. 17, 1533 (1982)
110. Harris, J. S., Ward, I. M.: J. Mater. Sci. 8, 1655 (1973)
111. Mills, N. J., Walker, N.: Polymer 17, 335 (1976)
112. Schirrer, R. et al.: Colloid and Polymer Sci. 259, 812 (1981)
113. Döll, W., Schinker, M. G., Könczöl, L.: in Deformation, Yield and Fracture of Polymers, p. 20.1, Cambridge/London, Plastics and Rubber Institute 1982
114. Döll, W., Könczöl, L., Schinker, M. G.: Polymer (in press)
115. Döll, W., Könczöl, L., Schinker, M. G.: in Vorträge der 14. Sitzung des Arbeitskreises Bruchvorgänge, p. 235, Mühlheim/Berlin, DVM Deutscher Verband für Materialprüfung e. V. 1982

116. Verheulpen-Heymans, N.: Polymer 20, 356 (1979)
117. Hertzberg, R. W., Skibo, M. D., Manson, J. A.: J. Mater. Sci. 14, 1754 (1979)
118. Williams, J. G.: J. Mater. Sci. 12, 2525 (1977)
119. Williams, J. G.: J. Mater. Sci. 14, 1758 (1979)
120. McEvily, A. J., jr., Boettner, R. C., Johnston, T. L., in: Fatigue — An Interdisciplinary Approach (ed.) Burke, J. J., Reed, N. L., Weiss, V., p. 95, Syracuse N.Y., Syracuse University Press 1964
121. Laird, C., Smith, G. C.: Phil. Mag. 7, 847 (1962)
122. Jacoby, G., Cramer, Ch.: Rheologica Acta 7, 23 (1968)
123. Jacoby, G. H., in: ASTM STP 453, p. 147, Philadelphia, ASTM 1969
124. McMillan, J. C., Pelloux, R. M., in: Fatigue crack propagation ASTM STP 415, p. 505, Philadelphia, ASTM 1967
125. Hertzberg, R. W.: ibidem, p. 205
126. Laird, C.: ibidem, p. 131
127. Feltner, C. E.: J. Appl. Phys. 38, 3576 (1967)
128. Elinck, J. P., Bauwens, J. C., Homès, G.: Int. J. Fracture Mech. 7, 277 (1971)
129. Hertzberg, R. W., Manson, J. A.: J. Mater. Sci. 8, 1554 (1973)
130. Skibo, M. D., Hertzberg, R. W., Manson, J. A.: J. Mater. Sci. 11, 479 (1976)
131. Sauer, J. A., McMaster, A. D., Morrow, D. R.: J. Macromolecular Sci. — Physic B 12, 535 (1976)
132. Skibo, M. D., et al.: J. Mater. Sci. 12, 531 (1977)
133. Mackay, M. E., Teng, T. G., Schultz. J. M.: J. Mater. Sci. 14, 221 (1979)
134. Takemori, M. T., Kambour, R. P.: J. Mater. Sci. 16, 1108 (1981)
135. Hertzberg, R. W., Skibo, M. D., Manson, J. A.: in Fatigue Mechanism ASTM STP 675 (ed.) Fong J. T., p. 471, Philadelphia, ASTM 1979
136. de Charentenay, F. X., Laghouati, F., Dewas, J.: in Deformation, Yield and Fracture of Polymers, p. 6.1, Cambridge/London, Plastics and Rubber Institute 1979
137. Könczöl, L., Schinker, M. G., Döll, W., in: Fatigue in Polymers, p. 4.1, London, Plastics and Rubber Institute 1983
138. Schinker, M. G., Könczöl, L., Döll, W.: J. Mater. Sci. Let. 1, 475 (1982)
139. McCrum, N. G., Read, B. E., Williams, G.: Anelastic and Dielectric Effects in Polymeric Solids, London-New York-Sidney, John Wiley a. Sons Ltd. 1967
140. Schinker, M. G., Könczöl, L., Döll, W.: to be published in Colloid and Polymer Sci.
141. Takahashi, K.: J. Macromol. Sci. — Phys. B 19, 695 (1981)
142. Sternstein, S. S., Meyers, F. A.: J. Macromol. Sci. — Phys. B 8, 539 (1973)
143. Brown, H. R.: J. Polym. Sci., Polym. Phys. Ed. 21, 483 (1983)

Received January 12, 1983
H. Kausch (editor)

Crazing and Fatigue Behavior in One- and Two-Phase Glassy Polymers

J. A. Sauer and C. C. Chen
Department of Mechanics and Materials Science, Rutgers University,
New Brunswick, NJ 08903, U.S.A.

Advances in Polymer Science 52/53
© Springer-Verlag Berlin Heidelberg 1983

List of Symbols

d	Root-mean square end-to-end distance of chain having entanglement molecular weight, M_e
E	Youngs modulus
E'	Storage modulus
E''	Loss modulus
K	Stress intensity factor
ΔK	Stress intensity factor range
l_e	Chain contour length between entanglements
M_e	Enganglement molecular weight
$\bar{M}_n$	Number average molecular weight
$\bar{M}_w$	Weight average molecular weight
N	Number of alternating cycles
N_{ci}	Cycles to onset of craze initiation
N_f	Cycles to fracture in fatigue
N_s	Cycles to onset of strain softening by shear
S	Stress amplitude
t	Time
tan δ	A measure of internal friction
T	Temperature
T_g	Glass transition temperature
β	Secondary transition
λ	Extension ratio
λ_{max}	Maximum extension ratio
σ_c	Craze intensification stress
σ_{ci}	Craze initiation stress
σ_T	Tensile strength
σ_y	Compressive yield stress
$\Delta\sigma_y$	Drop in σ_y after yielding
$\Delta\sigma_y/\sigma_y$	A measure of strain softening
ν	Test frequency

1 Introduction

Crazing is a mode of localized plastic deformation that occurs particularly in glassy polymers subject to tensile stresses. It involves orientation of molecular chain segments in the direction of the principal stress together with cavitation or voiding. It tends to develop at stress concentration sites such as surface flaws or internal inhomogeneities. Many of the macroscopic aspects of crazing, and of the influence of various variables such as stress state, time, temperature, and environment, were discussed in early investigations of this phenomenon by Sauer et al. [1, 2] and by Maxwell and Rahm [3]. The microstructure of crazes was studied by Bessonov and Kuvshinskii [4] and by Spurr and Niegisch [5]. During the ensuing years important contributions have been made by many investigators. Criteria for craze initiation and for shear yielding were proposed by Sternstein and coworkers [6, 7]. Excellent summaries of the research to 1973 have been given by Kambour [8] and by Bowden [9].

Important current concepts of the micromechanisms involved in crazing and environmental aspects of this subject have been developed by Kramer and co-workers [10, 11]; they are described in Chapter 1 of this volume.

The beneficial role of crazing as an important mode of plastic deformation in rubber modified polymers has received considerable attention. Due to the stress concentrating effect of the dispersed second-phase particles, plastic deformation, by matrix crazing throughout the volume of the specimen, is enhanced; and an appreciable increase occurs in impact strength and toughness. Discussion of the general principles of rubber toughening and of the effects of particulate fillers on yielding and fracture mechanisms have been given by Bucknall [12, 13].

Crazing can also have a deleterious effect on the performance of glassy polymers. Under conditions conducive to brittle fracture, such as the presence of notches, alternating stresses, and aggressive environmental media, crazes act as preferred low energy paths for crack propagation. Application of linear elastic fracture mechanics to the study of craze and crack growth is detailed in Chapter 3, it has been described earlier by Williams [14]; fracture phenomena, including molecular aspects, have been reviewed by Kausch [15] and by Andrews and Reed [16].

The behavior of amorphous and crystalline polymers under alternating stresses has received an increasing degree of attention within recent years, partly as a result of a greater use of polymers in various engineering applications. Crazing frequently plays a vital role in fatigue fracture of polymers and it leads to some interesting new phenomena such as the presence of concentric discontinuous crack growth (DCG) bands on the fracture surface. A book by Hertzberg and Manson [17], and recent reviews by Sauer and Richardson [18] and by Radon [19] summarize the current state of our knowledge relative to the general nature of fatigue fracture, the influence of various material and environmental variables, and the laws governing fatigue crack growth.

One purpose of the present review is to investigate and elucidate deformation and fracture mechanisms in glassy polymer systems subject to alternating stresses and to compare and contrast their behavior with that observed under monotonic loading. Craze initiation, strain softening due to shear deformation, and fatigue life and fracture surface morphology are studied as a function of applied stress and test frequency. A second purpose is to investigate the deformation and fracture response, under static and dynamic conditions, of several rubber modified glassy polymers and to compare their behavior with that of corresponding unmodified polymers. Two different polymers system are considered preferentially: one is polystyrene (PS) and rubber modified versions of this polymer, such as medium impact polystyrene (MIPS) and high impact polystyrene (HIPS); and the other includes copolymers of styrene-acrylonitrile (SAN), of varying acrylonitrile content and of varying molecular weight, together with a rubber modified version of this polymer, viz. acrylonitrile-butadienestyrene, (ABS). Their composition and sample preparation techniques will be described in Section 2.

Some studies of fatigue behavior in rubber modified polymers, such as ABS, HIPS and rubber reinforced PVC, have been reported [20-29]. However, there appear to be no detailed studies of the influence of stress state and frequency on deformation

modes, fatigue performance, and fracture surface morphology for unnotched specimens of glassy polymers or of polymer systems incorporating an elastomeric second
phase. Much more attention has been paid to the influence of a rubber component
on impact strength. However, many current day applications of rubber modified
plastics demand resistance both to impact loading and to various conditions of
alternating stressing. It is essential, therefore, for engineering and design reasons,
as well as for enhancing scientific understanding, that we acquire knowledge of the
deformation and fracture mechanisms involved in fatigue and of the influence of
material and test variables on craze and crack initiation, on subsequent growth and
on fracture surface morphology of polymeric specimens subject to dynamic, as well
as static, loading.

2 Materials and Experimental Procedure

The PS used in this study was a Foster Grant heat resistant grade supplied in the
form of extruded rod. GPC studies indicated it had an $\bar{M}_w$ of 2.74×10^5 and an $\bar{M}_n$ of
1×10^5 [25]. Two rubber modified versions of this polymer were also investigated.
One, obtained in granule form was a medium impact grade, MIPS, containing a
small amount of rubber. A high impact grade, HIPS, available both as extruded rod
and in pellet form, contained a higher rubber content. From SEM examination of
fracture surfaces, it has been estimated that the range of particle sizes in HIPS is
about 0.5 to 5 μm and that the average particle size lies in the range 2.5 to 3.0 μ [26].
Also, from measurement of the tensile yield stress of HIPS and comparison with
published information [13], it is estimated that the volume fraction of 2nd phase
particles is about 30 %. As recently discussed by Turley and Keskkula, it is the volume
fraction of the rubber phase and not the weight percent of the added rubber that
controls mechanical properties [27].

A small amount of orientation is present in the extruded samples but previous
tensile and fatigue measurements made on PS have shown essentially no change in
properties after annealing for 24 hrs at 90 °C [28]. Hence, most tests were carried out
on the samples as received, after suitable machining and polishing. The polishing
was done by hand using successively finer grades of emery paper accompanied by
repeated washings in water followed by drying for 6 hrs in a vacuum oven at 50 °C.

SAN materials, containing 25 %, 30 % and 51.2 % of acrylonitrile (AN), were
obtained from Dow in the form of pellets. For the composition containing 30 %
of acrylonitrile, three different molecular weight grades, with $\bar{M}_w$ values of 1.15×10^5,
1.30×10^5 and 1.85×10^5, were studied. Test samples were prepared by compression
molding into rectangular bars from which specimens were machined after annealing
for two hrs at 96 °C. The ABS used in this study was a Borg Warner general purpose
grade supplied in the form of extruded rod. Samples for testing were machined from
the rod and carefully polished in the manner indicated above.

Tensile tests, at controlled strain rates, were performed on an Instron tensile tester.
Samples were rectangular, 6.35 mm by 3.17 mm in cross-section, or cylindrical, with
a diameter of 5.1 mm. Both types had a gauge length of 12.7 mm. Fatigue tests were
carried out on similar cylindrical samples, or on rectangular specimens, 5.1 mm by
3.17 mm in cross-section, at various selected stress amplitudes and at frequencies

ranging from 0.02 to 21 Hz. These latter measurements were made on a servohydraulic Instron apparatus under a sinusoidal loading pattern. In most cases, tests were carried out on four separate samples and the recorded average cycles to fracture, N_f, was obtained by averaging the individual log N_f values. Compression tests were run on cylindrical samples, 2.54 cm high and 1.27 cm in diameter to monitor yield stress and to note the dependence of material rigidity and of strain softening on composition.

For transparent polymers, such as PS and SAN, the initiation and subsequent development of crazing in either tension or tension-compression fatigue can be detected visually. However, more accurate values of the craze initiation stress, σ_{ci}, or, in the fatigue studies, of the cycles to craze initiation N_{ci}, can be determined by use of a light reflection method [3, 28]. A newly formed craze acts as a mirror. Hence, by monitoring the intensity of the reflected light from a light source shining at an acute angle to the axis of the transparent specimen, the stress, or time to onset of crazing for a tension test, or cycles to craze initiation for a fatigue test, can be inferred from the first rise in intensity above background.

For the non-transparent HIPS and ABS materials, it is possible to obtain information about fatigue-induced strain softening, due to shear or to onset of crazing, by observation of hysteresis loops. From detailed study of each loop as a function of N, the number of cycles, one can obtain an estimate of the cycles to onset of strain softening due to shear, N_s, and of the cycles for onset of craze initiation, N_{ci}. N_s is taken as the first cycle at which a change in loop width of equal magnitude on both the tension and compression side is observable. N_{ci} is estimated from the first cycle at which there is a detectable change in the area, or in loop width at some particular stress value, on the tensile side of the hysteresis loop but with essentially no change on the compression side of the loop.

For the unmodified polymers, with low internal friction, our observations show that there is only a small temperature rise in the specimen during the fatigue test even at the highest stress amplitudes and test frequencies employed in this study. For the more viscoelastic rubber modified polymers, the rise in specimen temperature during the test was assessed by means of a Barnes Infrared Temperature sensor, focused on the sample during the fatigue testing. For HIPS, tested at 17.3 MPa and at 21 Hz, a maximum temperature rise of about 7 °C was recorded. For ABS tested at 27.6 MPa and at 21 Hz, the maximum recorded temperature rise was ~20 °C. This falls to about 2 °C at 2 Hz; at lower test frequencies the specimen temperature rise is not significant at any of the stress amplitudes employed.

Dynamic mechanical relaxation tests were carried out by use of a Piezotron apparatus. Thin reedlike samples were subject to alternating tension at low applied stress and at a fixed frequency of 3 Hz. From the recorded observations, made as a function of temperature at a controlled rate of 2 °C/min, the storage modulus, E', the loss modulus, E'', and the loss tangent, $\tan \delta = E''/E'$ were determined.

The glass transition temperature, T_g, was also determined for some of the test materials. Values given in this paper are somewhat higher than reported literature values as they were obtained from the temperature location of the principal maximum in the loss modulus vs. temperature data from tests made at 3 Hz.

Fracture surfaces were examined optically and by scanning electron microscopy (SEM) using an Etec apparatus. Prior to SEM investigation, the fracture surfaces were coated with a thin layer of gold-palladium.

3 Polystyrene and Rubber Reinforced Polystyrenes

3.1 Deformational Response under Monotonic and Alternating Loading

The tensile stress-strain response of the homopolymer, and of two rubber modified grades of polystyrene, is shown in Fig. 1. The principal mode of deformation is crazing and all three materials exhibit a craze yield stress. However, there is no evidence of localized necking in any of the three materials. The craze yield stress decreases and the elongation to fracture, and the toughness, increase significantly with increase in rubber content.

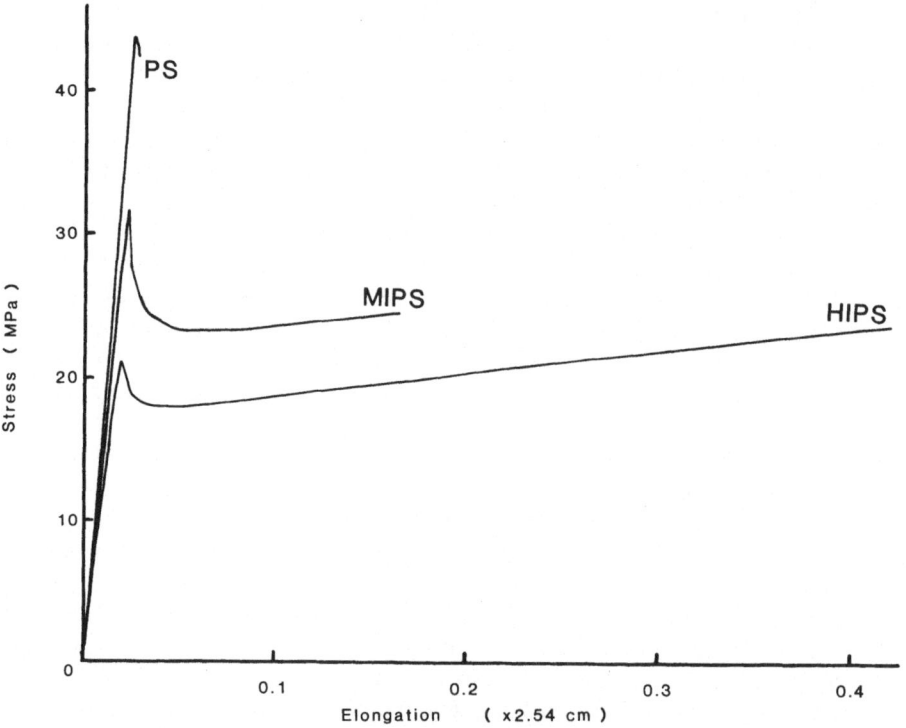

Fig. 1. Tensile stress-strain curves of PS, MIPS and HIPS obtained at a displacement rate of 12.7×10^{-2} cm/min

For the transparent homopolymer, crazes are seen, by eye, to develop from surface sources at a stress of about three-fourths of the yield or fracture stress. As the stress is further increased, new crazes nucleate and other crazes grow in size. At the craze yield stress maximum, the plastic strain rate due to crazing reaches the value of the imposed deformation rate. At these high stresses, a microcrack develops, frequently from an internal inhomogeneity and quickly propagates through the existing craze planes to produce brittle type fracture. The tensile fracture surface, as shown in Fig. 2a, is remarkably smooth, as the planar crazes, which have developed in transverse directions to the applied stress, serve as precursors and channels for crack propagation.

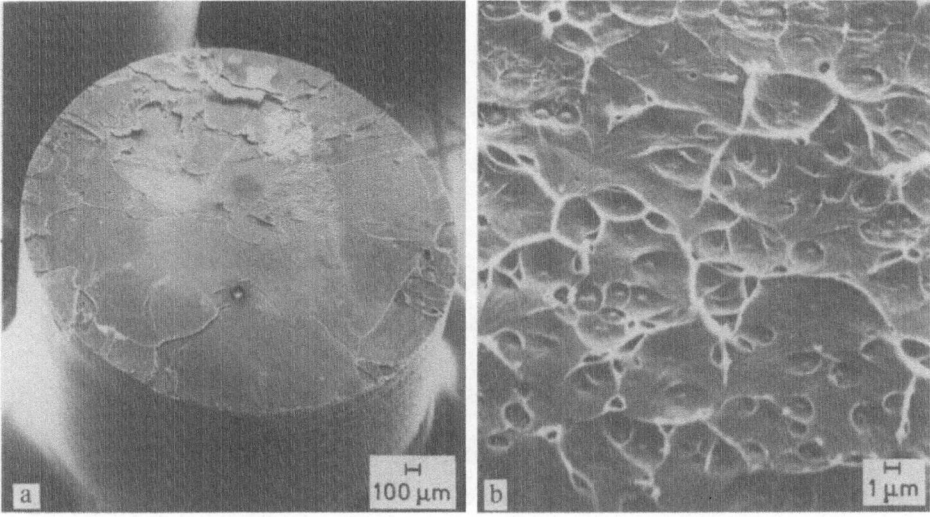

Fig. 2a. Typical tensile fracture surface of PS; **b** high magnification of a portion of tensile fracture surface of PS

The crack frequently initiates from the breakdown of a craze that formed at an internal defect, as a void or impurity particle. Then, as shown by various investigators, as crack speed increases, the crack jumps rapidly from one craze bulk interface to another and from one craze to another. This can lead to a so-called 'mackerel' type pattern on the fracture surface or to a 'craze island' type structure [29, 30], see also Chapter 1 and 3. As crack length increases and local stress rises, numerous secondary fractures, as shown in Fig. 2b, are generated ahead of the crack front.

In the rubber modified polystyrenes, the principal function of the dispersed second phase particles is to generate an increased amount of matrix crazing throughout the volume of the specimen. Crazing develops at a much lower applied stress due to the stress concentrating effect of the dispersed particles. As Fig. 1 illustrates, much greater plastic deformation is produced in the rubber reinforced polymers prior to catastrophic fracture. Development and growth of crazes throughout the specimen lead to the phenomena of 'stress whitening' [12]. This is first noticed before the craze yield stress is reached but it intensifies as deformation is increased beyond yield. Stress whitening is evident on the fracture surface as well and it its more intense for HIPS than for MIPS.

Figure 3a is an SEM scan of the tensile fracture surface of a HIPS sample. The morphology is quite different from that of a PS sample but similar to that reported [26]. The fracture surface shows two distinct regions. In the first slow growth region fracture appears to develop from breakdown of crazes at, or near the surface. The crack propagates by incremental jumps from one craze plane to another nearby one and the surface is finely textured. Fracture becomes catastrophic only when the crack reaches a certain critical size. The slow and fast crack zones are clearly depicted on the fracture surface. The fast zone exhibits a rougher surface, probably arising from the development of additional crazes above and below the crack tip and from crack bifurcation. A higher magnification view of a portion of the tensile fracture surface

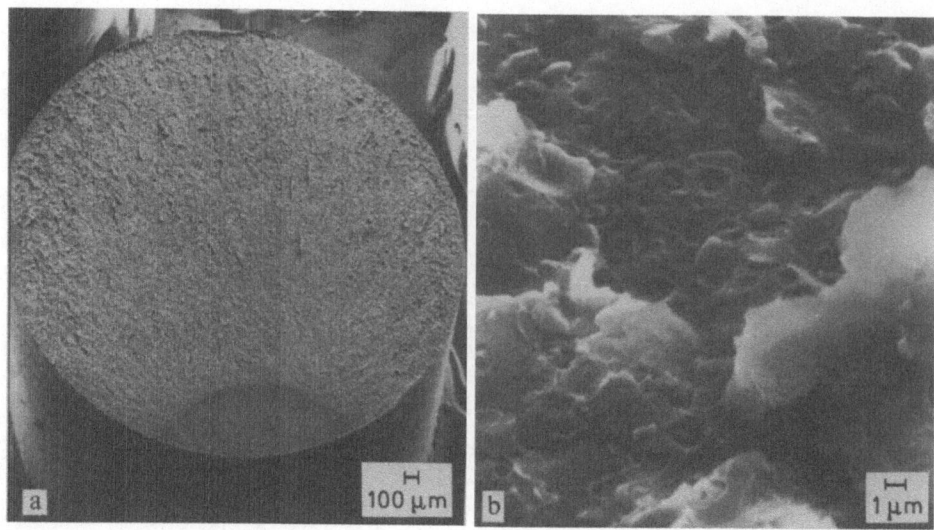

Fig. 3a. Typical tensile fracture surface of HIPS; **b** high magnification of a portion of the tensile fracture surface of HIPS

of HIPS is shown in Fig. 3b. Because of the high elongation that occurs in this rubber modified polymer before fracture, the surface is not smooth and there is evidence that the particles have been extensively deformed. Some of the larger particles have fractured while others show extensive voiding and cavitation. Many of the smaller particles, below 1 μm or so in size have been essentially bypassed by the fracture crack and, in some cases, both the particle and the resulting cavity are visible.

When similar test samples of PS, and of rubber modified PS, are subject to an alternating tension-compression stress instead of a monotonically increasing tensile stress, a brittle type fracture, with little or no macroscopic plastic deformation is produced, provided only that the alternating stress amplitude be above a certain critical value [26]. Deformation in the form of crazing still occurs but the extent of crazing is greatly restricted. This is illustrated, for a transparent PS sample that has been cycled without fracture for 80% of its expected life, in Fig. 4 [28]. Unlike the situation in simple tension, where crazes develop at various points throughout the reduced gauge length, there is only one visible plastic deformation or damage zone. Thus, one of the characteristic features of fatigue stressing of glassy polymers is that plastic deformation, in the form of crazes, becomes highly localized and tends to persist in the plane where it originated.

A similar situation prevails for the rubber modified polymers except that it is more difficult to detect the localized crazing or damage zone prior to fracture due to the lack of transparency. Nevertheless, one observes that stress whitening, which in simple tension occurred uniformly throughout the gauge length, is generally not present under fatigue loading. Only if alternating stress values are very high does one see any evidence of stress whitening and then only in the immediate vicinity of the flaw site. Once a fatigue-induced craze nucleates, from a flaw or a stress concentrating rubber particle, it becomes unstable under additional cycling, and a microcrack or void develops within the craze by fracture or viscous rupture of the craze fibrils.

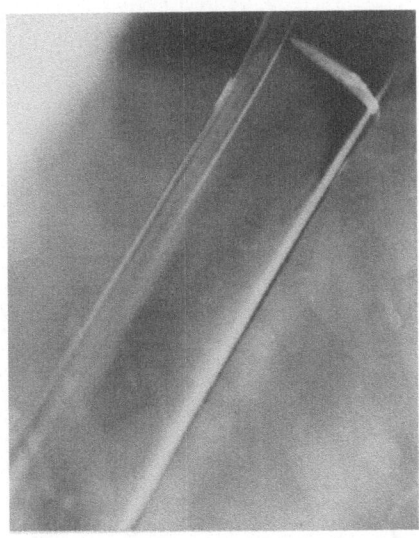

Fig. 4. Photograph of PS fatigue specimen cycled for 80% of its expected fatigue life at a stress amplitude of 17.2 MPa and at 21 Hz

Fatigue fracture thus involves an initiation phase of localized damage, as well as a subsequent crack propagation phase. Material variables and test variables may affect each of these phases quite differently. To illustrate, molecular weight is not considered to be an important variable from the point of view of craze initiation [8, 31], but it is known that fatigue lifetime, and fatigue crack propagation (FCP) rate, can be greatly extended by increasing molecular weight and thereby increasing entanglement density, craze strength and craze stability [25, 32–34]. Other variables, such as the presence of aggressive environments or of surface coatings may have, depending on the circumstances (stress state, frequency, temperature) a more dominant effect on the initiation phase. To illustrate, nitrile rubber coatings can extend the fatigue lifetime of PS specimens by a decade and this is attributed primarily to inhibition of surface crazing [18].

Several methods, as described in the preceding section, have been used to acquire information relative to the onset of the initial localized plastic deformation under alternating loading. Figure 5 is a plot of reflected light intensity vs. cycles for a transparent PS sample tested at 21 Hz at a stress amplitude of 17.2 MPa [28]. The number of cycles, N_{ci}, to initiate the craze, as determined from the first jump in intensity over background, is about 5,000 cycles and the cycles to fracture, N_f, is about 11,000. In later sections, the ratio of N_{ci} to N_f, which in this example is about 0.45, is shown to be a function of both stress amplitude and frequency.

For non-transparent specimens, as shown by Bucknall and Stevens [24], useful information relative to the deformation mode can be obtained by recording hysteresis loops as a function of cycles. Figure 6 shows hysteresis loops obtained at 0.2 Hz at various N values for PS tested at a stress amplitude of 24.1 MPa and Fig. 7 for HIPS tested at 17.2 MPa. For PS, with $N_f = 1,451$ cycles, there is no detectable change in loop area at this stress amplitude up to the final cycle. This illustrates the highly localized nature of the fatigue-induced damage zone in PS and indicates that, for this polymer, hysteresis loop observations are not an effective method for detecting craze

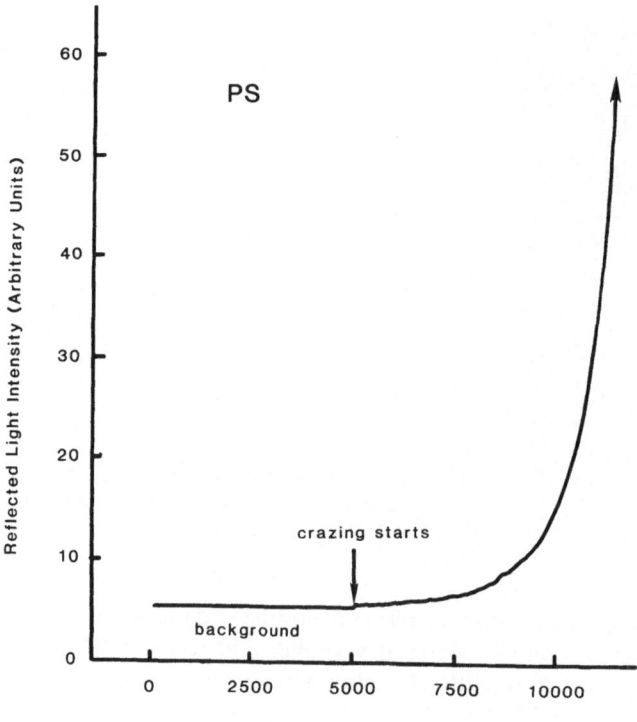

Fig. 5. Reflected light intensity vs. cycles for a PS sample, tested at a stress amplitude of 17.2 MPa and at 21 Hz

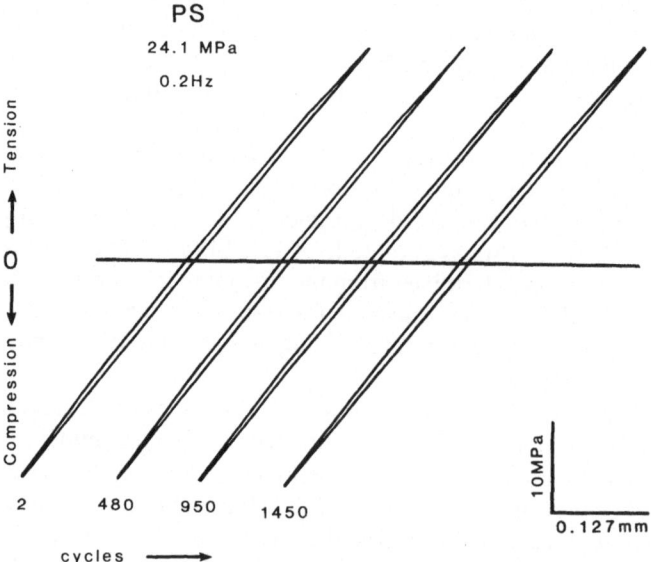

Fig. 6. Hysteresis loops for PS vs. number of cycles for specimen tested at a stress amplitude of 24.1 MPa and at 0.2 Hz

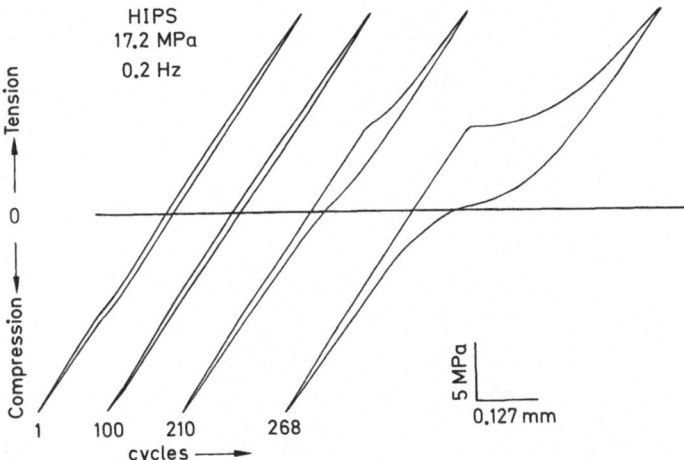

Fig. 7. Hysteresis loops for HIPS vs. number of cycles for specimen tested at a stress amplitude of 17.2 MPa and at 0.2 Hz

initiation. The light reflection method is a more sensitive indicator but, unfortunately, it is of use only with transparent polymers. For HIPS, the situation is different and Fig. 7 shows that the tensile part of the hysteresis loop changes in form and increases in area with increased cycling. This is an expected consequence of craze type yielding. For the stated conditions of Fig. 7, craze initiation occurs in HIPS prior to 25 cycles and craze and crack growth intensify upon continued cycling.

Fracture surface morphology is markedly different for fatigue fracture than for tensile fracture [17, 18, 35]. Figure 8a shows an SEM scan of a typical fatigue fracture surface for PS tested at 0.2 Hz. This scan may be compared with that of the tensile fracture surface shown in Fig. 2a. In all samples of PS investigated to date, fatigue fracture developed from a surface source. This illustrates the great importance of surface finish and test environment in alternating load applications. It also indicates why fatigue durability can be enhanced by appropriate surface treatments and coatings [36, 37].

The fracture surface shows four regions of different morphological features. The source is surrounded by a smooth, semi-circular mirror region (RI) indicative of relatively slow crack growth through essentially a planar craze zone. Within this region, one sees a series of concentric crack growth bands, Fig. 8b. Then, as the stress intensity at the crack tip increase, a second region (RII) arises that is still comparatively smooth. Due to formation of additional crazes ahead of the crack front but out of the crack plane this region may exhibit radial lines and tear ridges. Upon further cycling, the surface roughens and a third region (RIII) develops wherein the fast moving crack frequently jumps from one craze-bulk interface to another or from one craze plane to another. Finally, catastrophic fracture occurs on the last half cycle and this portion of the fracture surface (RIV) region is very rough.

Under some circumstances, as when unnotched samples of PS are stressed from a maximum to a minimum value in the tensile mode [35], or when notched samples are tested in a similar mode [17], the smooth region near the fracture source show

Fig. 8 a and b. Fatigue fracture surface of PS tested at a stress amplitude of 17.2 MPa and at 0.2 Hz; **a** low magnification scan; **b** higher magnification of region near fracture source

a series of discontinuous crack growth bands [17,18]. Fig. 8a shows that DCG bands can also occur under conditions of alternating tension and compression when the test frequency is low. Additional examples of such bands will be shown later. They have a characteristic morphology which results from the continuous nature of craze development with cycling but the discontinuous nature of crack growth [17].

Figure 9a shows an SEM micrograph of a typical fatigue fracture surface of a HIPS sample fractured under similar conditions to Fig. 7. The surface is unlike that of a similar HIPS sample tested at 21 Hz [26] and also unlike that of the tensile fracture surface of Fig. 3 or the PS fatigue fracture surfaces of Fig. 8. Here there are no clearly defined slow and fast growth regions as in the case of tensile fracture of MIPS or HIPS; and there is no smooth, mirror-like portion, as in fatigue fracture of PS. Instead, the entire surface is rough and multiple fracture cracks have apparently developed from many surface sites. Because of the stress concentrating effect of the dispersed particles, crazes initiate much sooner, and craze breakdown to form a crack, or multiple microcracks, also occurs sooner. One result is that the average fatigue lifetime of unnotched samples of HIPS has been found to be less than that of PS, for tests made at comparable stress amplitude and frequency [28].

A high magnification picture of a portion of the fracture surface, Fig. 9b, shows that the fatigue crack has cut through the second phase particles. The matrix surrounding the rubber particles appears smoother and the rubber particles are better delineated than for the case of tensile fracture (Fig. 3b). The small white spots are believed to be the remnants of broken and retracted craze fibrils. The fracture has passed through craze planes that develop around the equator of the particles. Somewhat similar features in fatigue fracture of notched samples of HIPS have been noted by Manson and Hertzberg [23]. The composite nature of the particles with essentially spherical

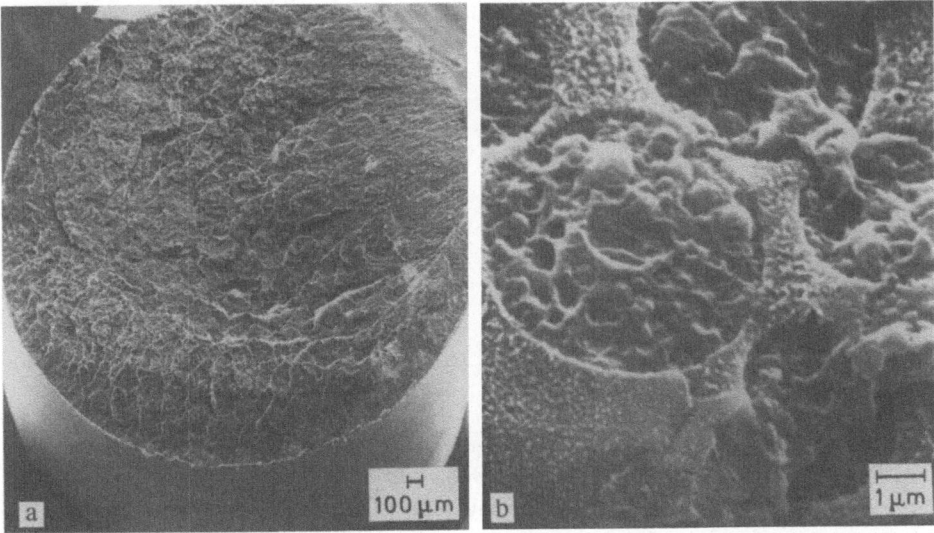

Fig. 9a and b. Fatigue fracture surface of HIPS tested at a stress amplitude of 17.2 MPa and at 0.2 Hz; **a** low magnification scan; **b** high magnification of a region near the source

sub-inclusions of PS, is evident. It is this feature which causes the volume concentration of second phase particles to be much higher than the weight percent of added poly-butadiene. For the large occluded particles, there does not appear to be any extensive cavitation such as that observed in solid rubber particles by Donald and Kramer from transmission electron microscopy (TEM) studies of thin films of HIPS extended in tension [38]. However, small voids separated by ligaments are visible along the outside edges of some particles. The fracture surface morphology shown in Figs. 9a and 9b is characteristic for the stress and frequency conditions stated. However, the morphology is altered, as discussed in following sections, by changes in either of these variables.

3.2 Effects of Stress Amplitude and Stress State on Fatigue Behavior

The influence of stress amplitude on cycles to fracture for unnotched samples of PS has been studied by several investigators [32, 39]. For fracture in the region of 10^3 to 10^5 cycles, there appears to be a linear variation between stress and log N with N increasing as the stress decreases. However, a stress amplitude is further reduced, the S–N curve flattens and approaches a minimum stress value, the endurance strength, below which no failure occurs, at least up to 10^7 cycles. The endurance strength is itself a function of molecular weight; for the commercial grade PS of this study tested at 21 Hz it is about 9.6 MPa; while for anionic PS samples of 2,000,000 molecular weight it is about 13.8 MPa [25]. The higher endurance strength value is a result of greater craze stability with increase of molecular weight and with increase in the number of entanglements between randomly coiled molecular chains.

For a series of stress amplitudes the average number of cycles to craze initiation, as determined by the light reflection method, and the average number of cycles to

Table 1. Effects of Stress Amplitude on Cycles to Craze Initiation and Fracture

Material	Stress Amplitude	Cycles to Initiation	Cycles to Fracture	N_{ci}/N_f
	(MPa)	N_{ci}	N_f	(%)
PS	13.8	42,200	62,800	67.2
	17.2	9,230	20,500	45.0
	20.7	2,050	9,120	22.5
HIPS	10.3	9,050	9,390	96.4
	13.8	1,150	2,030	56.7
	17.2	10	183	5.4
	20.7	1	23	4.3

failure for 4–5 specimens was determined at a frequency of 21 Hz. The results obtained are presented in Table 1 and S–N_{ci} and S–N_f curves are displayed in Fig. 10. At the lowest stress amplitude of 13.8 MPa about two-thirds of the total fatigue life of PS specimens is spent in craze initiation. As stress amplitude increases, the number of cycles for initiating plastic deformation by localized crazing decreases more rapidly than does the number of cycles for fracture, or the number for craze and crack propagation. Hence, the percentage of time to initiate the first potential damage zone decreases with increasing stress amplitude. From Fig. 10 it appears that as the applied stress approaches the endurance strength of the material, the S–N_f curves and the S–N_{ci} curves tend to approach one another. Thus, at low stresses most of the fatigue lifetime is spent in initiation of a surface craze while at high stresses most of the lifetime is spent in craze and crack propagation.

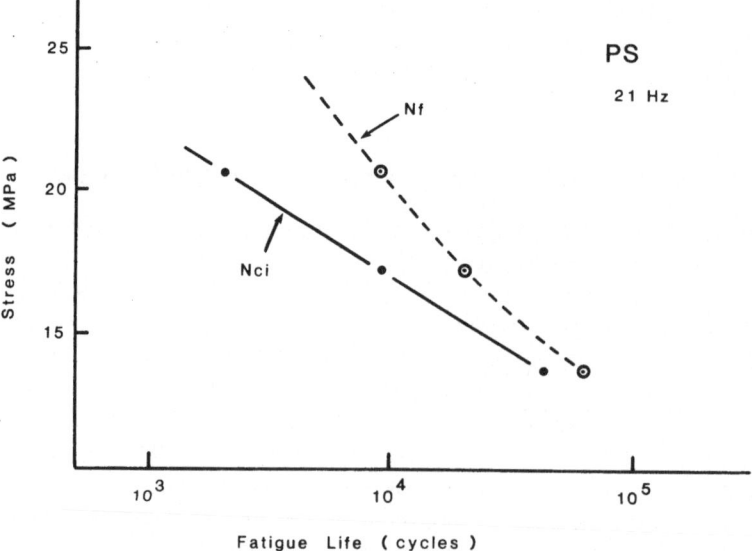

Fig. 10. Cycles to craze initiation, N_{ci}, and cycles to fracture, N_f, vs. stress amplitude for PS tested at 21 Hz

For the rubber modified polymer, the effect of stress amplitude on cycles to craze initiation and fracture for $v = 0.2$ Hz is also shown in Table 1. In this case, however, the estimate of N_{ci} is less accurate as it was made from small observed changes in recorded hysteresis loops. At any given stress level, the number of cycles for first observation of craze yielding and the number of cycles for fracture are significantly less for the rubber modified polymer than for PS. In HIPS, tested at a low stress amplitude of 10.3 MPa, craze nucleation sufficient to produce a detectable change in the stress-strain response occurs only after 9,000 cycles or after 96% of the total cycles to fracture. As the stress amplitude is increased to 13.8 MPa, the percent of the fatigue lifetime spent in craze initiation decreases to about 57%. At still higher stresses craze initiation occurs very early, usually within the first 10 cycles. Thus in HIPS, as for PS, most of the fatigue lifetime in unnotched specimens at high applied stresses is spent in craze and crack growth and, at low applied stresses, in craze initiation.

For a stress amplitude of 17.2 MPa, Fig. 7 showed the changes that occur in the dynamic stress-strain response of HIPS at various N values. By monitoring such hysteresis loops, one can determine the specific dependence on N of properties such as the secant modulus [24]. Or one can detect onset ot strain softening by measuring the width of the hysteresis loop, taken at a given value of the tension or compression stress, and note how this changes with N. Such a plot is shown in Fig. 11 for a HIPS sample that fractured at 202 cycles.

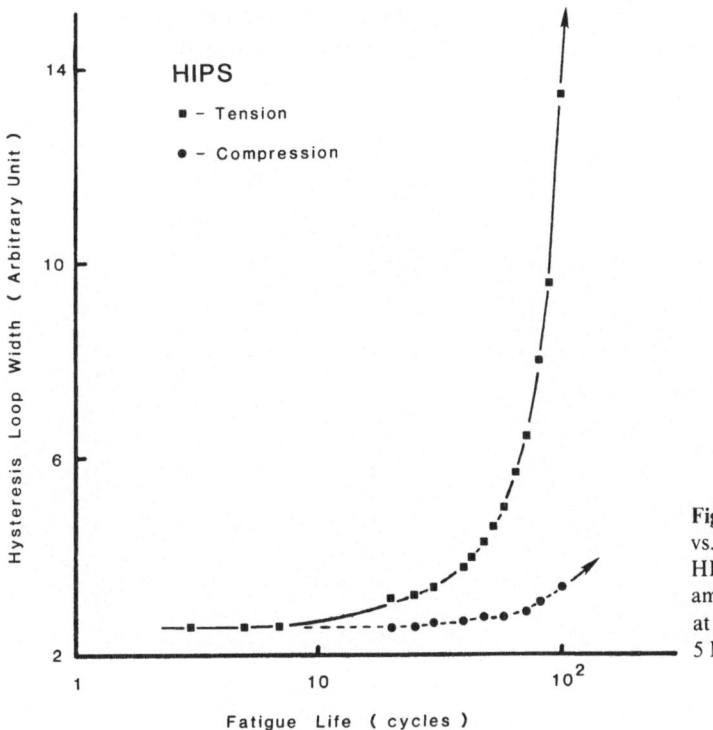

Fig. 11. Hysteresis loop width vs. number of cycles for a HIPS sample tested at a stress amplitude of 17.2 MPa and at 0.2 Hz. Width noted at 5 MPa

On the tension side, the loop width begins to rise at about 7 cycles. This is a result of strain softening due to craze initiation. With continued cycling, the plastic deformation increases continuously and rapidly with increase of N. On the compression half cycle, no changes are observed for about 30 cycles and only very small changes are noted thereafter. This comparative behavior indicates clearly that crazing, as noted also by Bucknall and Stevens [24], is by far the dominant mode of deformation in HIPS samples subject to alternating stresses.

The stress amplitude in fully reversed cycling has a definite effect on the morphology of the fatigue fracture surface. Figure 12a shows an SEM micrograph of the fracture surface of PS tested in fatigue at 0.2 Hz under a stress amplitude of 24.1 MPa. This may be compared with Fig. 8a which shows a fracture surface for a sample tested at 17.2 MPa. At the higher stress, there are only three distinct regions instead of four. One is the small smooth region near the fracture source containing DCG bands. A second region occupying only a small part of the fracture surface is shown at higher magnification in Fig. 12b. Here we see numerous secondary fracture features somewhat similar to those noted in tensile fracture (Fig. 2b). The third region, occupying most of the fracture surface, displays a series of rough broad bands. Such wide bands can arise in tensile fracture, and result apparently from some from of instability [39]. Another effect of high stress amplitude, which is particularly noted when tests are made at a high mean stress, is for fatigue cracks to develop from more than one surface fracture source. [35]

Figure 13 shows a typical fatigue fracture surface for a HIPS sample tested at 0.2 Hz at a low stress amplitude of 10.3 MPa. To note the effect of stress amplitude, these pictures should be compared with those of Fig. 9 obtained at a higher stress amplitude of 17.2 MPa. At the lower stress amplitude the fracture surface, Fig. 13a, is markedly different from that of Fig. 9a. Here fracture developed from a surface

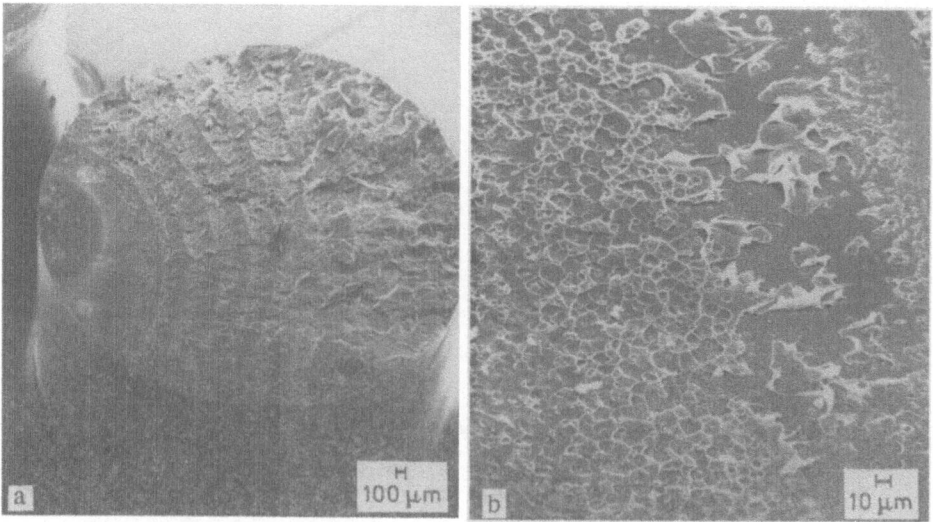

Fig. 12a and b. Fatigue fracture surfaces of PS tested at a stress amplitude of 24.1 MPa and at 0.2 Hz; **a** Low magnification scan; **b** High manification of region beyond DCG bands

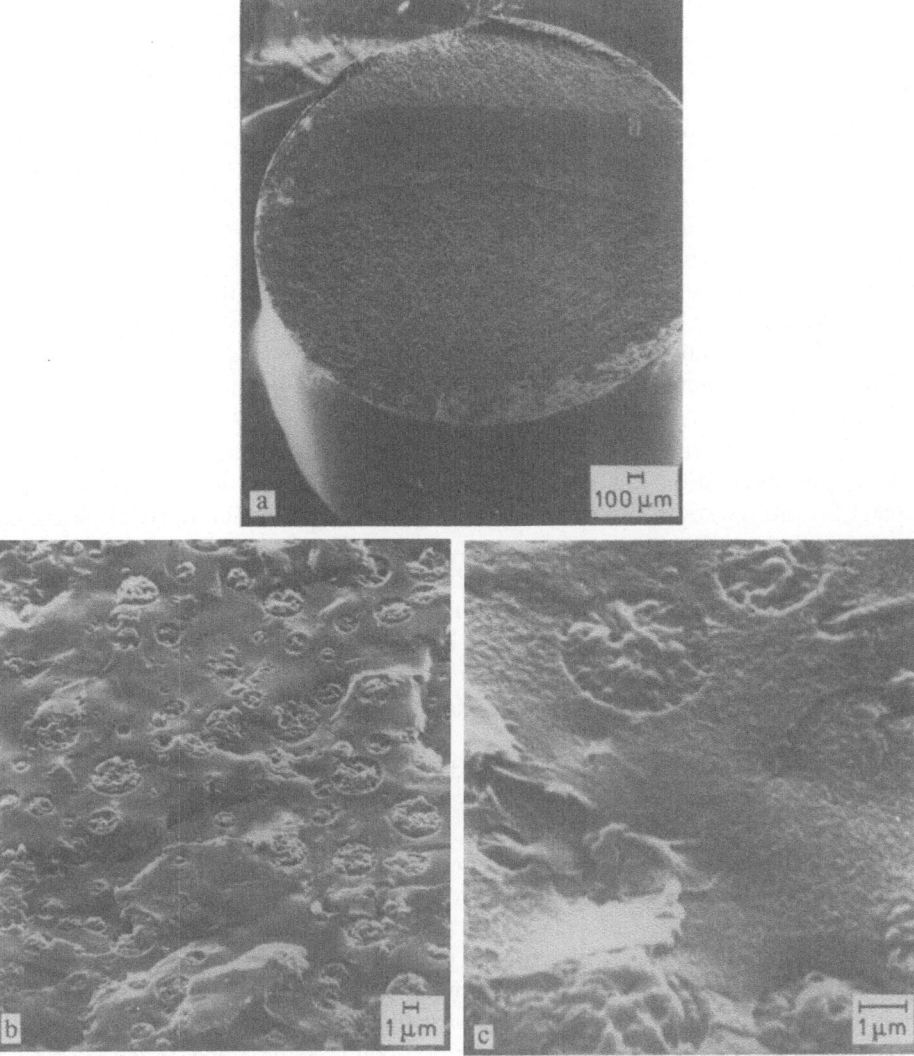

Fig. 13a—c. Fatigue fracture surfaces of HIPS tested at a low stress amplitude of 10.3 MPa and at 0.2 Hz; **a** Low magnification scan; **b** Higher magnification near source; **c** High magnification near transition region

source and, with continued cycling, grew radially outward until one-half or more of the area was covered. At this point, fracture speed increased and the transition between the fatigue propagation region and the fast growth tensile region is clearly revealed. In the tensile fracture portion, the morphology is similar to that noted in fracture under monotonic loading and displayed in Fig. 3, i.e., there is a comparatively smooth, fine-textured region followed by a rougher region, indicative of greater crack bifurcation and instability.

Figure 13b is a higher magnification view near to the fracture source. Most of the particles have been fractured while some smaller ones have been bypassed. Note the well-defined fractured second phase particles and the comparatively smooth nature of the matrix surrounding the particles. This behavior is indicative of fracture through crazes generated at the particle matrix interface. At still higher magnification the matrix is seen to be covered with small white dimples. These dimples, arising from disentanglement or fracture of the molecular chain bundles within the craze fibrils, are rounded and uniformly dispersed. However, farther out on the fracture surface and closer to the transition region to last cycle tensile fracture, the dimples tend to become more tongue-shaped with tufts pointing generally in the direction of crack propagation. Figure 13c is a high magnification view of a portion of this region. The fibril tips are about 0.1 μm in size and their appearance is somewhat similar to the hairy features noted by Doyle et al. in fast tensile fracture and attributed to viscous rupture of chainmolecule bundles [40].

Although fatigue fracture of HIPS specimens usually develop from surface sources, one sample, tested at the low stress amplitude of 9.5 MPa showed an internal origin. Except for this, however, the morphological features were similar to those described above for the sample tested at a stress amplitude of 10.3 MPa. It is concluded that at 0.2 Hz and at low stress amplitudes, the fatigue fracture surface is well defined with slow radial growth occurring from a surface flaw or occasionally from an internal source, until the crack size becomes large enough to cause catastrophic fracture. In contrast, at high stress amplitudes close to the tensile yield stress of HIPS the fatigue fracture surface (Fig. 9a) is ill defined and multiple fracture sources appear to be present.

Additional fatigue tests have been carried out in a fatigue mode that did not involve reversed stressing [41]. In these tests, samples of HIPS were cycled at 21 Hz from a minimum tensile stress of 3.4 MPa to a maximum tensile stress of 17.2 MPa. In another investigation, samples of PS were cycled under similar stress levels but at 26.7 Hz [35]. When these results are compared with results obtained at comparable frequency but under a reversed stress of 17.2 MPa, the average lifetime to fracture is found to increase by a factor of four to five for the samples tested purely in the tensile mode. This applies both to PS and to the rubber modified polystyrene. This finding suggests that, for a given maximum stress, cycling craze fibrils purely in a tensile mode is much less damaging than cycling in a reversed tension-compression mode. In the latter case fibrils are alternately stretched and then compressed. It is, apparently, the alternate buckling and crimping of the craze fibrils on the compression side of the cycle which is so damaging to craze stability. Also, if the maximum stress is retained at 17.2 MPa but the range of stress reduced by raising the minimum stress to 8.8 MPa, then no fatigue failure occurs even after 10^7 cycles; and, under these test conditions, which are a mixture of creep and fatigue, it is possible to generate multiple crazes along the gauge length without craze breakdown or crack development [35].

When smooth specimens of PS are cycled at 26.7 Hz in a tensile mode at a maximum stress of 17.2 MPa discontinuous crack growth bands similar to those shown in Fig. 8 are seen in the mirror region near the fracture source. These bands are also similar in morphology to those observed in testing notched PS specimens at a minimum stress that is maintained at one-tenth the maximum stress [34]. However, DCG bands were not seen when tests are carried out under a fully reversed stress of 17.2 MPa at comparable frequency [35]. Even under these conditions, a recent study has shown that concentric bands may be seen in the mirror region near the source if tests are carried out in a silicone oil environment [42].

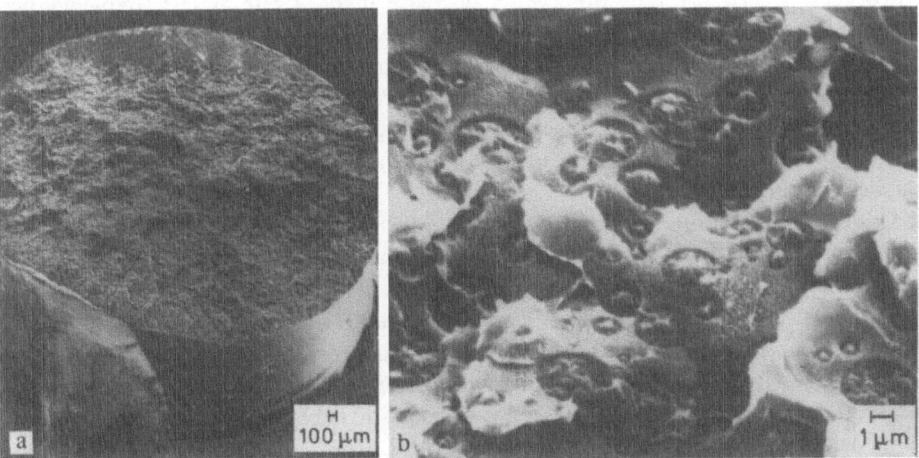

Fig. 14a and b. Fatigue fracture surfaces of HIPS tested at 21 Hz in a tensile mode. The maximum stress was 17.2 MPa and the minimum stress was 3.4 MPa; **a** Low magnification scan; **b** High magnification scan

Figure 14 shows fracture surface of HIPS specimens tested in cyclic tension at a maximum stress of 17.2 MPa and a minimum stress of 3.4 MPa. The surface features are distinctly different than those reported for similar samples cycled at the same frequency of 21 Hz but under a reversed stress of 10.3 MPa [26,41]. One difference is that upon cycling in a tensile mode to a high maximum stress of 17.2 MPa the fracture surface, Fig. 14a, is generally very rough and more than one fracture source may be present. This was also noted when fracture occurred under a reversed stress of 17.2 MPa at 0.2 Hz (Fig. 8). A high magnification view (Fig. 14b) shows the presence of a grainy structure in the matrix surrounding the many fractured second-phase particles. Similar halo-like features surrounding particles were noted in notched HIPS samples tested and fractured in a tensile mode [23]. The grainy features, 0.1 μm or less in size, are thought to respresent ruptured and retracted craze fibrils that result when the fatigue fracture crack propagates through existing craze planes. Their apparent absence in tests under fully reversed cycling at 21 Hz [41] is probably a result of the damaging effects on surface morphology and on craze lifetime of alternate extension and buckling of craze fibrils. Examination of Fig. 14b, and of similar SEM scans, shows that the small particles appear to cavitate to a much greater extent than the larger more occluded particles and remnants of the cavitated and fractured particle are clearly visible in the center of the small cavities. This finding is in accord with TEM studies of deformed thin films of HIPS [38]. Smaller particles, which tend to be more homogeneous, are also more susceptible to cavitation under applied tensile stress.

3.3 Effects of Test Frequency on Fatigue Response and Fracture Morphology

The influence of test frequency on fatigue crack propagation of notched specimens has been studied by several investigators, but mostly by Hertzberg, Manson and coworkers [17]. For PS, they observed a decrease in FCP rate as test frequency was increased from 1 Hz to 10 Hz to 100 Hz [43, 44]. No results were presented for HIPS but, for a blend of HIPS and PPO, the FCP rate also decreased with increase in frequency [45]. Similar changes were observed in other glassy polymers, such as PMMA, and it was noted that the frequency sensitivity was greatest when the test frequency was comparable to the frequency of molecular motions associated with the β visco-elastic damping peak [46].

Under cyclic loading it is possible to generate in viscoelastic materials significant temperature rises in test specimens. Various examples illustrating the possible effects of stress amplitude and frequency on specimen temperature have been presented [17, 18]. Our interest here is rather in the effects of test frequency on total fatigue lifetime of unnotched specimens so that both initiation and propagation phases are involved in the fatigue fracture process. For PS, thermal effects appear to be insignificant at all frequencies investigated; for the more viscoelastic rubber modified HIPS material, some degree of specimen temperature rise is to be expected, particularly at the higher stress amplitudes and the higher test frequencies.

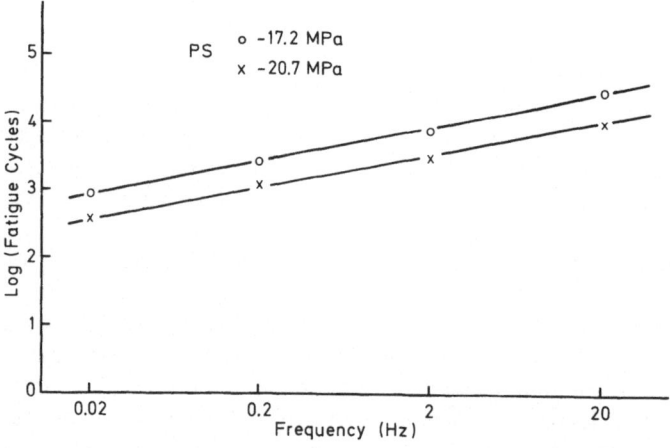

Fig. 15. Effect of frequency on cycles to fracture for PS specimens tested at stress amplitudes of 17.2 MPa and 20.7 MPa

Figure 15 shows how the average fatigue lifetime of PS depends on frequency for two different stress amplitudes. The variation appears to be a linear one on this log-log plot, with the number of cycles to fracture increasing with increase of frequency, and at essentially the same rate for both stress amplitudes. For the rubber modified HIPS the fatigue endurance is plotted as a function of frequency in Fig. 16. Here too the lifetime increases with increase of test frequency and again the variation is a linear one on a log-lot plot. The slope of these curves is also essentially independent of stress

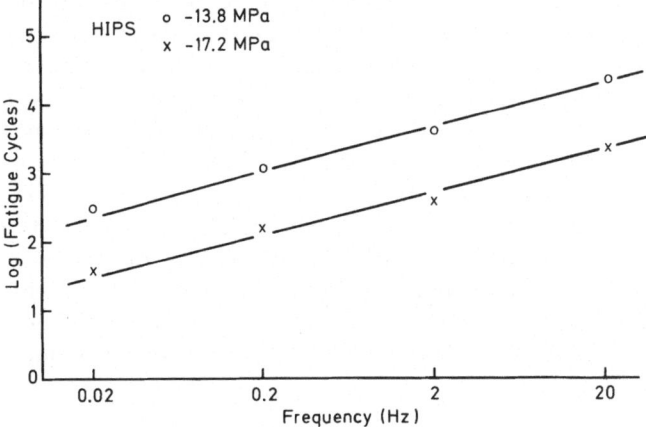

Fig. 16. Effect of frequency on cycles to fracture for HIPS specimens tested at stress amplitudes of 13.8 MPa and 17.2 MPa

amplitude. Increasing the stress amplitude from 13.8 MPa to 17.2 MPa decreases the fatigue lifetime of HIPS about one decade at all frequencies.

It is interesting to compare the test results for the homopolymer and the rubber modified polymer. This comparison, for a stress amplitude of 17.2 MPa, is shown in Fig. 17. It is evident that, when test samples are tested at the same stress amplitude, HIPS is inferior to PS in fatigue endurance at all frequencies. Yet it has been observed from fracture mechanics type experiments on notched samples that, for a given stress intensity factor range, ΔK, HIPS is superior to PS in fatigue performance [17]. This comparison emphasizes the important role played by craze and crack initiation in the fatigue fracture process of unnotched specimens. In the rubber modified polymer, despite a greater resistance to crack propagation once a crack has nucleated, the overall fatigue lifetime is less than that of PS. This is a direct result of the fact that the dispersed particles cause craze nucleation to occur much earlier in time; or for a given constant rate of loading, to occur at a much lower stress as seen in Fig. 1.

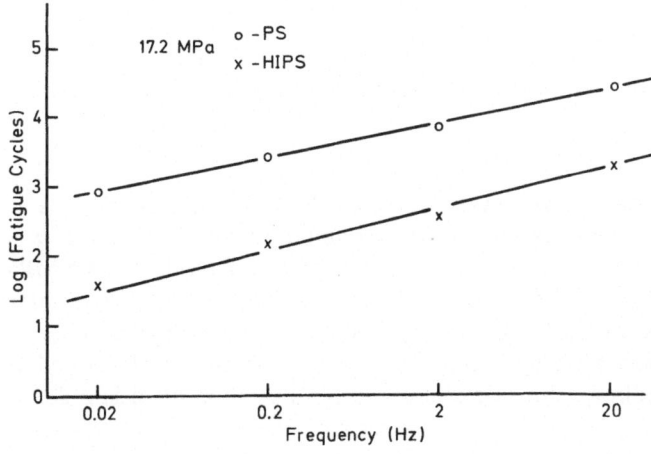

Fig. 17. Cycles to fracture vs. frequency for samples of PS and HIPS tested at a stress amplitude of 17.2 MPa

The current test results covering a wide frequency range are in accord with the findings of a prior study made on a different test apparatus at a single frequency of 1,250 cpm [28, 47]. They conclusively show that the difference in fatigue behavior due to incorporation of a rubber phase is not a result of greater thermal effects in HIPS. Even at the lowest frequencies, where there is no significant rise in specimen temperature, the rubber modified polymer shows consistently poorer fatigue performance than the unmodified polymer.

The influence of frequency on the S–N curve is shown for HIPS in Fig. 18. The general form of the S–N curve appears to change little with change of frequency. The fatigue data indicate that for PS, on reducing frequency by two decades, the average lifetime to failure is reduced by about a decade while for HIPS, with the same reduction in frequency of two decades, average lifetime is reduced about a decade and a half.

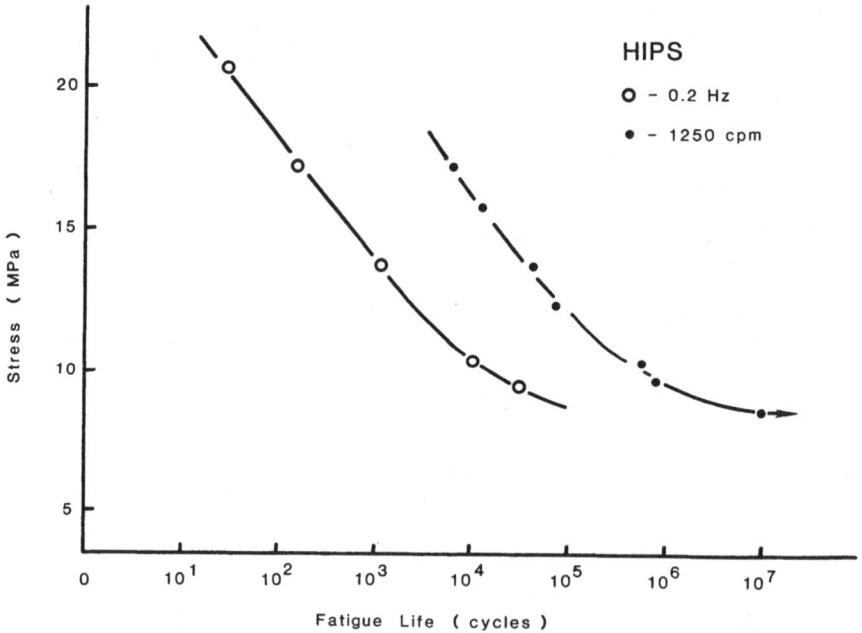

Fig. 18. Stress amplitude vs. cycles to fracture of HIPS specimens tested at 0.2 Hz and 21 Hz

Hysteresis loop observations have been made for both PS and HIPS at two frequencies, viz. 0.2 Hz and 0.02 Hz and at a stress amplitude of 17.2 MPa. It was noted in Fig. 6 that, for PS tested at 0.2 Hz and at a stress amplitude of 24.1 MPa there was no detectable change in loop area and hence no indication of appreciable crazing up to 1,450 cycles, the last cycle before fracture. Additional tests made at 0.02 Hz and 17.2 MPa show a similar behavior. For one sample that fractured at 803 cycles, loops were monitored for 660 cycles and no indication of plastic deformation was detected.

For a HIPS sample tested at a stress amplitude of 17.2 MPa and a frequency of 0.2 Hz, hysteresis loops taken at various cycles (Fig. 7) indicated that craze initiation was first observed for this sample after about 20 cycles, while 283 cycles were required to fracture. For similar fatigue tests carried out at the lower frequency of 0.02 Hz, the cycles to fracture were decreased (from 283 to 64) and loop asymmetry and craze formation began sooner, at about 1–2 cycles. The changes produced in hysteresis loops with cycling are shown in Fig. 19. With decrease of test frequency N_f reduces, the entire S–N curve shifts to the left as shown by Fig. 18, and, because of the increased time for each cycle, fatigue induced craze initiation occurs earlier in the specimen lifetime.

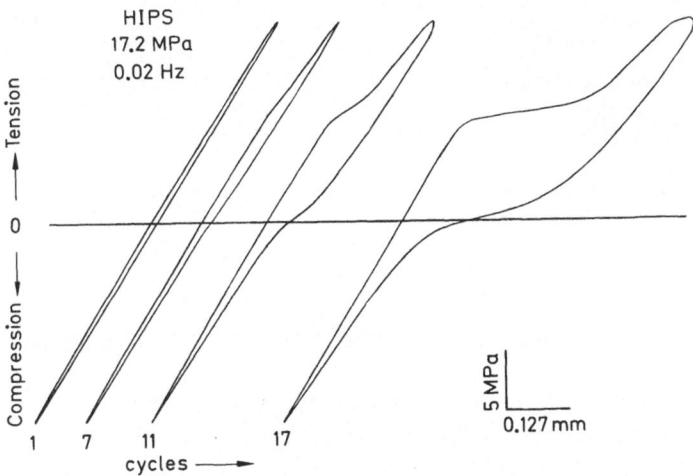

Fig. 19. Hysteresis loops of HIPS vs. number of cycles for specimen tested at a stress amplitude of 17.2 MPa and at 0.02 Hz

Test frequency also has a significant effect on the fatigue fracture surface morphology. Let us first consider the unmodified polymer. For a sample tested at 2 Hz, Fig. 20a shows the entire fracture surface and Fig. 20b depicts the region surrounding the fracture source. As noted with reference to Fig. 8, for a sample tested at 0.2 Hz at the same stress amplitude, four regions of different morphology are present. Discontinuous crack growth bands are seen, in region RI near the fracture source, in the form of some 20 separate concentric rings which increase in spacing as we move out from source. Such DCG bands are not seen, however, in tests at the same stress but at the higher frequency of 26.7 Hz [18]. A higher magnification view of the 11th and 12th bands is given in Fig. 21a. These bands have the characteristic DCG band micromorphology. The dark portion is the crack arrest zone. The craze grows continuously from the crack tip and many cycles may elapse before craze breakdown occurs and the crack jumps through the existing craze and then again arrests sometime before it meets uncrazed material. The process was described earlier by Hertzberg and Manson [17]; the most recent and quite dramatic findings of Döll are given in detail in Chapter 3; similar characteristic features have been noted on the fatigue

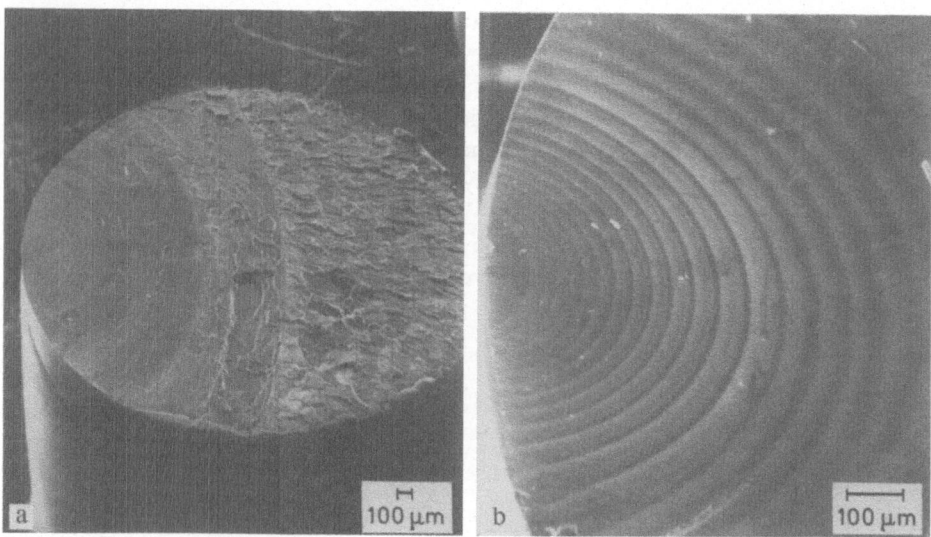

Fig. 20a and b. Fatigue fracture surface of a PS specimen tested at 17.2 MPa and 2 Hz; **a** Low magnification scan; **b** Higher magnification near source

fracture surface of notched specimens of PS and of other polymers tested in alternating tension.

When the crack jumps incrementally through the existing craze, it initially moves through the central region of the craze. A high magnification view of this portion of the 12th band is shown in Fig. 21b and of the rear portion of the same band in Fig. 21c. The dimpled structure of Fig. 21b results from crack penetration through the thicker part of the craze. Figure 21c depicts the characteristic craze island type structure that results when crack propagation becomes unstable and the crack moves along various craze bulk interfaces in the thinner part of the craze. Why the discontinuous growth bands are seen clearly at 0.2 Hz and at 2 Hz and are seldom seen in unnotched specimens tested at 21 Hz is not clear; perhaps at the slower test speeds there is more time for viscoelastic recovery processes to operate and less damage is done to the craze fibril structure during the compressive part of the cycle.

In fatigue tests on notched specimens, many materials exhibit not only DCG bands in the smooth region near the fracture source but also show at higher values of ΔK, a series of parallel striations, with each striation spacing now corresponding to the amount of crack growth per cycle [17]; see also Chapter 3. These striations tend to occur on the fracture surface in the region farther away from the source. They have also been observed in unnotched samples of PS tested in tension-compression fatigue at 26.7 Hz [35]. Figure 22 shows that similar striation markings appear on the fracture surface of the PS samples tested at 2 Hz. This SEM micrograph is of a portion of the fracture surface just beyond the last DCG band. The striation spacings (~ 1 μm) are much smaller than the DCG band spacings and the effect on their growth of secondary fractures, developing from the presence of various heterogeneities, is clearly visible.

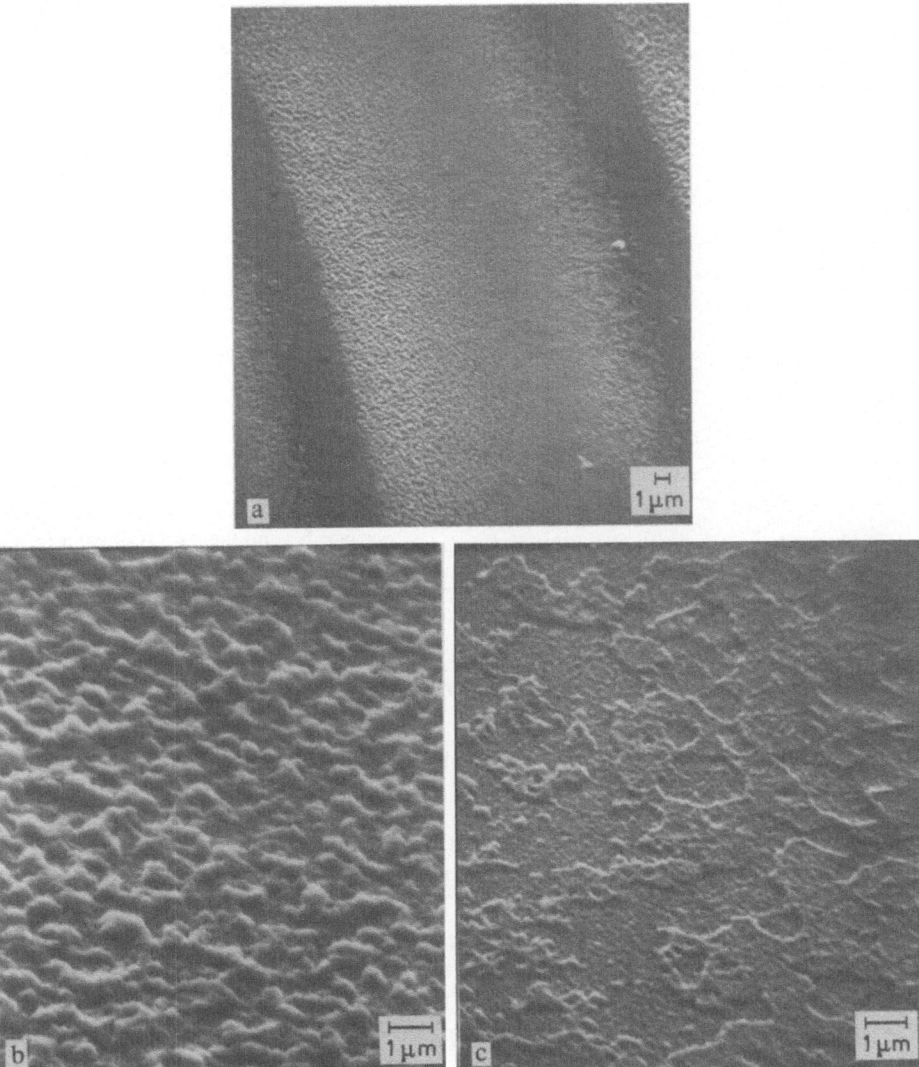

Fig. 21 a—c. Portions of the fatigue fracture surface of the PS sample of Fig. 20; **a** Region between the 11th and 12th bands; **b** Leading edge of the 12th band; **c** Trailing edge of the 12th band

To see the possible effects of test frequency on fatigue fracture surface morphology of HIPS we can compare results obtained at a stress amplitude of 10.3 MPa and at a frequency of 0.2 Hz (Fig. 13) with published results obtained at the same stress but at 21 Hz [26]. In neither case are discontinuous crack growth bands seen and both samples show fatigue fracture developing from a surface source. The fine textured stable crack growth region that follows is somewhat smoother and the transition line separating the slow and fast growth regions is more definite for the sample tested at low frequency.

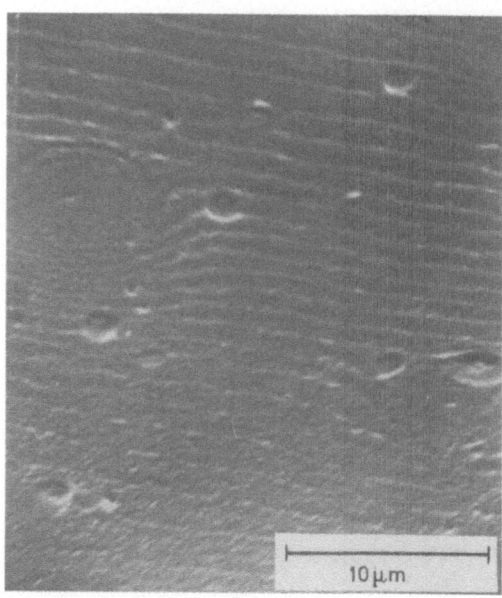

Fig. 22. A portion of the fatigue fracture surface beyond the last DCG band for the PS specimen of Fig. 20. Crack propagation direction from top to bottom

4 Styrene-Acrylonitrile Copolymers and ABS

4.1 Effects of Acrylonitrile Content and Molecular Weight on Mechanical and Relaxation Behavior of SAN Polymers

As discussed in preceding sections, one method for improving the toughness of glassy polymers is to reinforce them with a dispersed, and preferably grafted, rubber component. However, this is only achieved, as Fig. 1 shows, by a sacrifice in tensile yield strength; and, as demonstrated also by Fig. 17, by a sacrifice in fatigue performance. One method of raising the yield and fracture stress is to copolymerize styrene with acrylonitrile. A rubber modified version of this copolymer, as ABS, should then show a relatively high tensile yield strength, compared to HIPS, along with good toughness and impact resistance.

Many investigations of the mechanical properties of ABS and of the nature of rubber toughening in this material have been made and these findings are reviewed and discussed by Bucknall[12] and by Mann and Williamson[48]. There have been fewer studies of fatigue performance, and most of these are concerned only with crack propagation in notched specimens[17, 18]. Here we are concerned with the effects of stress state and frequency on craze initiation, on softening due to shear deformation, and on fatigue fracture surface morphology. Also, the comparative performance of the rubber reinforced ABS, under both monotonic and cyclic loading, relative to the unmodified SAN copolymer and relative to rubber reinforced polystyrene is of interest. In this section we concentrate on mechanical behavior under monotonic loading and on relaxation properties of unmodified copolymers; and, in the next section, we discuss how these properties and the tensile fracture surface morphology are altered by incorporation of a rubber phase.

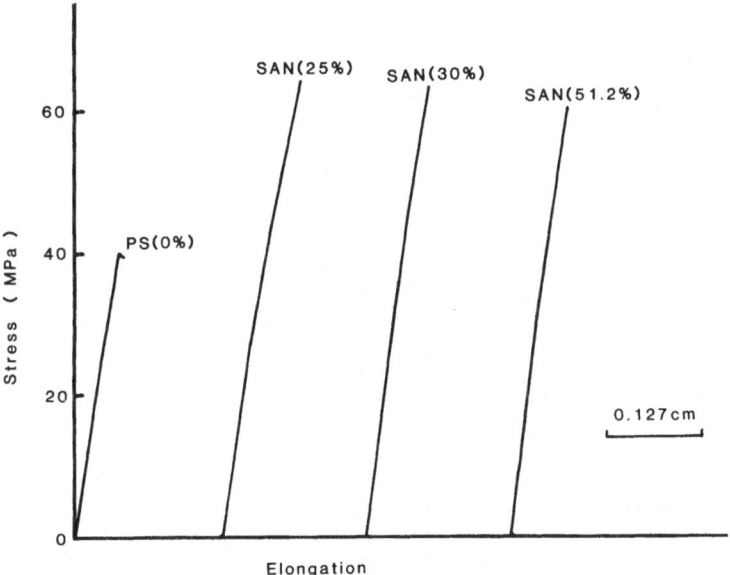

Fig. 23. Stress-strain curves of PS and of SAN copolymers with varying AN content obtained at a displacement rate of 5.08×10^{-2} cm/min

Figure 23 shows a typical stress-strain curve of PS and indicates how this curve is modified by increase of acrylonitrile (AN) content. Although PS, and all three of the SAN copolymers show the typical brittle behavior of many glassy polymers, it is evident that the tensile strength has been increased significantly by the presence of acrylonitrile. A linear increase has been reported for AN contents ranging from 0 to 20 % [49] but this trend clearly does not continue for higher AN contents. Copolymerization of styrene with acrylonitrile also has a significant effect on the craze stress as we shall show.

Since the SAN specimens are transparent, it is possible to use the light reflection method to detect onset and growth of crazing. A typical plot of stress and of reflected light intensity vs. time (or strain, as the tests were run at a constant crosshead speed) is shown in Fig. 24 for the SAN resin containing 30 % of acrylonitrile. The stress where the first rise in light intensity occurs, σ_{ci}, is essentially the stress for initiation of the first surface craze. Since the magnitude of this stress is affected by the surface condition, and by how well the surface is polished, we also record another stress, designated σ_c, and called the craze intensification stress, which is the stress at which the rate of crazing increases significantly. This stress, taken from the graph as indicated, is approximately equal to the stress at which the first clear visual indication of crazing occurs.

In Fig. 25, we show how σ_{ci}, σ_c and the tensile strength, σ_T, depend on acrylonitrile content. The experimental points are average values from separate tests on three or more samples, except for the highest AN content, where only one sample was tested. The craze intensification stress increases linearly with AN content; the stress for first craze initiation follows a somewhat similar pattern but there was more scatter in the

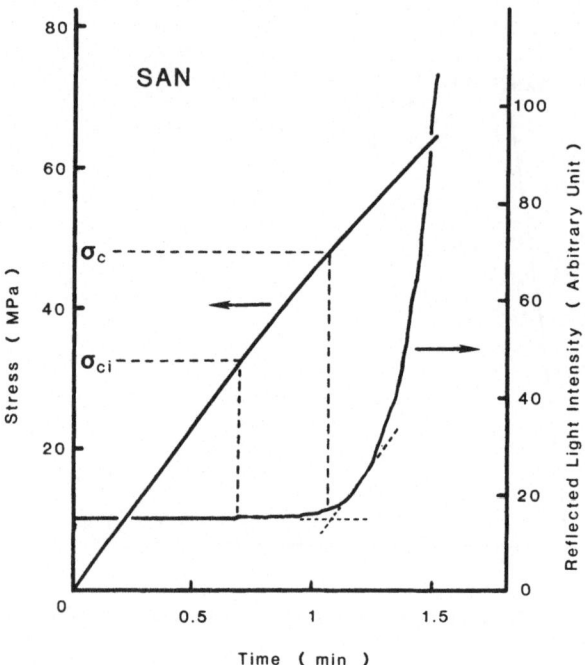

Fig. 24. Reflected light intensity and tensile stress vs. time for an SAN sample with 30% AN tested at a displacement rate of 5.08 $\times 10^{-2}$ cm/min

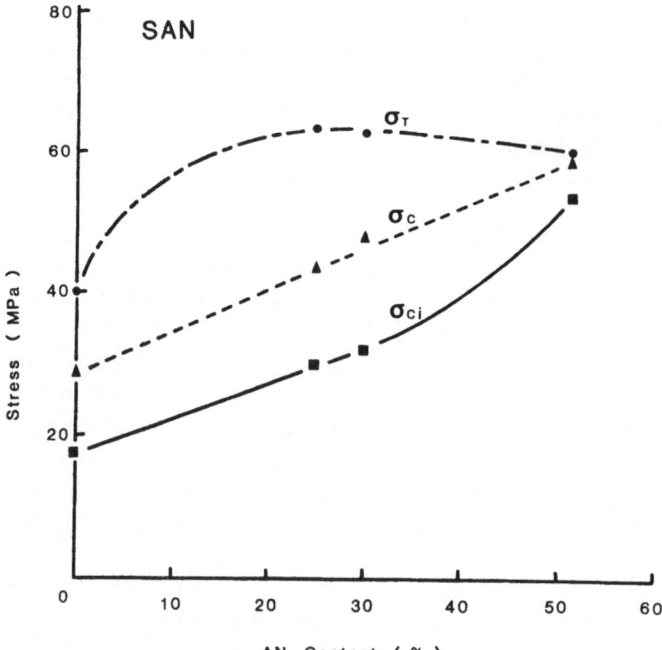

Fig. 25. Variation of craze initiation-stress, σ_{ci}, craze intensification stress, σ_c, and tensile fracture stress, σ_T, with AN content

data for the various samples tested. The above result is in general accord with published findings which show that SAN resins have greater resistance to crazing under creep conditions [50], and have much better resistance to environmental crazing than does PS [51]. Figure 25 indicates that the tensile fracture stress, unlike the craze stress, reaches a maximum value in the region of 25 to 30% AN and thereafter decreases slightly. A similar effect has been noted by Ziemba [51].

Stress-strain curves have also been determined for SAN samples of 30% AN content, having three different molecular weights. The tensile strength was found to increase with increasing molecular weight and there was also an increase in the strain to fracture. These changes are a result of an increased degree of molecular entanglement and a greater resistance to craze breakdown. Similar effects have been noted in the homopolymer PS [25].

Using the light reflection method, we have investigated the effects of molecular weight on the craze initiation stress and on the craze intensification stress. For the craze initiation stress, σ_{ci}, there was considerable scatter in the data and no obvious trend. The variation of the craze intensification stress, and of the tensile strength, with molecular weight is shown in Fig. 26. Also shown is the variation with $\bar{M}_w$ of the glass transition temperature, T_g, as determined from dynamic mechanical measurements made at 3 Hz. The craze intensification stress appears to increase rapidly with molecular weight at low molecular weights and then more slowly as $\bar{M}_w$ increases. T_g follows a similar trend. However, the average tensile strength increases linearly with $\bar{M}_w$ over the range investigated. At the highest molecular weight, 1.85×10^5, it is about 25% higher than the value for the samples having an $\bar{M}_w$ of 1.15×10^5.

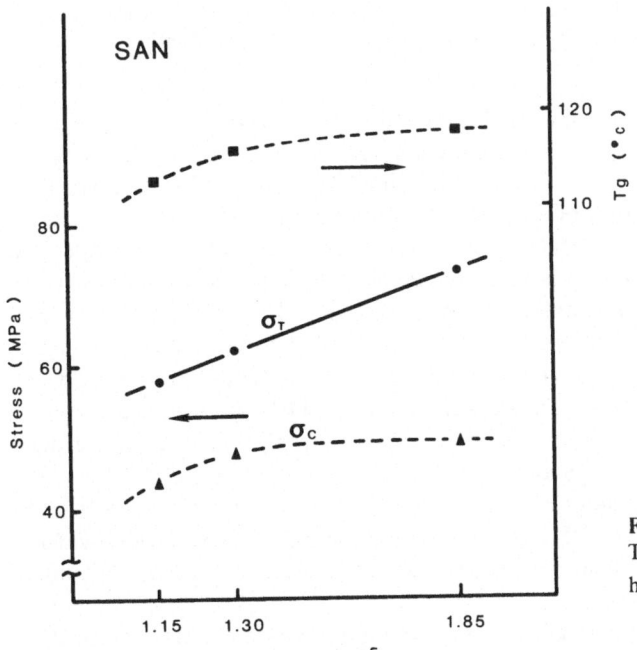

Fig. 26. Variation of σ_c, σ_T and T_g with $\bar{M}_w$ for SAN samples having 30% AN

Thus, by using AN contents of 30 % and high $\bar{M}_w$ values, the tensile strength of the unmodified polymer can be significantly increased from about 40 MPa to over 73 MPa. With increase of molecular weight, as shown from studies of thin films of PS by Donald and Kramer, the craze extension ratio is reduced and the increased entanglement density then results in greater craze stability and greater resistance to fracture [52].

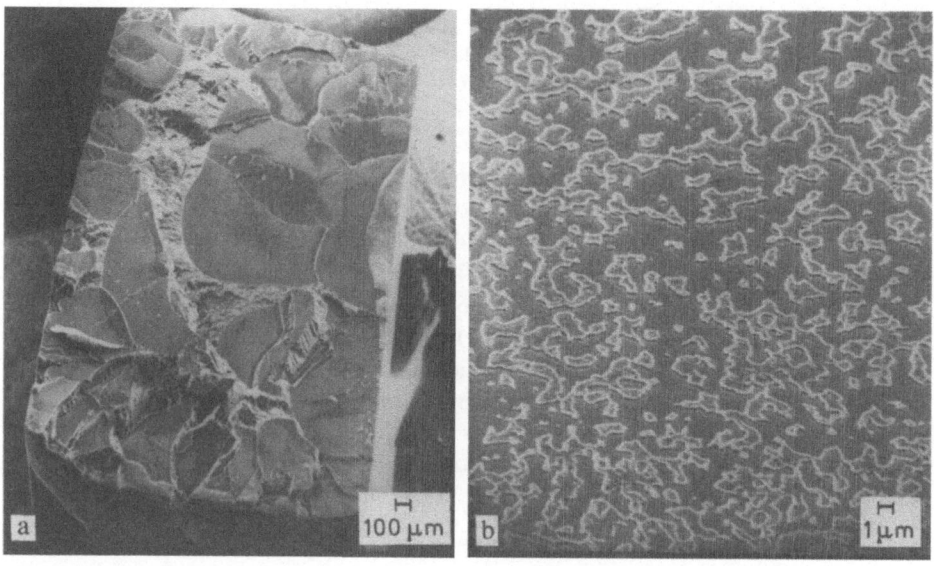

Fig. 27a and b. Tensile fracture surface of an SAN sample containing 25 % AN and tested at a displacement rate of 5.08×10^{-2} cm/min; **a** Low magnification scan; **b** High magnification scan

The tensile fracture surface morphology of the SAN resins differs somewhat from that of PS shown in Fig. 2. A typical fracture surface of an SAN rectangular specimen is shown in Fig. 27a for a resin containing 25 % AN and Fig. 27b is a high magnification view of a portion of a fractured craze near the center of the specimen. The island type structure is characteristic of crack propagation through the thin part of an existing craze. Many smooth craze planes at different elevations are visible in Fig. 27a and the general morphology is similar to that of PS except that the overall surface is not as flat and there is some indication of shear deformation between craze planes. This findung is in general accord with the results of a study by Donald and Kramer on thin films of SAN [53]. They concluded, from TEM studies, that the craze microstructure, including a midrib section of higher extension ratio than the main part of the craze, is similar for PS and SAN. They also noted that, in the thin films, localized shear deformation zones occurred, as well as localized crazes, and that the tendency for crazing reduced as AN content increased. Similarly, we note less evidence of craze planes and more of shear deformation on the fracture surface as AN content rises.

The SEM scans of Figs. 28a and 28b show that molecular weight also has an effect on fracture surface morphology. For the lowest molecular weight sample with

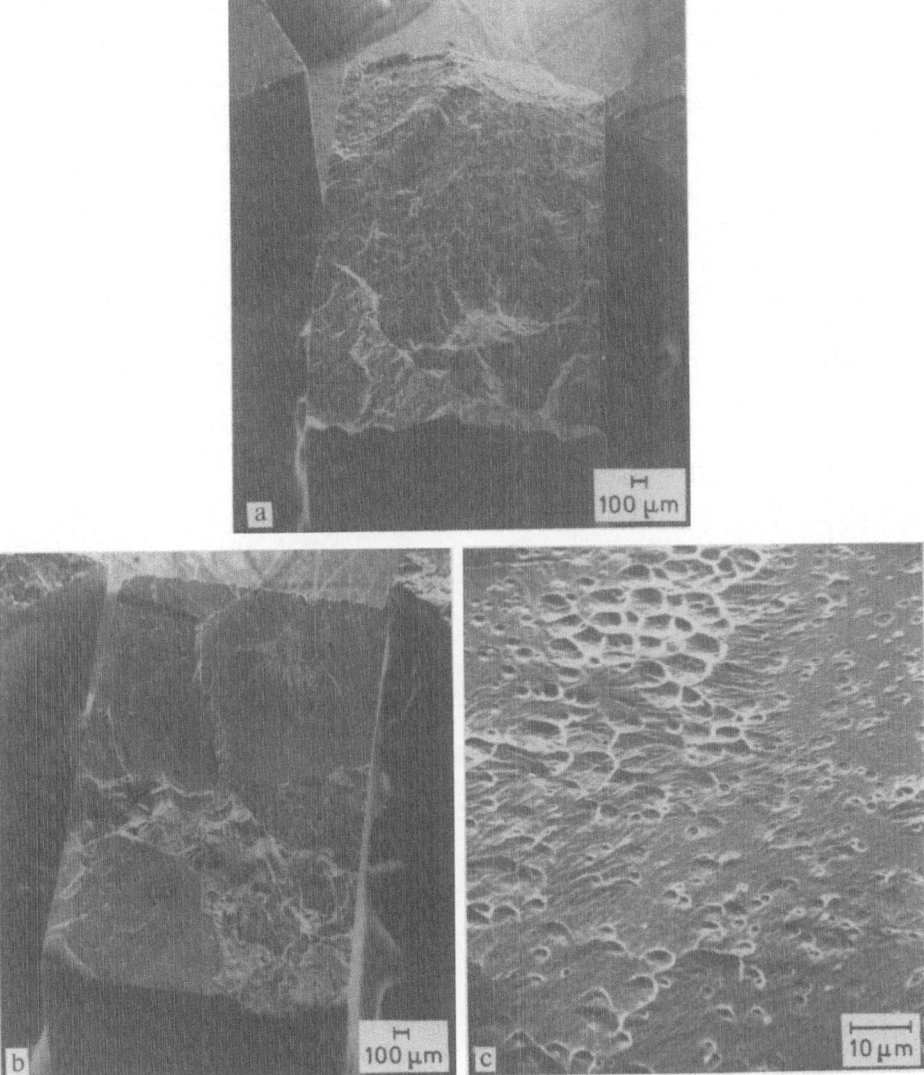

Fig. 28a—c. Tensile fracture surfaces of SAN specimens with 30% AN and varying $\bar{M}_w$; **a** $\bar{M}_w$ = 1.15×10^5; **b** $\bar{M}_w$ = 1.85×10^5; **c** High magnification near end of smooth crazed portion of (a)

$\bar{M}_w$ = 1.15×10^5 craze breakdown occurred early before extensive craze formation and growth and the fracture surface shows only a small smooth portion, characteristic of fracture through a pre-existing craze. For the higher molecular weight sample with and growth and the fracture shows only a small smooth portion, characteristic of fracture through a pre-existing craze. For the higher molecular weight sample with $\bar{M}_w$ = 1.85×10^5 crazes are more stable and most of the fracture surface shows fracture through pre-existing craze planes. The greater craze stability for the higher $\bar{M}_w$ sample permits more matrix deformation to occur prior to final fracture as it raises the craze breakdown stress. The greater strength and ductility of the high molecular weight sample are evident from comparison of the stress-strain curves of samples of different molecular weight. In SAN, as in other glassy polymers, internal heterogeneities can lead to secondary fractures developing ahead of the crack front. Many such secondary fractures are visible in Fig. 28c, which is a high magnification view of a portion of the smooth part of the fracture surface of Fig. 28a. Note the general similarly of these secondary fractures with those shown in Fig. 2b for PS.

Compression tests, as well as tension tests, were carried out on the samples of varying AN content. All samples exhibited a compressive yield stress, σ_y; and with further increase of strain, a subsequent drop in stress, $\Delta\sigma_y$, occurred before the stress again started to rise. From the test data, the elastic modulus and the yield strength were determined as a function of nitrile content and the results are shown in Fig. 29.

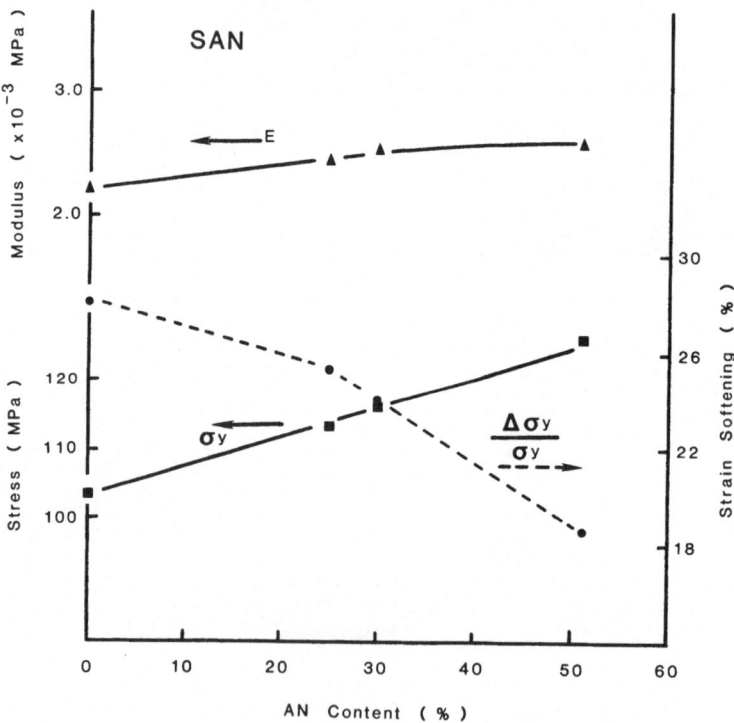

Fig. 29. Elastic modulus, E, compressive yield strength, σ_y, and strain softening ($\Delta\sigma_y/\sigma_y$) vs. AN content for various SAN resins

The compressive yield strengths are much higher than the tensile ones as crazing and early fracture are no longer present and the deformation mode is entirely shear yielding. It is evident that the resistance to shear yielding increases significantly with AN content and in essentially a linear fashion. The modulus values are not accurate measurements of the stiffness as they have been estimated only from axial displacement measurements. They do show, however, that stiffness increases with AN content but at a somewhat lower rate than yield strength.

The drop in stress after yielding is a measure of strain softening and all of the SAN copolymers exhibit this feature. It has been suggested by Haward [54] that materials with a strong propensity for crazing, such as PS and PMMA, also show a pronounced degree of strain softening in a compression stress. In Fig. 29 the degree of strain softening, defined as $\Delta\sigma_y/\sigma_y$, is seen to decrease with increasing AN content. Following Haward, this implies less tendency for localized deformation. The tension test results (Fig. 25), and the fracture surface morphology, both imply that the extent of crazing is reduced with increase in the percentage of acrylonitrile.

Our test results show that the glass transition temperature, T_g, increases with increasing AN content as well as with increasing $\bar{M}_w$. This is evident from Fig. 30, where the storage modulus, E', is plotted vs. temperature for samples of SAN containing 25%, 30% and 51.2% AN. For these copolymers, the trend of E' vs. T is similar to that of PS with E' decreasing slowly with increasing T, due to ordinary thermal expansion effects, and then dropping precipitously as the glass transition region is approached.

The variation of the dynamic modulus, measured at 24 °C, and of T_g with AN content, for all samples tested, is shown in Fig. 31. Both properties increase with increase of AN content, with a rise in modulus of about 20% and a rise in T_g of about

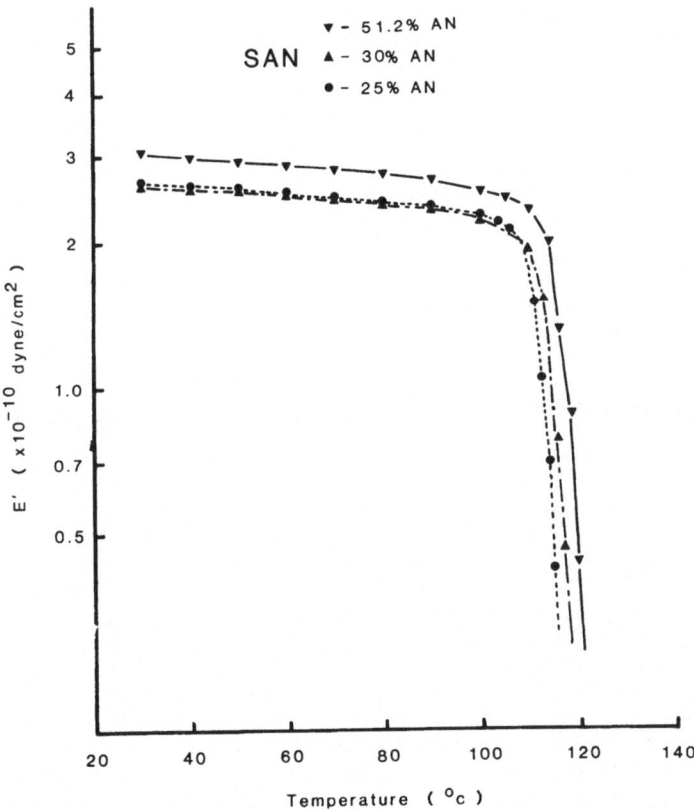

Fig. 30. Storage modulus, E', vs. temperature for SAN of varying AN content

6 °C on proceeding from 0 % AN to 51.2 % AN. Thus we conclude that increasing AN content leads to increases in modulus, in T_g, and in resistance to crazing and shear yielding. However, a study of copolymers having a wider range of AN contents, shows that T_g reaches a maximum value near 50 % AN and then drops slightly with further increases in acrylonitrile content [55].

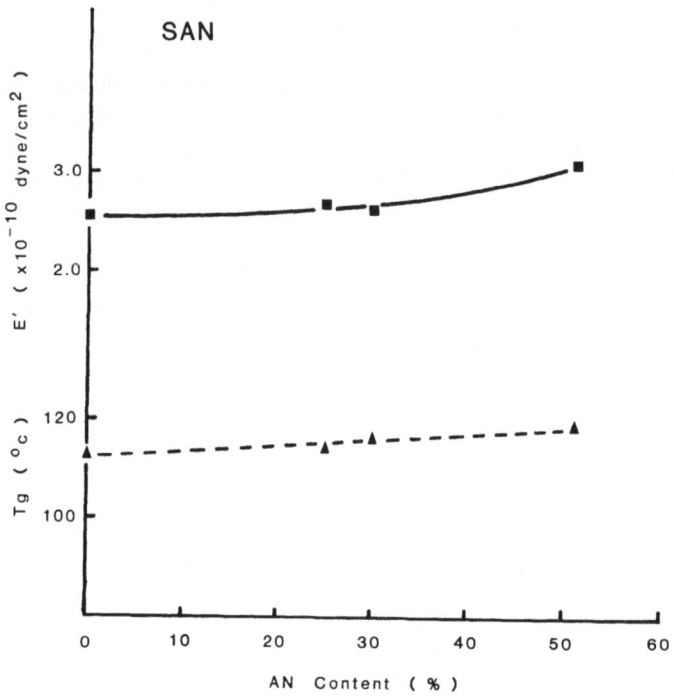

Fig. 31. E' and T_g vs. AN content for various SAN polymers

4.2 Mechanical and Relaxation Behavior of ABS and Comparison with SAN and HIPS

Typical tensile stress-strain curves for ABS and HIPS are shown in Fig. 32. On comparing that of ABS with the stress-strain response for a rubber free SAN copolymer (Fig. 23) we see that the effect of the added rubber is to reduce the tensile strength and modulus while the ductility and toughness, as measured by the area under the stress-strain curve, increase. ABS is much stronger than HIPS and has a yield strength comparable to that of PS. Its elongation to fracture, however, while much greater than that of PS and SAN, is not as great as that of HIPS. This is because the presence of inhomogeneous shear deformation in ABS produces necking and this localizes the plastic deformation and the stress whitening to the vicinity of the necked region. If, however, crazing is accentuated by introduction of glass beads, then elongation to fracture of ABS increases significantly [56]. Elongation to fracture also increases at

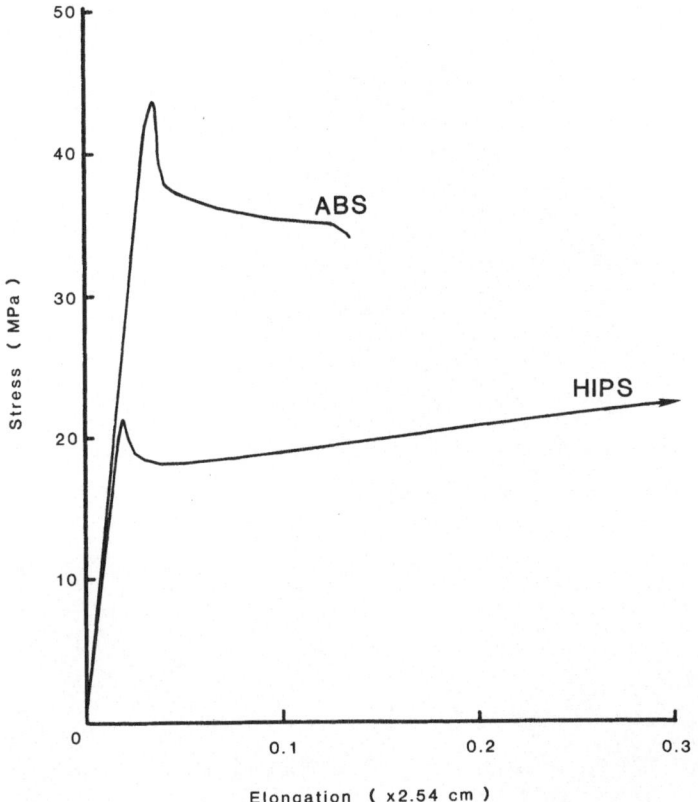

Fig. 32. Stress-strain curves of ABS and HIPS; displacement rate 12.7×10^{-2} cm/min

higher speeds of testing as necking is then suppressed [13]. In HIPS, in contrast to ABS, crazing is the dominant mode of deformation, the entire gauge length is stress whitened, there is essentially no necking, and there is no change in cross-sectional area. That inhomogeneous shear deformation is involved in ABS, as well as crazing, has been demonstrated by transmission electron microscopy pictures of deformed and etched samples by Bucknell and coworkers [57] and by observations of deformed thin films by Donald and Kramer [58].

Fracture initiated in the tensile tested ABS samples, as noted also by Truss and Chadwick [59], from either surface flaws or from internal flaws. Figure 33a shows an SEM picture of the tensile fracture surface of a sample broken at a comparatively high deformation rate of 12.7 cm/min. The fracture surface is unlike that of SAN (Fig. 27a) or that of rubber modified polystyrene (Fig. 3a). Fracture, for this specimen, has developed from both a surface source and from an internal source; and fine radial flow lines emanate from both sources. The slow growth region adjacent to the source tends to develop a conical shape as has been noted [59]. This is probably a result of localized shear formation. In ABS specimens subject to creep deformation at low values of stress, the creep strain is found to be due almost entirely to shear but, at higher stresses, shear is accompanied by crazing [12, 60]. Crazes can also be induced

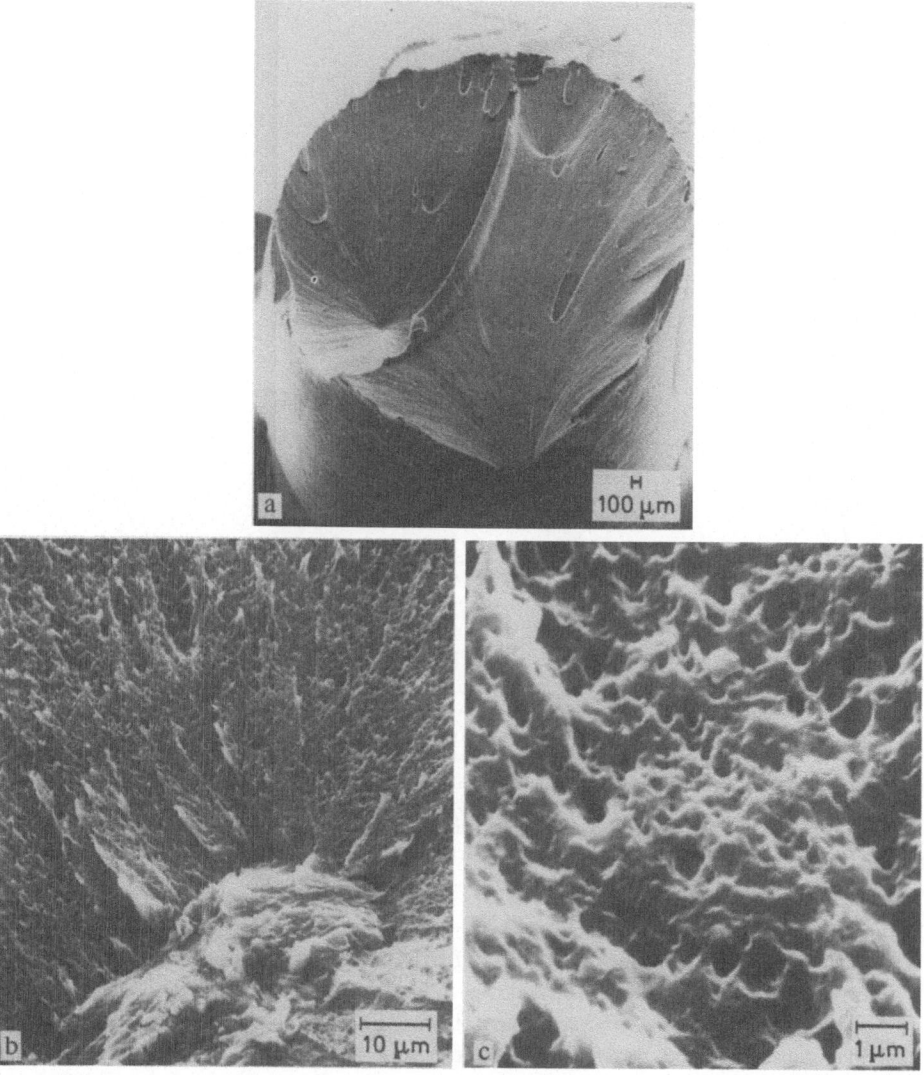

Fig. 33a—c. Tensile fracture surfaces of ABS specimens tested at a displacement rate of 12.7 cm/min; **a** Low magnification scan; **b** Higher magnification near internal source; **c** High magnification near external source

in ABS even in very short loading times ($\sim 10^{-6}$ sec) if the loading is impulsive and the induced stress well above the static yield stress [61].

Figure 33b is a high magnification view of the region adjacent to the internal source. The source appears to be a large foreign particle or aggregate some 40 μm long and 20 μm wide. There is evidence of considerable localized drawing in the surrounding matrix and the drawn material tends to align along the crack propagation direction. A still higher magnification view of a region near the surface source is shown in Fig. 33c. A somewhat similar 'honeycomb' type morphology has been observed in ABS specimens fractured in a methanol environment [62]. Voids, with a size range from 0.1 to about 0.8 μm appear over most of the fracture surface and the matrix material is distorted and drawn. The voids probably result from cavitation or desocketing of dispersed rubber particles which in ABS cover this size range. TEM studies of thin sections of deformed ABS show some matrix crazing has occurred and some holes have formed in the second phase particles [63]. Also, in TEM studies of deformed thin films Donald and Kramer have observed cavitation in both small rubber particles, which tend to be homogeneous, and also in larger particles, which are of the occluded type [58]. This intense cavitation encourages shear deformation, as it reduces constraints. The average particle size is much less in ABS than in HIPS (compare Fig. 33c with Fig. 13c) and since small particles are ineffective as craze initiators, this too encourages shear deformation.

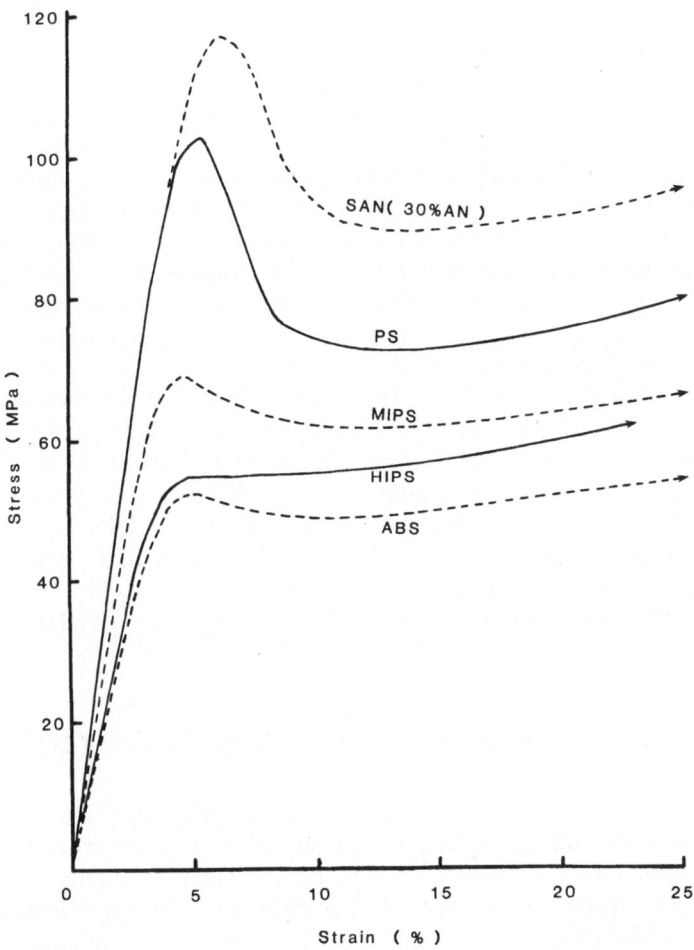

Fig. 34. Stress-strain curves in compression for SAN, PS, ABS, MIPS and HIPS; displacement rate of 5.08×10^{-2} cm/min

Tension tests have been made on two samples of ABS at a slower speed, viz. 12.7×10^{-2} cm/min. One sample failed by development of a crack from an internal flaw while the second sample developed cracks from both a surface flaw and an internal flaw. At the slower speed of testing, fracture stress was reduced and there was somewhat less evidence of secondary fractures ahead of the crack front, such as those shown in Fig. 33a. Neither parabolic secondary fractures, nor conical regions surrounding the fracture source, were observed on the fracture surfaces of HIPS (Fig. 3a).

The results of compression tests on specimens of SAN, PS, MIPS, HIPS and ABS are shown in Fig. 34. As noted earlier, SAN has a higher yield stress but a lesser degree of strain softening than PS. The effect of rubber inclusions is to greatly reduce the compressive yield stress of the SAN copolymer and to reduce also the degree of strain softening. Hence, less localization of deformation is to be expected in ABS than in SAN and this is in accord with results from tension tests which show the development of a region of stress whitening in ABS but not in SAN. Similarly, MIPS and HIPS also show a lower compressive yield strength and less strain softening than the unmodified PS and they too, unlike PS, undergo a uniform stress whitening over the entire gauge length.

On comparing ABS with HIPS we see, unlike the situation for tensile yield, that the compressive yield stress is higher for the HIPS material. Also, the compressive and tensile yield strengths in HIPS are significantly different as in tension the apparent yield stress is low because of extensive crazing. In ABS, however, shear deformation is the dominant mode in the yielding process for both tension and compression. The slightly higher value for the compression yield stress in ABS, compared to its tensile yield stress, is simply a natural consequence of the change in the hydrostatic component of stress; and it can be accounted for by either a Mohr-Coulomb yield criterion or a pressure modified Von Mises criterion [64, 65].

Relaxation tests (not shown) were made on an ABS sample over a temperature range from -140 °C to 120 °C. Little change was noted in the position of the main T_g transition compared to that observed in SAN but a loss peak and a modulus drop were evident near -100 °C due to the T_g transition of the polybutadiene component. The measured value of the storage modulus at 20 °C was about 1.58 GPa for the ABS sample. This compares with a value for SAN of about 2.68 GPa. Hence, rubber modification has reduced the modulus of the copolymer about 40%. For HIPS compared to PS, the corresponding reduction, for the samples we tested, was about 25%.

4.3 Effects of Stress on Fatigue Behavior and Deformation Modes of SAN and ABS

In this investigation, we have studied the fatigue behavior of smooth specimens of SAN copolymers containing 25% and 30% of AN. For the samples containing 30% of AN, fatigue data have also been acquired at three different molecular weights. Since the specimens are transparent, the average number of cycles to craze initiation, N_{ci}, was determined usually for two separate specimens of each test group. The test results, along with values of cycles to fracture, N_f, obtained from tension-compression, fatigue tests made at a stress amplitude of 27.6 MPa and at a frequency of 21 Hz, are presented in Table 2. Comparison of the fracture lifetime data with that obtained for PS (Fig. 10 and Table 1) shows that the SAN copolymers have appreciable greater fatigue endurance than does PS. To illustrate, PS specimens subject to a cyclic stress of 20.7 MPa developed an average fatigue lifetime before fracture of about 9100 cycles; in comparison, SAN types A, B, and C developed a longer fatigue lifetime even though cycled at a much higher stress amplitude. The average fatigue lifetime also increases with increase of AN content. This is illustrated by the higher N_f value for group B, with 30% AN, than for Group A, with 25% AN; although the difference indicated in N_f values is not large, the increase has occurred despite a somewhat lower molecular weight for the B group specimens.

Table 2. Effects of Molecular Weight and AN Content on Fatigue of SAN

Group Designation	Content (%)	$\bar{M}_w$	No. Specs. Tested	N_{ci} (cycles)	N_f (cycles)	N_{ci}/N_f (%)
SAN A	25	1.45×10^5	4	11,900	13,500	88
SAN D	30	1.15×10^5	4	1,630	4,310	38
SAN B	30	1.30×10^5	3	12,600	14,800	85
SAN C	30	1.85×10^5	4	17,500	22,900	76

Molecular weight itself has a strong influence. Figure 35 shows how the cycles to fracture and cycles to craze initiation, for samples containing 30% AN, varies with the weight average molecular weight. At the lower $\bar{M}_w$ values, N_f rises rapidly with increase in molecular weight and then more slowly at the higher $\bar{M}_w$ values. The increase in fatigue endurance with molecular weight is attributed to a greater craze stability as a result of a higher entanglement density and a reduced number of chain end defects. Increases in lifetime, or decreases in fatigue crack propagation rate, with increase of molecular weight have also been noted in other glassy polymers such as PS [25], PC [66], PVC [21] and PMMA [33]. Also, with increase of molecular weight, the chain contour length, l_e, between entanglements is reduced to a certain extent; and, if the entanglement points are considered as crosslinks then, as pointed out in the earlier Chapters the maximum extension ratio for the craze is given by

$$\lambda_{max} = l_e/d$$

where d is root-mean-square end-to-end distance of a chain having an entanglement molecular weight of M_e. Observed craze extension ratios for SAN are about 0.8 of λ_{max}. With reduction in l_e, the entanglement density is increased and, as a result, a longer lifetime to craze breakdown would be anticipated. In addition, it has been shown that polymers with low values of l_e, such as polycarbonate, or high AN content SAN, have a much grater tendency to deform by shear rather than by crazing, whereas polymers with high l_e values, like PS, deform entirely by crazing under tensile loading [67].

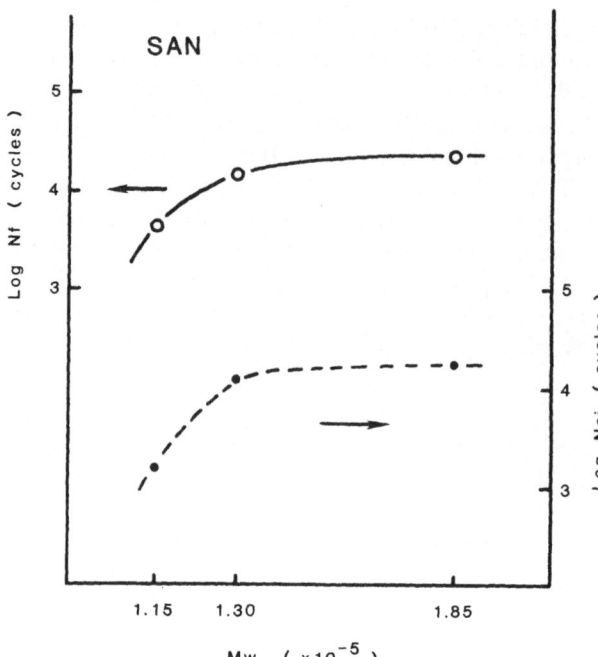

Fig. 35. Variation of N_f and N_{ci} with $\bar{M}_w$ for SAN samples containing 30% AN tested in tension-compression fatigue at 27.6 MPa and at 21 Hz

It is evident from Table 2, and from comparison with data of Table 1, that the resistance to fatigue-induced crazing, as well as to fracture, is much greater for the SAN copolymMrs than for PS. For a given applied stress, crazing initiates at a much later time in SAN than in PS. Craze initiation in SAN also occurs, except for the lowest molecular weight material, very late in the fatigue process. Hence, fatigue failure is "initiation controlled" as most of fatigue lifetime is spent in nucleating the first craze, or potential damage zone, while only a small portion of the lifetime (10–25%) is spent in craze and crack growth. The test results also show that there is a molecular weight dependence of craze nucleation as well as of fracture. The average number of cycles for craze initiation increases with increasing molecular weight of the SAN polymer. Hitherto, it has been generally assumed that the craze initiation phase is essentially independent of molecular weight [31], but there is an increasing amount of evidence that this assumption may not be correct [68].

All of the SAN samples, whether of 25% or 30% AN tested in fatigue at 21 Hz, displayed a fracture surface morphology that was in some respects akin to that of PS and in other respects quite different. Like PS, fracture developed from a surface source and the region surrounding the source was comparatively smooth but, unlike PS tested at this frequency, discontinuous growth bands were seen in this region.

We present the results of our observations here for the samples containing 30% acrylonitrile. Figure 36a shows the fatigue fracture surface for a specimen of Group D having the lowest $\bar{M}_w$ value and the shortest fatigue lifetime. There is a comparatively smooth and radially lined slow growth region emanating apparently from a surface defect near to the corner of the specimen. This region is followed by a rough, textured region, indicative of fast crack propagation through previously uncrazed material.

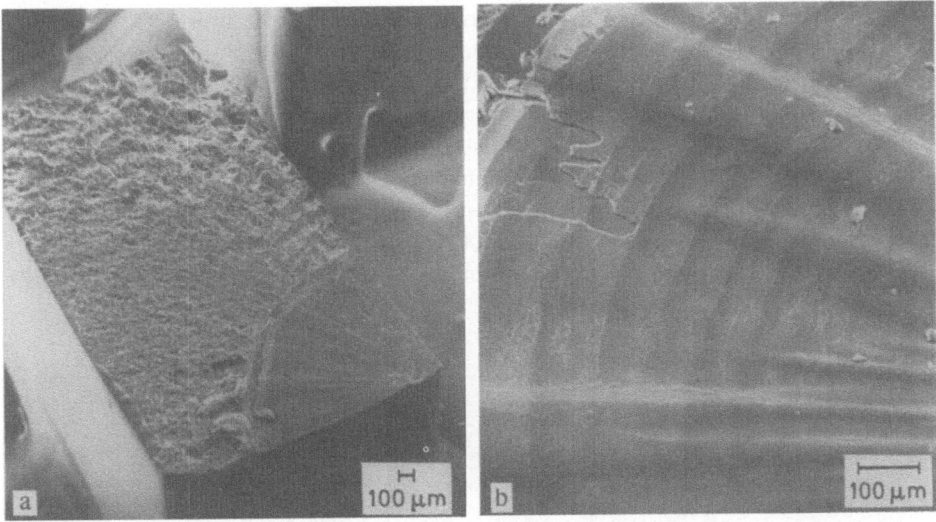

Fig. 36a and b. Fatigue fracture surface of an SAN sample (30% AN-Group D) tested in tension-compression fatigue at 27.6 MPa and at 21 Hz; **a** Low magnification scan; **b** Higher magnification of banded region

The concentric bands surrounding the source in the comparatively smooth region are shown at higher magnification in Fig. 36b. They increase in spacing with increase in crack length and they extend from one side surface of the specimen to another. However, they are not as sharply defined as the DCG bands observed in PS at lower frequencies (Fig. 8b and Fig. 20b). There is also another unusual feature which is brought out in higher magnification SEM scans, such as those shown in Fig. 37.

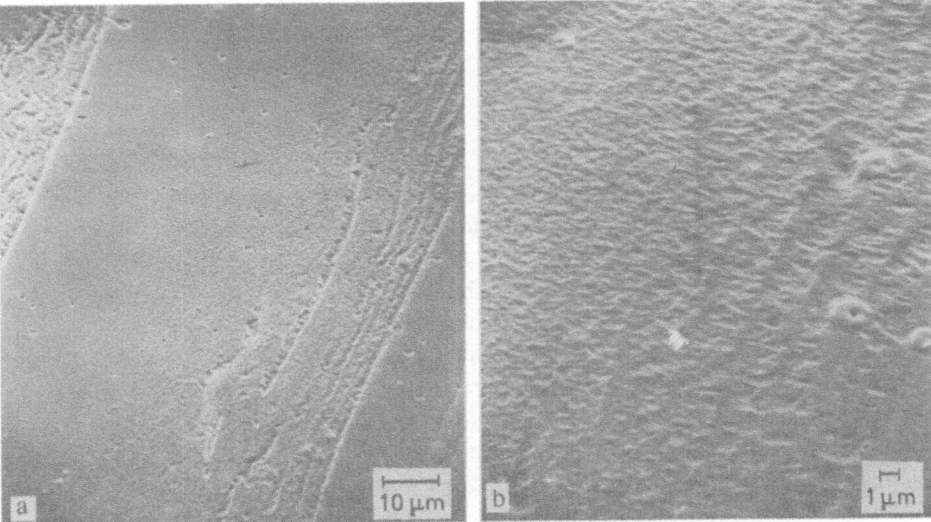

Fig. 37a and b. Views of a portion of the fatigue fracture surface of the SAN sample of Fig. 36; **a** Region of 4th DCG band; **b** Beginning of 3rd DCG band

Fig. 37a shows in greater detail the micromorphology of the fourth DCG band of Fig. 36b and Fig. 37b shows a higher magnification view of a portion of the third band. Both pictures indicate that within the broad bands of 50 to 80 µm in width, there are an additional series of concentric narrow bands or striations with widths of the order 1 to 2 µm. Somewhat similar irregular striations lying between broad bands have also been noted in polycarbonate by Mackay et al. [69].

Upon increasing molecular weight, several changes are noted in the fracture surface morphology. The first is that the size of the comparatively smooth slow growth region relative to the fast growth region increases with increase in molecular weight.

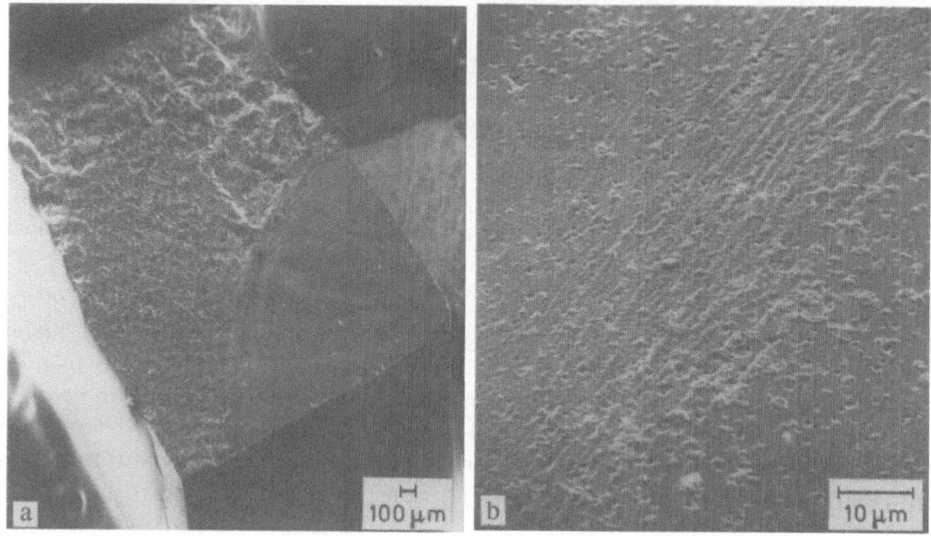

Fig. 38a and b. Fatigue fracture surfaces of two SAN samples (30% AN-Group B) tested in tension-compression fatigue at 27.6 MPa and at 21 Hz; **a** Low magnification scan; **b** Detail of edge of 2nd DCG band

This is evident from comparison of Fig. 38a, showing a specimen of the B group with that of Fig. 36a. For the highest molecular material, Group C, this region is still larger and encompasses almost 50% of the total aea rather than 25% or so for the low molecular weight samples. These changes are another indication of the greater number of entanglements per coiled molecular chain with increase of molecular weight and the consequent greater stability of the crazes to breakdown. A second change with increase in $\bar{M}_w$ is that the crack arrest portion of the band appears to decrease in size and the region at the beginning of the band, Fig. 38b, shows rather illdefined striations as well as many small hillocks and corresponding cavities. Perhaps the higher molecular weight and greater craze strength leads to more tearing at the craze-matrix interface rather than viscous rupture of the craze fibrils [15].

With regard to AN content, no marked changes were noted between fractured surfaces of Group A specimens and of Group B specimens. There are, however, significant changes produced in the fatigue properties and fracture surface by incorporation in SAN of a rubber phase, as will now be discussed.

For the rubber modified copolymer, ABS, we have carried out fatigue studies over a range of stress amplitudes and frequencies under reversed tension-compression cycling; in addition, some tests have been made under cyclic tension. Results will be presented first for tests carried out under different applied amplitudes and comparisons will be made between ABS and SAN, between ABS and HIPS, and between our data on ABS and that of others [17, 20, 24].

The influence of stress amplitude on fatigue life of ABS specimens tested at 0.2 Hz is demonstrated by the S–N curve of Fig. 39. The curve is similar in form to that for HIPS (Fig. 18) with log N_f increasing linearly with decreasing stress amplitude in the intermediate cycle range. At lower stresses, the S–N curve flattens and appears to be approaching a horizontal plateau, or endurance limit, below which no fracture occurs. Some S–N data for ABS have been reported. Under square wave loading, Bucknall and Stevens tested ABS over a limited range of stress amplitude and they did not observe the flattening tendency [24]. The lifetime values given in Fig. 39 are comparable to those obtained in rotating bending tests at 0.5 Hz for intermediate values of stress amplitude but they are higher than the reported values for low stress amplitudes [70].

When ABS is compared to HIPS at comparable stress amplitudes, fatigue lifetimes are several decades higher. Several factors are involved. First, tensile strength and modulus of the SAN matrix are increased compared to values for the PS matrix. Secondly, the SAN matrix is more resistant to crazing and thirdly, the primary mode of deformation for ABS appears to be shear yielding rather than crazing.

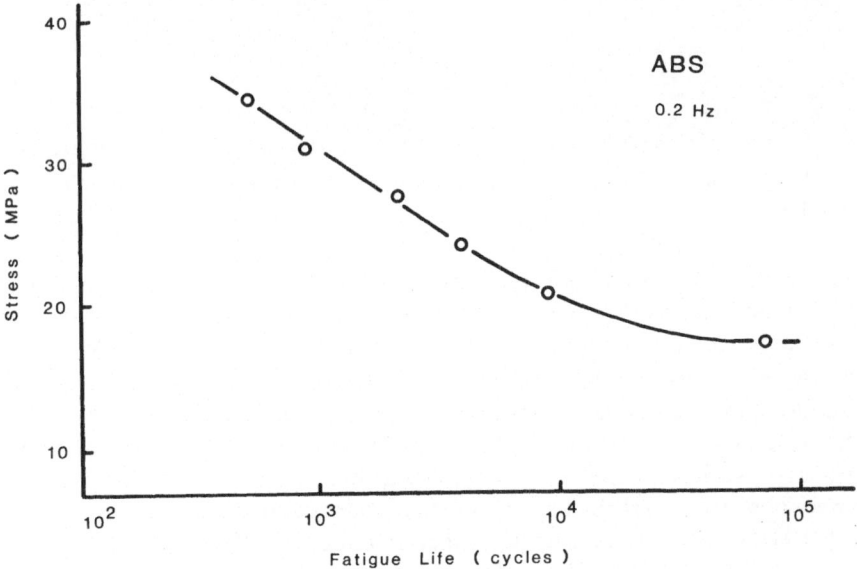

Fig. 39. Stress amplitude vs. cycles to failure for ABS samples tested at 0.2 Hz

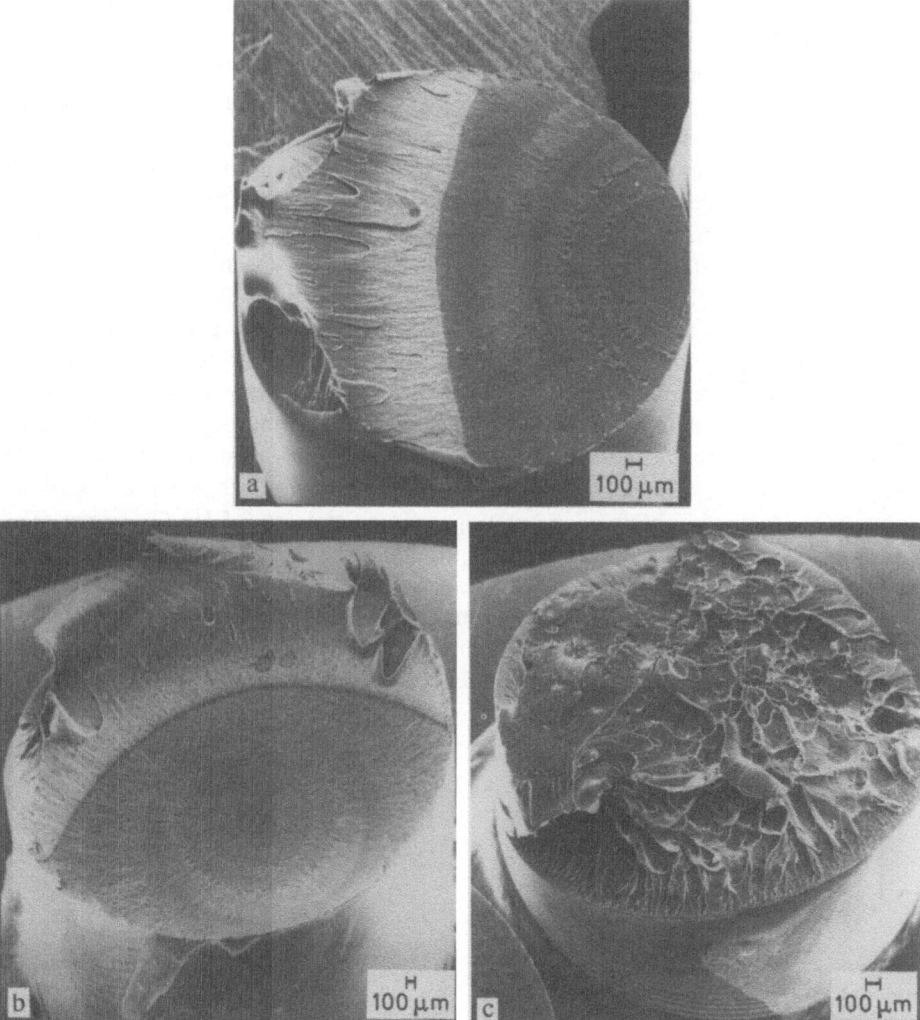

Fig. 40a—c. Fatigue fracture surfaces of ABS samples tested at different stress amplitudes at 0.2 Hz; **a** 17.2 MPa; **b** 20.7 MPa; **c** 34.5 MPa

The fatigue fracture surface morphology of ABS varies with stress amplitude and it is quite different from that observed in tensile fracture. SEM scans of typical fracture surfaces for stress amplitudes of 17.2 MPa, 20.7 MPa and 34.5 MPa are shown in Fig. 40. At the lowest stress, Fig. 40a, it is seen that fracture has developed from a surface flaw and that the fracture surface is quite smooth. Several concentric bands surrounding the source indicate that crack growth was probably discontinuous in this region. No such bands and no smooth, slow growth region on the fatigue fracture surface were observed in HIPS specimens tested under comparable conditions (Fig. 19a). At the higher stress of 20.7 MPa, fatigue fracture in ABS developed from an internal source (Fig. 40b) and the surface exhibited an "eye-shaped" slow growth region, again with DCG bands, followed by a fast growth region on an inclined plane.

Secondary fractures, that develop ahead of the main crack, are seen in both Figs. 40a and 40b and this part of the fracture surface is more characteristic of tensile fracture (Fig. 33a).

Increasing the stress amplitude to higher values tends to reduce the size of the slow growth region and to produce a more irregular fracture surface with multiple internal and external sources. This is shown, for a stress amplitude of 34.5 MPa in Fig. 40c. Some of the internal sources, and more are seen at lower stress amplitudes of 27.6 and 31.0 MPa, shown an apparent flat region essentially perpendicular to the stress direction, indicative of crack growth through pre-existing crazes. Subsequent fracture then appears to occur on inclined surfaces, possibly along angled shear bands that develop from the tip of the craze. However, as stress amplitude rises, the size of these small flat craze zones decrease and the entire fracture surface becomes very rough as Fig. 40c shows.

The DCG bands near the source, for the sample tested at the low stress of 17.2 MPa, are shown in greater detail in Fig. 41a. The bands, with a spacing of the order of 100–200 μm, increase in size as the crack extends inward and they run continuously from one surface to another. These, and the gradations in void size and local texture across the width of each band, are characteristic of DCG bands [17]. At high magnification (Fig. 41b), the matrix shows evidence of considerable localized plastic deformation and many small cavities, varying in size from 0.1 to about 0.7 μm are visible. These may represent cavitated rubber particles or they may be holes in the matrix where the dispersed rubber particles have separated from the matrix. Somewhat similar DCG bands, with comparable spacing, have been observed upon testing notched ABS specimens in alternating tension [34]. These bands were considered to represent the plastic zone size ahead of the crack tip. According to the Dugdale

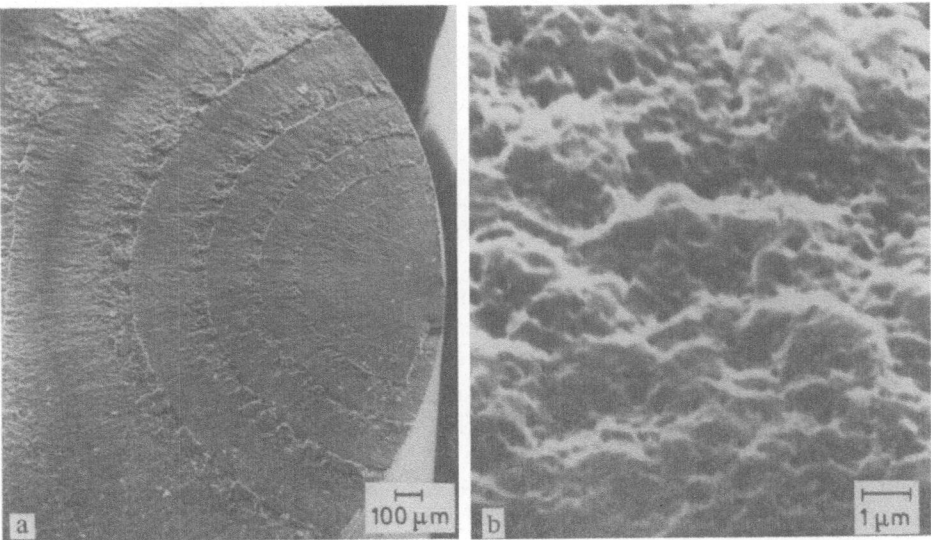

Fig. 41a and b. Fatigue fracture surfaces of an ABS specimen tested at 17.2 MPa and at 0.2 Hz; **a** Region near source showing DCG bands; **b** High magnification of region close to source

model [71], see also Chapter 3, the plastic zone size should vary as the second-power of the stress intensity factor, and this was observed for the DCG band spacing [17, 72].

Additional information relative to the deformation modes present in ABS, and of the effect of stress amplitude on these, has been obtained from analysis of hysteresis loop measurements made as a function of cycles. For an imposed stress amplitude of 34.5 MPa hysteresis loops, taken at various values of N, are shown in Fig. 42 for a sample that failed after 313 cycles. At this high stress, shear deformation, as indicated by an increase in the width or area of both tensile and compressive portions of the hysteresis loop, occurs within the first two cycles, while close to 100 cycles are required before sufficient localized crazing has developed so that the tensile portion of the loop is increased relative to the compression portion. Crazing becomes somewhat more pronounced within the last few cycles before fracture.

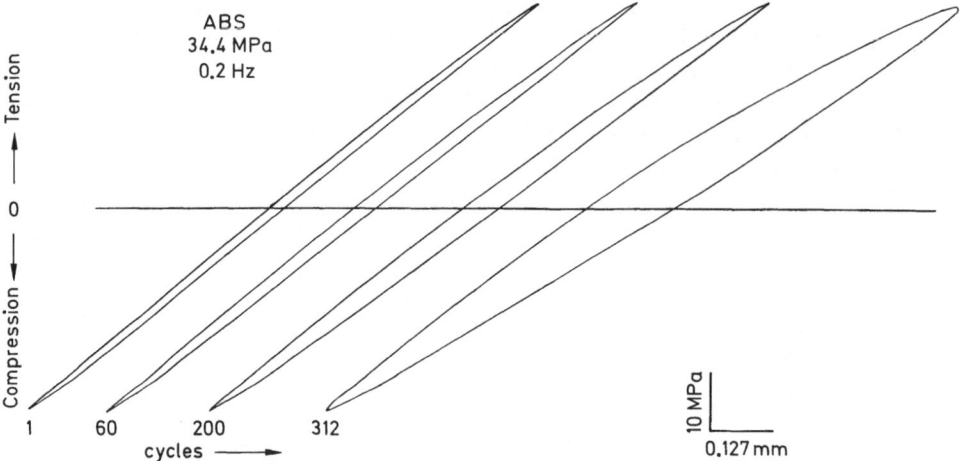

Fig. 42. Hysteresis loops for ABS vs. number of cycles for specimen tested at a stress amplitude of 34.5 MPa and at 0.2 Hz

The variation in loop width (at a nominal stress of 10 MPa) for both the tension and compression side of an ABS sample, tested at a stress amplitude of 31 MPa, is shown in Fig. 43. In this case strain softening initiates early and develops rapidly within the first 100 cycles or so. It then continues to increase with increased cycling, at essentially a steady rate on both the tension and compression side of the hysteresis loops. As crazing becomes a significant mode of deformation, which here occurs at about 650 cycles, the loop width increases more rapidly on the tensile side. From this type of graph, we can estimate, for various stress amplitudes, the number of cycles required to initiate shear type softening, N_s, and the number of cycles required to initiate craze yielding, N_{ci}.

In Table 3, this information is presented for five ABS specimens, tested at stress amplitudes varying from 20.7 MPa to 34.5 MPa. For comparison purposes, similar type data on HIPS is also shown. The superior craze and fatigue resistance of ABS is immediately evident. The data also show that except possibly for the lowest stress

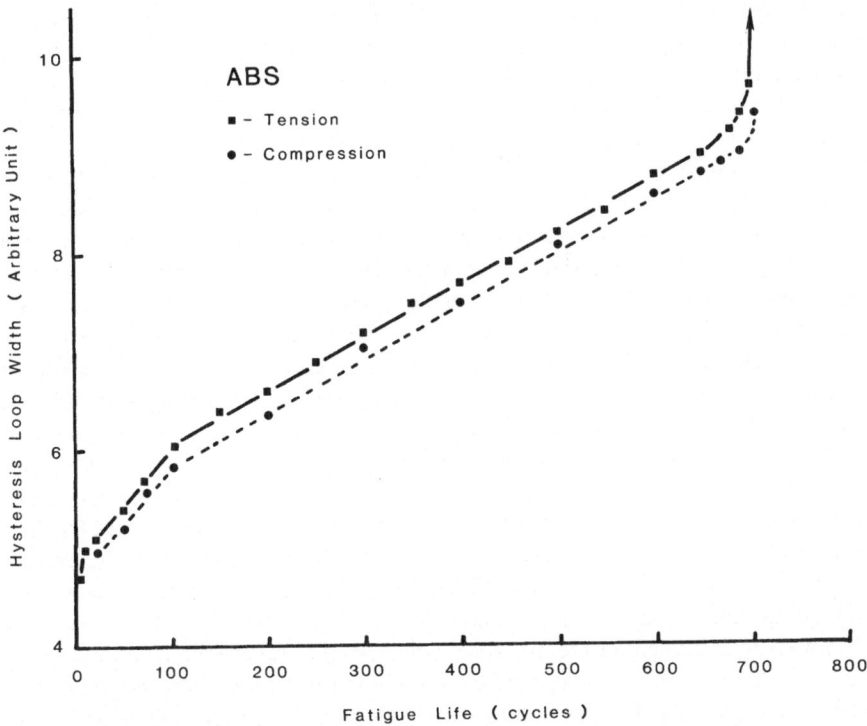

Fig. 43. Hysteresis loop width measured at a tension, or compression, stress of 10 MPa vs. number of cycles for an ABS sample tested at a stress amplitude of 31 MPa and at 0.2 Hz

Table 3. Cycles to Strain Softening (N_s) and Cycles to Craze Initiation (N_{ci})

Polymer	Stress Amplitude (MPa)	N_s (cycles)	N_{ci} (cycles)	N_f (cycles)
HIPS	20.7	—	1	23
	17.2	—	10	183
	13.8	—	1,150	2,030
	10.3	—	9,050	9,390
ABS	34.5	2	80	313
	31.0	2	650	711
	27.6	10	1,500	1,630
	24.1	90	4,100	4,150
	20.7	—	10,800	10,900

(and longest time to fracture) the dominant deformation mode in ABS is shear. Shear deformation, as noted earlier, is favored by the presence of many small and cavitated rubber particles. At high stresses, crazing, induced by stress concentration at the larger, occluded second phase particles, develops quite early in the lifetime cycle. Then, as the stress is lowered, many more cycles, and a much higher proportion of the total number, is required before craze deformation begins to play a significant role. This behavior is significantly different than that of HIPS as in fatigue of rubber modified polystyrene shear yielding does not play an active role and the localized deformation is associated with crazing. This finding is consistent with the differences already referred to relative to fracture of the two rubber modified polymers.

Some fatigue tests have been made on ABS, and on HIPS, under square wave loading, at a frequency of 0.033 Hz by Bucknall and Stevens [24]. Our graph of hysteresis loop width vs. cycles (Fig. 43) shows similar trends with cycling as their graphs of tension and compression secant modulus vs. cycles. They too found that most of the fatigue-induced damage in ABS is a result of shear deformation while in HIPS it is a result of crazing. Our findings relative to the influence of shear deformation and craze deformation in fatigue appear to be in accord with conclusions drawn from creep studies [57]. There too it was found that at low stresses, crazing only became significant in ABS at long times but that with increase of stress, crazing tended to become a more significant contributor to the total deformation even at shorter times.

To investigate the possible effects of stress state, and in particular to determine the influence of the compressive component of the stress on fracture lifetime and fracture surface morphology, some fatigue tests were made on ABS specimens in cyclic tension. In these tests the maximum stress was 27.6 MPa and the minimum stress was zero. The average lifetime for specimens, tested at 21 Hz, was 12,700 cycles. This can be compared with an average lifetime, for specimens tested in reversed tension-compression at 27.6 MPa of about 5,050 cycles (Fig. 46). From this we see, as noted previously for HIPS specimens, that fatigue damage is accelerated by stress cycling in a compressive mode, even though, in the case of ABS, the principal deformation mode is shear and not crazing.

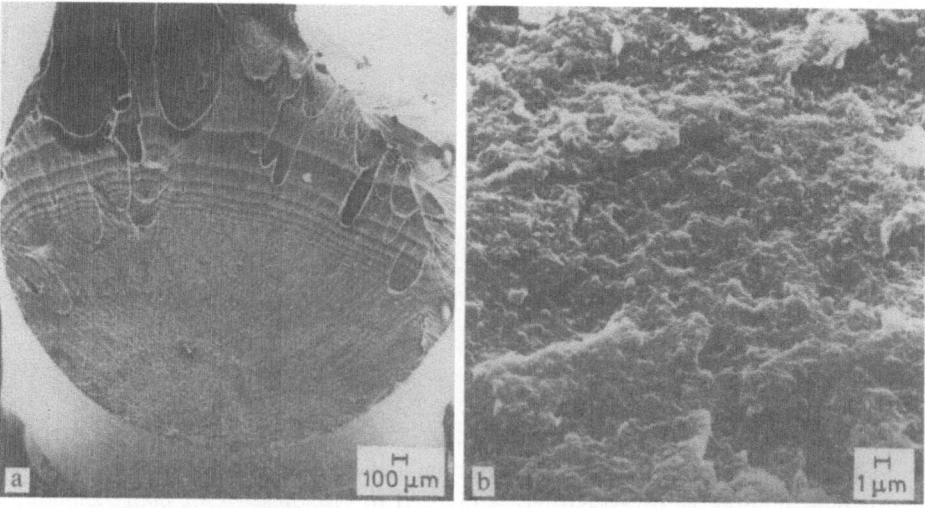

Fig. 44a and b. Fatigue fracture surface of an ABS sample tested at 21 Hz in an alternating tensile mode; at a maximum stress of 27.5 MPa and a minimum stress of zero; **a** Low magnification scan; **b** Higher magnification of region near source

The fatigue fracture surface, under tensile cycling, shows some interesting features. As noted in the low magnification picture of Fig. 44a, the fracture has developed from a surface, or near surface, source and there appear to be two sets of concentric bands on the fracture surface. One set, surrounding the origin is similar in general appearance, and in band spacing, to that shown previously in Fig. 41a. These bands are considered to be discontinuous crack growth bands, although they are not as well-defined — perhaps because of the higher frequency — as those of Fig. 41a.

At high magnification, the slow growth region near the source, Fig. 44b, shows rather rounded features as well as numerous small cavities. The cavities, more visible at still higher magnification, are generally well below 1 μm in size. The other set of concentric lines visible near the center of the fracture surface are thought to be striations corresponding to crack growth per cycle. Such striations usually develop in the fast growth region where secondary parabolic features are already present. A higher magnification view of a portion of this region is shown in Fig. 45a. Some 10 different ridges, with their spacing increasing rapidly with increasing crack size from perhaps 10 μm to 200 μm, are visible. In addition, numerous drawn fibrillar elements, lying in the crack propagation direction, overlay the parallel ridges. The highly drawn nature of the ABS material in this portion of the fracture surface is clear from Fig. 45b showing, in greater detail, a small region lying between the sixth and seventh striations. The fibrillar elements, aligned in the crack propagation direction, show evidence of extensive plastic deformation. Compare this picture, taken at the same magnification, with that of Fig. 44b, to see the difference in fracture surface morphology between the slow growth region near the source and the fast growth region away from the source.

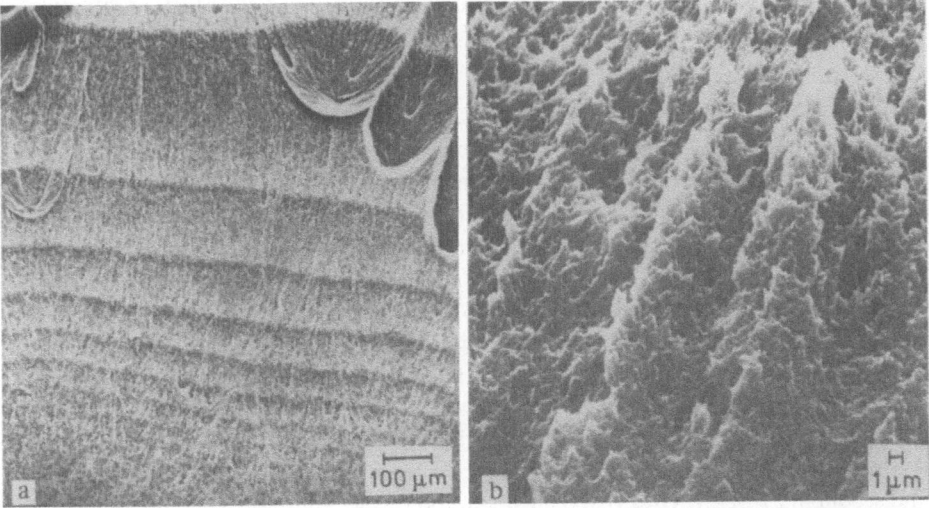

Fig. 45a and b. Fatigue fracture morphology of a region near center of specimen of the ABS sample of Fig. 44; **a** Portion of fast growth region showing fatigue striations and secondary fractures; **b** High magnification of a region between striations

4.4 Effects of Test Frequency on Fatigue Response and Fracture Morphology

Fatigues tests have been run on a series of ABS samples under reversed tension-compression at a stress amplitude of 27.6 MPa, at four different frequencies, viz. 0.02, 0.2, 2, and 21 Hz. The average fatigue lifetime as a function of frequency is shown in Fig. 46 and, for comparison purposes, the data for HIPS obtained at 17.2 MPa is also shown. The average fatigue life-time of ABS increases in a linear manner with frequency on this log-log plot. The rate of increase, however, is reduced compared to that of HIPS, or of PS (Fig. 15). The reduced frequency sensitivity is perhaps a result of a reduced magnitude of the β-transition in the SAN copolymer compared to that in PS. A reduced frequency sensitivity of FCP rate in ABS vs. a HIPS-modified PPO has also been noted [45].

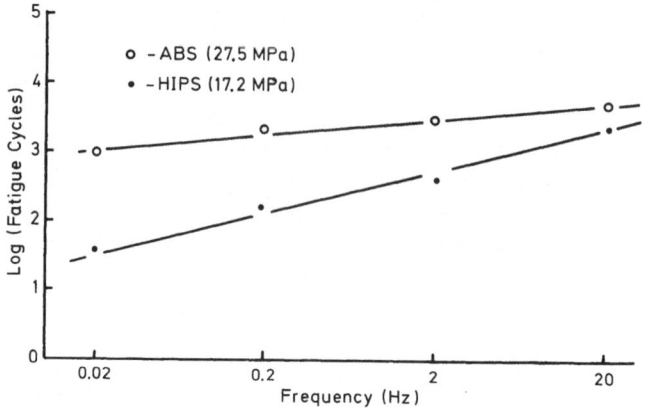

Fig. 46. Cycles to fracture vs. frequency for ABS samples tested at a stress amplitude of 27.5 MPa and HIPS samples tested at a stress amplitude of 17.2 MPa

Test frequency has some effect on fracture surface morphology. Figure 47 shows the fatigue fracture surfaces of ABS specimens tested at a stress amplitude of 31 MPa at two different frequencies. At the lower speed, 0.02 Hz, there is a relatively flat slow growth region that developed from an internal flaw. Radial lines emanate from this flaw in all directions. The slow growth region is followed by an inclined fast growth region with many indications of secondary fractures, some of which have the conical shape, associated with tensile fracture at high stress (Fig. 33). At the higher frequency, multiple surface sources and various internal sources have developed and the entire fracture surface, Fig. 47b, is very rough.

The effect of frequency for a lower applied stress of 27.5 MPa is shown in Fig. 48. At the lowest frequency, fracture developed from an internal near-surface source, and secondary fractures are visible near the end of the comparatively flat slow-growth region and, in greater concentration, in the faster growth regions of the fracture surface. The morphology is, in general, similar to that of Fig. 47a for a sample tested at the same frequency but at a slightly higher stress.

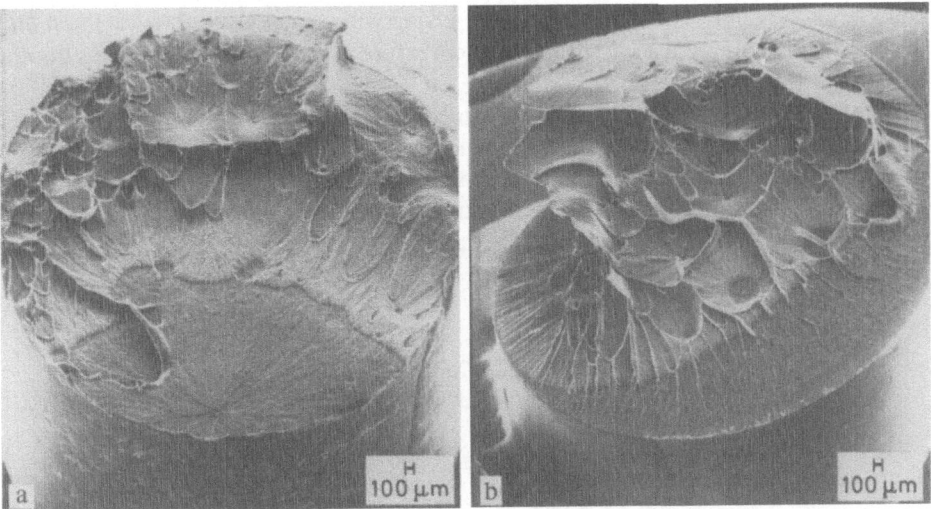

Fig. 47a and b. Fracture surfaces of ABS tested at two different frequencies at a stress amplitude of 31 MPa; **a** 0.02 Hz; **b** 0.2 Hz

The fracture surface of the sample fatigued at a higher frequency, 2 Hz, and at the same stress amplitude of 27.5 MPa, is shown in Fig. 48b. In this sample, fracture originated from two separate surface sources and these are separated by a torn ridge of material in between. The size of the slow growth region, here in two parts, is somewhat larger than for the sample tested at 0.02 Hz. There is some evidence, on both portions of the smooth part of the fracture surface of Fig. 48b, of discontinuous

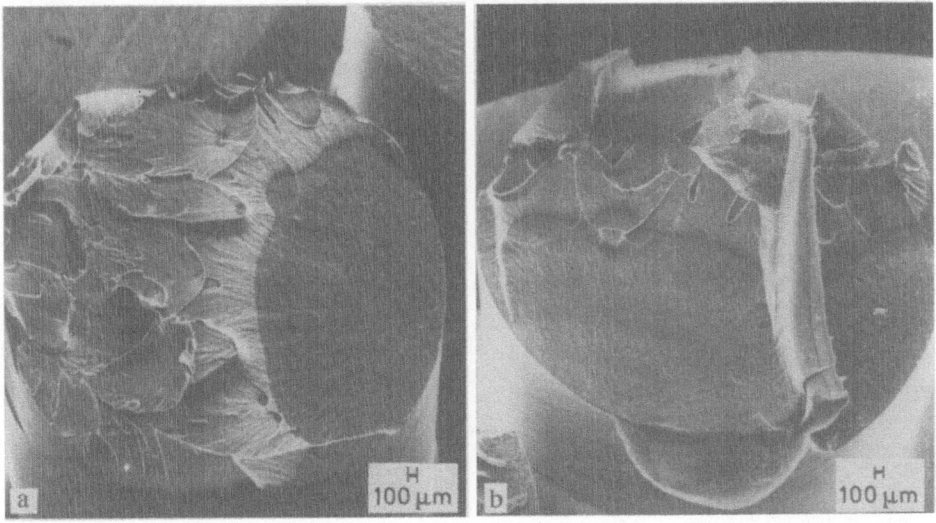

Fig. 48a and b. Fracture surfaces of ABS tested at two different frequencies at a stress amplitude of 27.6 MPa; **a** 0.02 Hz; **b** 2 Hz

crack growth bands. The bands are rather broad with spacings of about 200 μm and the band morphology vaóries, with a rougher texture at the leading edge and a smoother texture at the trailing edge.

A typical fatigue fracture surface obtained at the same stress, 27.5 MPa, and at a still higher frequency of 21 Hz is shown in Fig. 49a. At this test speed the entire fracture surface is much smoother than those obtained at lower frequencies and, at higher magnification, Fig. 49b, the fracture source appears to be a rather large irregular cavity, possibly generated from a small heterogeneity, that can be seen at the center of the opening. Although there are some gradations in micromorphology

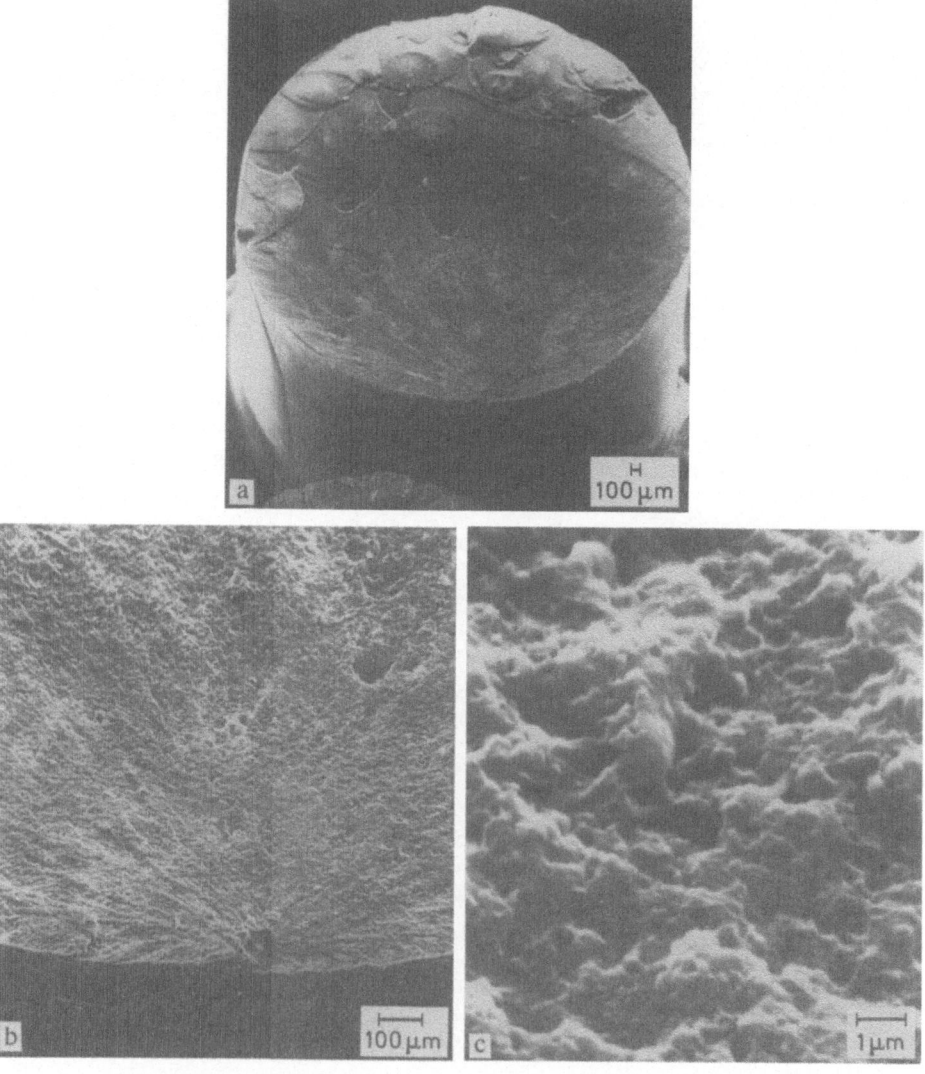

Fig. 49a—c. Fracture surfaces of an ABS specimen tested at 27.6 MPa at 21 Hz; **a** Low magnification scan; **b** Higher magnification of source region; **c** High magnification of region near source

in this region, there are no distinct DCG bands. Similarly in PS, on changing from low frequencies of 0.2 Hz or 2 Hz, DCG bands disappeared from the fracture surface. A higher magnification view of the region near the source, Fig. 49c, shows that, while the fracture surface is comparatively flat at low magnification, considerable localized plastic deformation is present in the matrix; and many cavities as well as small rounded features, probably representative of non-cavitated rubber particles still attached to the matrix, are visible.

Farther out from the fracture source near the center of the specimen faint striation lines are present. Though not seen in Fig. 49a, they are clearly visible in the higher magnification micrograph of Fig. 50a, along with the typical secondary fracture parabolic features generally observed in the fast growth region. A high magnification view of the central portion of the specimen near to the striation region is shown in Fig. 50b. The micromorphology is somewhat similar to that shown in Fig. 49b for a region near the fracture source except that here the plastically deformed material has many of its drawn elements oriented in the crack propagation direction. Again it appears that some of the larger rubber particles, after extensive localized deformation, have cavitated while some of the smaller ones have been bypassed by the fatigue crack. The matrix is not extensively fibrillated in the manner shown in Fig. 45b for specimens tested purely in a cyclic tension mode. Nevertheless, it is much more distorted than the matrix of fatigued HIPS specimens tested in tension-compression at the same frequency [26] or of HIPS specimens tested in a cyclic tensile mode (Fig. 14). For HIPS, where many of the rubber particles are in the range from 1–5 µm, the surrounding matrix deforms only by crazing; as a result, the fracture surface even at high magnification is very smooth and almost featureless, except for small dimples arising from broken and retracted craze fibrils. Also in HIPS, the occluded and deformed second phase particles are simply severed by the advancing fatigue crack.

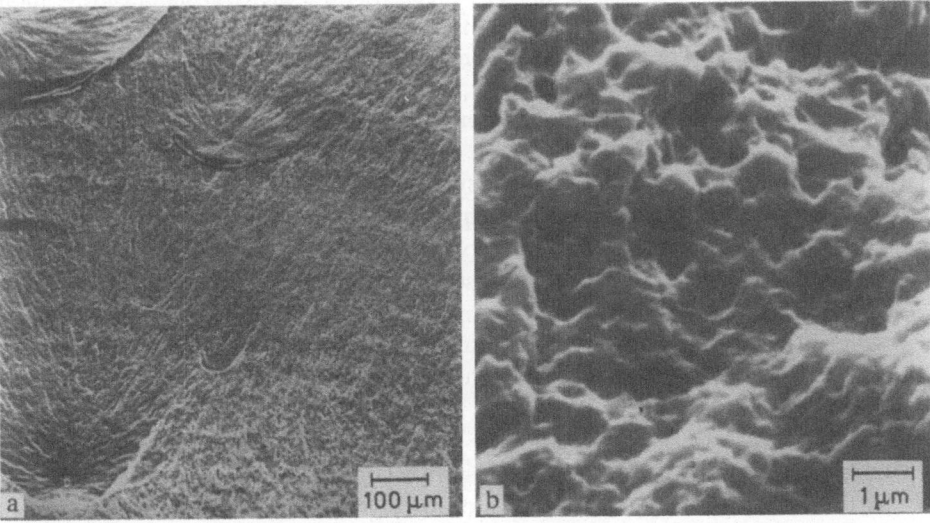

Fig. 50a and b. Fracture surfaces of ABS specimen of Fig. 49: **a** Region near center of specimen; **b** High magnification of region near striations

In contrast, in ABS, where particle size is much smaller and cavitation and shear deformation are present, as well as some crazing, the fracture surface is more extensively deformed and fractured occluded particles are rarely seen on the ruptured surface. The role of small particles in initiating shear has been discussed [73].

It has been reported that the effects of test frequency on fatigue behavior can be altered by the particular loading conditions observed. For example, in plasticized PMMA, it has been found that, at high loads, the fatigue crack propagation rate decreases with increase of frequency but at low loads. FCP is insensitive to cycle duration [74]. In our tests on ABS frequency effects were investigated at only two different stress amplitudes. In both cases, the average fatigue life increased with increase of test frequency; and similar results, but with greater frequency sensitivity, have been found for both PS (Fig. 15) and HIPS (Fig. 16).

5 Conclusion

1) Failure can develop under alternating loading, in both homogeneous and rubber modified glassy polymers, at stress values well below those required for fracture in simple tension. Also, in failure under cyclic stress, plastic deformation is limited and highly localized.

2) For high applied stress amplitudes, crazing initiates early in the fatigue process and cycles to fracture are low. As the stress amplitude is reduced, an increasing percentage of the fatigue lifetime is spent in craze and crack initiation. Fatigue lifetime at a given stress can be enhanced by delaying craze initiation (as by improving surface finish or applying suitable coatings) or by increasing craze stability (as by raising molecular weight and chain entanglement density).

3) Cycles to craze breakdown and to fatigue fracture increase significantly upon changing from complete stress reversal to cycling in a tensile mode at the same maximum stress. Buckling of the craze fibrils under compressive stress is conducive to early crack formation within the craze.

4) At the same stress amplitude, rubber modified polymers fail sooner in fatigue than do the unmodified polymers even though they have superior resistance to fatigue crack propagation. This is a result of much earlier initiation of crazing, localized plastic deformation, and subsequent crack development due to the stress concentrating effect of the dispersed second phase particles.

5) Fatigue lifetime increases with increase of test frequency, v, according to a power law relationship, for both PS and HIPS; and the slope of the lop N-log v plot appears independent of stress amplitude over the stress range investigated. Fatigue endurance also increases with frequency for ABS but its frequency sensitivity is less than that of HIPS.

6) SAN resins have greater resistance to craze initiation, to tensile fracture, to yielding in compression, to strain softening and to fatigue fracture than does PS. Properties vary with AN content and also, for a given AN content, with molecular weight. Increase of molecular weight raises both the craze and tensile stress and also the cycles to craze initiation and to fatigue fracture.

7) In simple tension and tension-compression fatigue, HIPS deforms by craze nucleation and growth while ABS deforms primarily by shear. Crazes develop in ABS prior to fracture but at a later stage than does shear deformation.

8) Fracture surface morphology varies with stress amplitude and test frequency as well as with composition and molecular weight. Even under fully reversed cycling, discontinuous crack growth bands have been observed on the fatigue fracture surface in PS, SAN and ABS but they have not been seen in HIPS.

9) In HIPS, most of the dispersed particles are fractured by the advancing fatigue-induced crack but particles smaller than 1 μm may be bypassed. The second phase nature of the particles is more clearly revealed by fatigue fracture rather than tension fracture and in the fatigue case, the surrounding matrix is remarkably smooth even at high magnification.

10) In ABS, where particle size is much smaller than in HIPS, the particles are less effective as craze initiators and the fatigue fracture surface shows evidence of considerable localized plastic deformation of the matrix polymer as well as of cavitation and/or loss of adhesion of the rubber particles.

Acknowledgments: We wish to express our appreciation to the National Science Foundation for their financial support under Grant No. DMR-78-27558. We also acknowledge the helpful discussions and counsel of A. M. Donald, R. W. Hertzberg, H. Karam, H. H. Kausch, E. J. Kramer, J. A. Manson, and J. S. Trent.

6 References

1. Sauer, J. A., Marin, J., Hsiao, C. C.: J. Appl. Phys. *20*, 507 (1949)
2. Hsiao, C. C., Sauer, J. A.: J. Appl. Phys. *21*, 1071 (1950)
3. Maxwell, B., Rahm, L. F.: Ind. Engrg. Chem. *41*, 1988 (1949)
4. Bessenov, M. I., Kuvshinskii, E. V.: Sov. Phys. Solid State *1*, 1321 (1960); *3*, 1947 (1962)
5. Spurr, O. K., Niegisch, W. D.: J. Appl. Poly. Sci. *6*, 585 (1962)
6. Sternstein, S. S., Ongchin, L.: ACS Poly. Prepr. *10*, 1117 (1969)
7. Sternstein, S. S., Myers, F. A.: J. Macromol. Sci., Phys. *B8*, 539 (1973)
8. Kambour, R. P.: J. Poly. Sci. *D7*, 1 (1973)
9. Bowden, P. B.: Chap. 5 The Physics of Glassy Polymers, Haward, R. N. (ed.), Applied Science 1973
10. Lauterwasser, B. D., Kramer, E. J.: Phil. Mag. *39A*, 469 (1979)
11. Kramer, E. J.: Chapter 1 in this volume; see also Chap. 3, in: Developments in Polymer Fracture-1, Andrews, E. H. (ed.), Applied Science 1979
12. Bucknall, C. P.: Toughened Plastics, Applied Science 1977
13. Bucknall, C. P.: Adv. Polym. Sci. *27*, 121 (1978)
14. Williams, J. G.: Adv. Polym. Sci. *27*, 67 (1978)
15. Kausch, H. H.: Polymer Fracture, Springer-Verlag Berlin, Heidelberg, New York 1978
16. Andrews, E. H., Reed, P. E.: Adv. Polym. Sci. *27*, 1 (1978)
17. Hertzberg, R. W., Manson, J. A.: Fatigue of Engineering Plastics, Academic Press 1980
18. Sauer, J. A., Richardson, G. C.: Int. J. Fract. *16*, 499 (1980)
19. Radon, J. C.: Int. J. Fract. *16*, 533 (1980)
20. Beardmore, P., Rabinowitz, S.: Appl. Polym. Symp. *24*, 25 (1974)
21. Skibo, M. D., Manson, J. A., Webler, S. M., Hertzberg, R. W., Collins, E. A.: ACS Symp. Series *95*, 311 (1979)
22. Hertzberg, R. W., Manson, J. A., Wu, W. C.: ASTM STP 536, p. 391 (1973)
23. Manson, J. A., Hertzberg, R. W.: J. Poly. Sci., Phys. Ed. *11*, 2483 (1973)
24. Bucknall, C. B., Stevens, W. W.: J. Mater. Sci. *15*, 2950 (1980)
25. Warty, S., Sauer, J. A., Charlesby, A.: Eur. Poly. J. *15*, 445 (1979)
26. Sauer, J. A., Habibullah, M., Chen, C. C.: J. Appl. Phys. *52*, 5970 (1981)
27. Turley, S. G., Keskhula, H.: Polymer *31*, 466 (1980)

28. Chen, C. C., Chheda, N., Sauer, J. A.: J. Macromol. Sci., Phys. *B19*, 565 (1981)
29. Hull, D.: J. Mater. Sci. *5*, 357 (1970)
30. Murray, J., Hull, D.: J. Poly. Sci. A-2, *8*, 583 (1970)
31. Fellers, J. F., Kee, B. F.: J. Appl. Poly. Sci. *18*, 2355 (1974)
32. Sauer, J. A., Foden, E., Morrow, D. R.: Poly. Engrg. Sci. *17*, 246 (1977)
33. Kim. S. L., Skibo, M., Manson, J. A., Hertzberg, R. W.: Poly. Engrg. Sci. *17*, 194 (1977)
34. Skibo, M., Hertzberg, R. W., Manson, J. A., Kim, S. L.: J. Mater. Sci. *12*, 531 (1977)
35. Sauer, J. A., McMaster, A. D., Morrow, D. R.: J. Macromol. Sci., Phys. *B12*, 535 (1976)
36. Warty, S., Morrow, D., Sauer, J. A.: Polymer *19*, 1465 (1978)
37. Chheda, N., Chen, C. C., Sauer, J. A., in: Advances in Materials Technology in the Americas-1980, V. I, p. 33, ASME 1980
38. Donald, A. M., Kramer, E. J.: J. Mater. Sci. *17*, 2351 (1982); J. Appl. Poly. Sci. *27*, 3729 (1982)
39. Rabinowitz, S., Krause, D. R., Beardmore, P.: J. Mater. Sci. *8*, 11 (1973)
40. Doyle, J. J., Maranci, A., Orowan, E., Stork, S. T.: Proc. R. Soc. *A-329*, 137 (1972)
41. Sauer, J. A., Woan, Der-jin, Chen, C. C., Habibullah, M.: 27th Intern. Symp. Macromol. V. II, 1045, 1981
42. Chen, C. C., Morrow, D. R., Sauer, J. A.: Poly. Engrg. Sci. *22*, 451 (1982)
43. Skibo, M. D., Hertzberg, R. W., Manson, J. A.: J. Mater. Sci. *11*, 479 (1976)
44. Hertzberg, R. W., Manson, J. A., Skibo, M. D.: Poly. Engrg. Sci. *15*, 252 (1975)
45. Skibo, M. D., Jariszewski, J., Kim, S. L., Hertzberg, R. W., Manson, J. A.: Proc. 36th Ann. Tech. Meeting, SPE, p. 304 (1978)
46. Hertzberg, R. W., Manson, J. A., Skibo, M. D.: Polymer *19*, 358 (1978)
47. Woan, Der-jin, Habibullah, M., Sauer, J. A.: Polymer *22*, 699 (1981)
48. Mann, J., Williamson, G. R.: Chap. 2, Physics of Glassy Polymers, Haward, R. N. (ed.), Applied Science 1973
49. Fletcher, K., Haward, R. N., Mann, J.: Chem. and Ind. *45*, 1854 (1965)
50. Moore, R. S., Gieniewski, G.: J. Appl. Poly. Sci. *14*, 2889 (1970)
51. Ziemba, G. P.: Encyl. Poly. Sci. Engrg. *V. 1*, p. 425, Wiley-Interscience 1970
52. Donald, A. M., Kramer, E. J.: Polymer *23*, 461 (1982)
53. Donald, A. M., Kramer, E. J.: J. Mater. Sci. *17*, 1871 (1982)
54. Haward, R. N.: Chap. 6, Physics of Glassy Polymers, Haward, R. N. (ed.), Applied Science 1973
55. Beevers, R. B., White, R. F. T.: Poly. Lett. *1*, 171 (1963)
56. Lavengood, R. E., Nicolais, L., Narkis, M.: J. Appl. Poly. Sci. *17*, 1173 (1973)
57. Bucknall, C. B., Clayton, D., Keast, W. E.: J. Mater. Sci. 7, 1443 (1972)
58. Donald, A. M., Kramer, E. J.: J. Mater. Sci. *17*, 1765 (1982)
59. Truss. R. W., Chadwick, G. A.: J. Mater. Sci. *12*, 583 (1977)
60. Bucknall, C. B., Drinkwater, I. C.: J. Mater. Sci. *8*, 1800 (1973)
61. Takahashi, K., in: Mechanical Behavior of Materials V. III, 556, Soc. of Mater. Sci., Japan 1972
62. Mai, Y. W.: J. Mater. Sci. *11*, 303 (1976)
63. Boyer, R. F., Keskhula, H.: Encyl. Poly. Sci. Engrg. *V. 13*, 375, Wiley-Interscience 1970
64. Sternstein, S. S., Ongchin, L., Silverman, A.: Appl. Poly. Symp. *7*, 175 (1968)
65. Sauer, J. A., Pae, K. D., Bhateja, S. K.: J. Macromol. Sci., Phys. *B8*, 649 (1973)
66. Pitman, G., Ward, I. M.: J. Mater. Sci. *15*, 635 (1980)
67. Donald, A. M., Kramer, E. J.: Polymer *23*, 1183 (1982)
68. Chen, C. C., Sauer, J. A.: to be publ.
69. Mackay, M. E., Teng, T. G., Schultz, J. M.: J. Mater. Sci. *14*, 221 (1979)
70. Bucknall, C. B., Gotham, K. V., Vincent, P. I.: Chap. 10, Polymer Science, Jenkins, A. D. (ed.), North-Holland Publ. Comp. 1972
71. Dugdale, D. S.: J. Mech. Phys. Solids 8, 100 (1960)
72. Hertzberg, R. W., Skibo, M. D., Manson, J. A.: J. Mater. Sci. *13*, 1038 (1978)
73. Argon, A. S.: Pure and Appl. Chem. *263*, 247 (1975)
74. Lebedeva, M. F., Aleshin, V. I., Aero, E. L., Kuvshinskii, E. V.: Int. J. Fract. *17*, 327 (1981)

Received August 11, 1982
H. Kausch (editor)

Crazes and Shear Bands
in Semi-Crystalline Thermoplastics

Klaus Friedrich
Arbeitsbereich Kunststoffe/Verbundwerkstoffe,
Techn. Universität Hamburg-Harburg,
D-2100 Hamburg 90, FRG

Advances in Polymer Science 52/53
© Springer-Verlag Berlin Heidelberg 1983

List of Symbols

A_{cr}	Sum of crazed area	mm^2
B	Specimen thickness	mm
D	Spherulite diameter.	μm
d	Thickness of a shear band	μm
E	Elastic modulus	MPa
f_s	Volume fraction of coarse spherulites	%
G	Shear modulus	MPa
H_K	Ball thrust hardness	MPa
K_{Ic}	Plane strain fracture toughness	$MPa \cdot m^{1/2}$
K_{IIc}	Mode II — fracture toughness	$MPa \cdot m^{1/2}$
L	Length	m
MFI	Melting flow index (230/2,16 by ASTM D 1238-65T)	$0.1 \cdot g \cdot min^{-1}$
n	Work hardening exponent	—
r_p	Plastic zone radius	μm
T	Temperature	°C or K
T_g	Glas transition temperature	°C or K
X_c	Degree of crystallinity	%
x	Shear displacement in a shear band	μm
α	Angle between the normal to the applied load and the crack direction	—
ε	Strain	—
ε_F	Strain at break	—
ε_i	Local strain inside of a shear band	—
$\dot{\varepsilon}$	Strain rate	min^{-1}
ν	Poisson's ratio	—
σ	Stress	MPa
σ_{BI}	Band initiation stress	MPa
σ_F	Fracture stress	MPa
σ_y	Yield stress	MPa
τ_{BI}	Shear stress for band initiation	MPa
θ	Adiabatic term	—
φ	Angle between shear band and loading direction	—

1 Introduction

1.1 General Aspects

One of the most important subjects of applied polymer science is, the understanding of the deformation mechanisms and the fracture properties of semi-crystalline polymers. At the same time, it is one of the most difficult to study, and the amount of research in this area is high (see e.g. [1-5]). One of the complications experienced with semi-crystalline polymers stems from the fact that they are composed of crystalline and amorphous phases, arranged in a diversity of microstructures. These are generally built up when a crystallizable polymer is cooled from the melt or is precipitated from a solvent. Their formation is seldom entirely under our control. Secondly there is a large variety of (time- and temperature-dependent) fracture conditions, ranging from fracture under static load to fatigue failure, from notch induced fracture to stress corrosion cracking. Each may be affected individually by a given microstructural feature (Fig. 1). The total problem, then, is of vast magnitude.

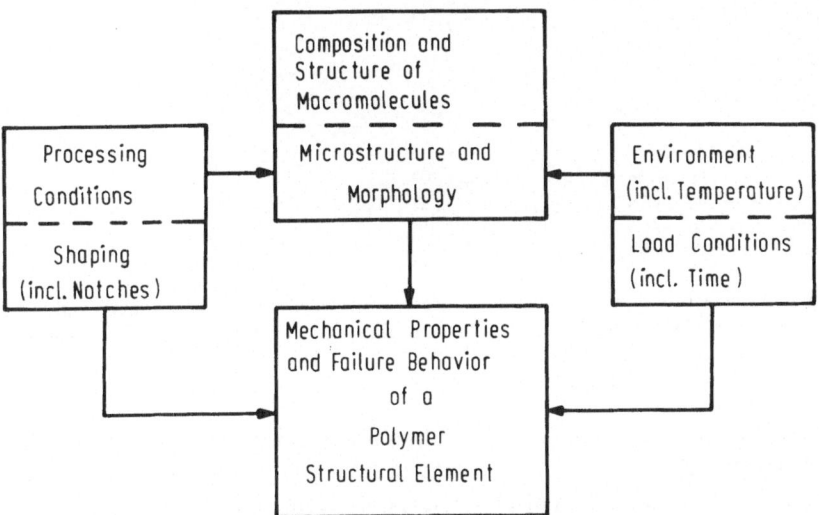

Fig. 1. Parameters which influence the properties of a polymer structural element (Courtesy G. Retting, Ludwigshafen)

1.2 Semi-Crystalline Microstructures

The manner in which crystalline and amorphous phases are arranged within a polymer solid depends strongly on the strain and thermal history. Unstrained polymers usually crystallize in a spherulitic manner, the spherulites being point-nucleated semi-crystalline entities which grow in a spherically symetrical form until their boundaries impinge. The spherulites are composed of radiating lamellar crystals, in which the molecular axis is perpendicular to the growth direction [5]. Only in some special cases are tangentially oriented crystal lamellae observed in spherulites (Fig. 2) [6-8]. The lamellae are folded-chain crystals somewhat similar in nature to the polymer single crystals which can be grown from dilute solutions. Special details as for instance the question of "adjacent reentry" will not be treated in this context (see e.g. [9]). As a rule the chains in the crystal lattice are oriented at 90° to the lamellar plane. The majority of them fold back at the surface of the lamellae. Quite a few of the chains, however, do not immediately reenter the crystal core of the same lamella by forming a tight loop but remain in the amorphous surface layer as free chain ends (cilia) or reenter, forming loose loops. Also, they may be partially included in the crystal lattice of another lamella, thus acting as an interlamellar tie molecule [9].

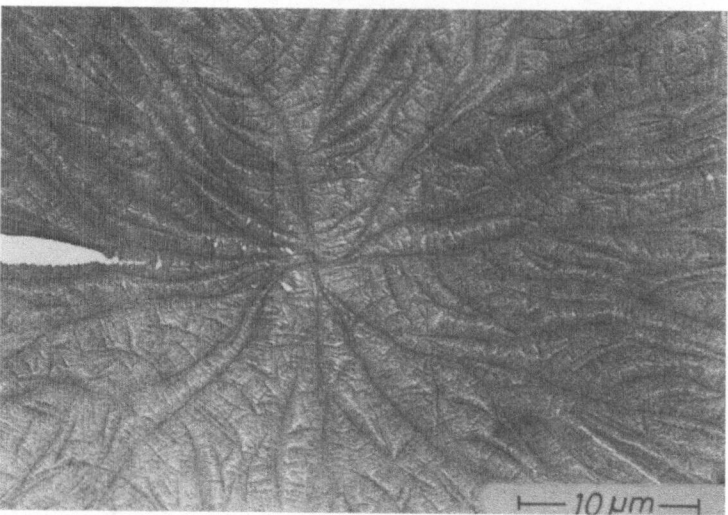

Fig. 2. Crosshatched structure of a spherulite in a solventproduced film of a 20 % isotactic and 80 % atactic PP mixture. The film exhibits a tendency to tear along the radial interfibrillar zones in the spherulite (defocussing contrast in TEM)

The degree of crystallinity within the spherulites is dependent on molecular weight, molecular structure and thermal and mechanical treatment. Thus, it varies widely, from a few percent to over 90 %, and differs not only from one polymer to another, but also within a given polymer [5]. Spherulite size, size distribution and microscopic morphological changes within the spherulites can be varied by thermal history and nucleating agents. Under special conditions, for example during the injection molding of thick walled parts, the different local cooling conditions can lead to differences in the morphology across the specimen thickness. In the case of rapidly cooled zones, a fine-spherulitic or even non-spherulitic microstructure can predominate. In the regions of slow undercooling, on the other hand, the morphology is coarsely spherulitic and contains individual voids or even holes [10].

Polymers which have been crystallized under high melt strains have a microstructure similar to that of a fiber cold drawn from an isotropic crystalline condition. Highly oriented semicrystalline fibers have a microfibrillar substructure. The microfibrils with lateral dimensions of 20 to 40 nm contain folded chain crystallites as the most important structural elements. In addition they contain amorphous regions with coiled chains and chain ends (ciliae). Loading such a microfibrillar structure in the direction of orientation results in the shear loading of the microfibrils, the dilatational loading of the amorphous regions, and the stretching of molecules connecting adjacent crystallites, the so-called inter- and intrafibrillar tie molecules [11-12].

1.3 Unsolved Problems in Failure Behavior

A summary of the processes which can take place when a thermoplastic material is subjected to a tensile stress is presented in Figure 3. It will be seen that one is dealing with two limiting tensile conditions, viz. plane strain and plane stress. In practice

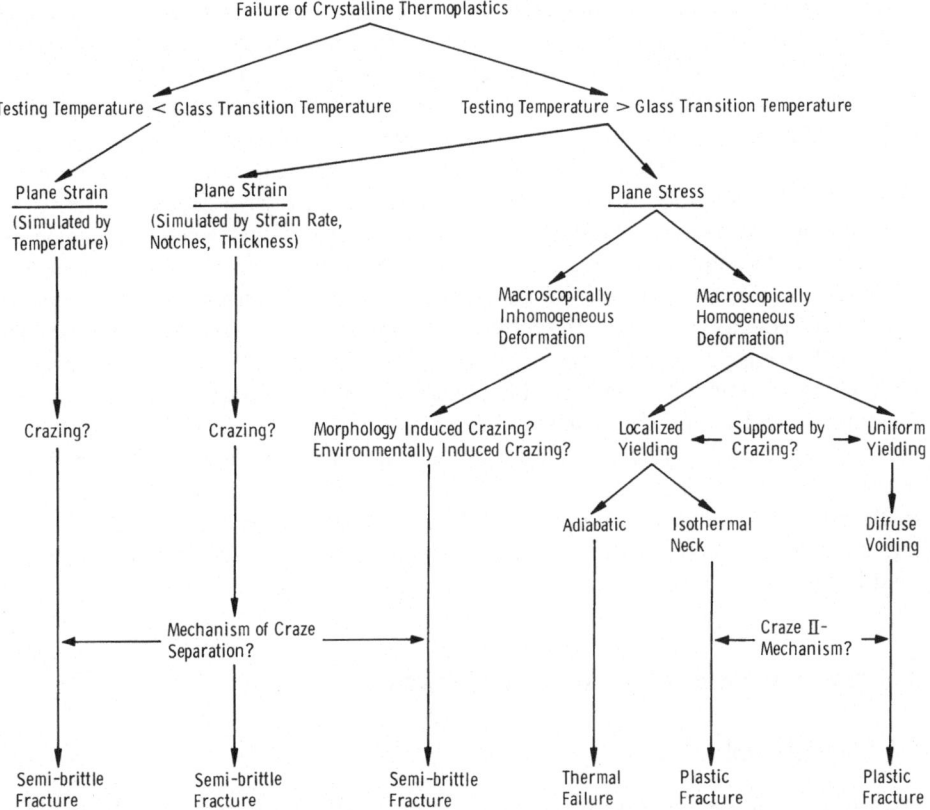

Fig. 3. Processes which may occur when a crystalline thermoplastic is subjected to tensile stresses

these two types of stress fields may occur together in different parts of the same test piece, but for the purpose of discussion it is convenient to distinguish between these two conditions which often lead to very different types of response [13].

There can be no doubt as to the importance of plane strain conditions for the fracture of plastics especially where sharp notches and thick sections are concerned. Such conditions nearly always lead to brittle or semi-brittle fracture. Vincent [14] has shown that the notch sensitivity in a braod range of amorphous and crystalline polymers is increased as the testing temperature is lowered and the loading rate is increased. Before fracture occurs, amorphous plastics often craze under these conditions. The complex questions of craze initiation, propagation and transformation into a crack have been treated extensively for amorphous polymers in the first three chapters of this book (see also [12−15]). The problem becomes more complicated when a crystalline-amorphous microstructure of the polymer has to be considered. It will have to be discussed in this chapter whether crazing is also a dominant failure mechanism, and what are the competing mechanisms?

One such competing mechanism is found under conditions of plane stress. In most of those cases the macroscopic deformation mechanism is usually that of a quasi-homogenous yielding process. It occurs either after localized yielding under formation of a neck which then spreads along the rest of the cross section leading to high amounts of strain at fracture. Or a uniform yielding of the complete gage length is observed, resulting in a typical plastic fracture. The process of yielding is well described and fairly well quantified for a large number of semi-crystalline polymers although there is not yet full agreement on the molecular processes involved [1-4, 12, 16].

Open questions also exist in the case of macroscopically inhomogeneous deformation, as it occurs for instance in the presence of aggressive environments [17-19]. Even less well known are those inhomogeneous deformation mechanisms which are induced by certain morphological features: crack-like defects in the spherulitic morphology, second phases, void or crack nucleating particles, or certain microstructural elements behaving differently such as spherulites and spherulite boundaries.

The following part of this report will be limited entirely to cases where craze-like features are observed in semi-crystalline materials. It is planned to show some of the phenomena and to discuss possible mechanisms.

In a final chapter a closely related phenomenon, the formation of shear bands in semi-crystalline polymers under compressive load will be described. It is attempted to discuss under which conditions shear bands are formed in semi-crystalline materials and how they interact with each other or with certain microstructural features, finally leading to crack initiation and shear fracture of the bulk polymer.

2 Crazes in Semi-Crystalline Polymers

2.1 Background

With respect to the general aspects of craze morphology reference will be made to the preceeding chapters of this volume (especially that of Kramer) even though they are dealing with amorphous polymers. The detection of crazes in semi-crystalline, opaque thermoplastics is harder to accomplish than with the transparent glassy polymers, though the existence of crazes in these materials can hardly be doubted [20-23]. Kambour [22] draws attention to some examples which may qualify as crazing in semi-crystalline polymers. Considering Geil's [24] experiences with polymer single crystals Kambour suggests that at the simplest level the stretching of polymer single crystals frequently causes fibrillation to occur. As a matter of fact, Morrow et al. [25] demonstrated that folded-chain crystals of PP exhibited periodic "(micro)-craze-cracks" with fibrils extending across these "cracks" when they were strained. Kojima [26] found with a similar polymer a fibrillation when deformation of the single crystals occured in the growth direction, and brittle fracture when the PP crystals were loaded perpendicular to the molecular fold plane. The results obtained by Gohil et al. [27] with single and multilayered crystals of POM revealed that the predominant mode of deformation of crystals involves a combination of micronecking, "brittle" fracture of lamellae and formation fibers, buckling in draw direction and shear along slip lines. Petermann and Gleiter [28-30] who studied the deformation of "substrate-free" PE single crystals suggested that the formation of fibers occurs by a two step process. The first step is the breaking off of single blocks of folded chains from the single crystals so that a "string-of-pearls" structure is obtained. In this "as drawn" state the molecules are roughly aligned parallel to the fiber axis (fiber texture). If the temperature is sufficiently high this process is followed by the thermally activated rearrangement (recrystallization) of the molecules in the drawn fibers so that a "bamboo" structure results. During this annealing process the alignment of the molecules in the crystalline regions becomes more regular and perfect (Fig. 4).

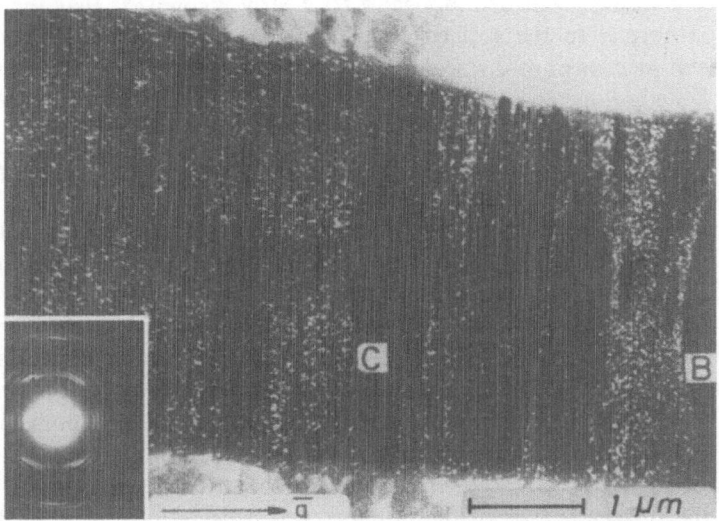

Fig. 4. Dark field electron micrograph of fibers and wide bundles (at B and C) that were anealed 1h at 383 K after drawing ("bamboo-structure"). The fibers consist of crystalline regions (bright areas) extended over the entire fiber cross section (Courtesy J. Petermann, Saarbrücken)

On a somewhat larger scale, several authors have shown [31-33] that the stretching of thin films consisting of planar analogs of spherulites causes the pulling-out of the spherulitic fronts into fibrillar bundles. In as much as this is accompanied by void formation it can be considered as a very interesting analogy to crazing. The well-known microfibrillation of many bulk semi-crystalline thermoplastics, notably linear PE and PP, produced on cold-drawing also has much in common with crazing. When conditions are suitable for cold-drawing the microfibrillation can be very well-developed, resulting in complete separation of the microfibrils [34]. The differences between a fibrillated cold drawn region and a craze reside, therefore, mainly in their macroscopic dimensions and many semi-crystalline polymers must consequently be regarded as intrinsically suitable for craze formation [20].

Craze-like features have been frequently observed in the past during the deformation of oriented semi-crystalline polymers, although they were sometimes named differently in the literature [23]: "transverse lines or lamellations" in cellulose fibers [35, 36], "fissures" and "strain bands" in fibers of PA, PETP, PE and PP [37-40], "cracks" in PA 6.6 [41] and "micro-cracks" by Zhurkov et al. [42, 43] for PP and PA 6. Also, the "pseudo-cracks" formed under strain at various temperatures in precracked PP und PB-1 films [44], as well as the "craze-cracks" observed by Flexman [45] during ductile impact fracture of PA 6.6 compositions, must be regarded as crazes in semi-crystalline polymers. In addition there can be no doubt that some of the "crack-structures" recently described for bulk PP [46], as well as the "micro-cracks and flow-zones" observed by Menges et al. [47-51] in PP, PE and POM, have a typical craze character.

Finally, solid evidence exists for crazes having formed in bulk, spherulitic PP under plane strain conditions. Van den Boogaart [52] clearly recognized that their planes were oriented normal to the applied stress and that they contained craze matter. Based on the displacement of surface scratches, he calculated the craze strain to be 100% using a wedge opening technique. Kambour [22] showed that even at room temperature craze bundles can be produced in PP in front of a sharp crack. The influence of morphology, gaseous environments and temperature on crazing and fracture in crystalline, isotactic PP films was intensively discussed by Olf et al. [23]. For microcrystalline, non-spherulitic PP the authors found a crazing behavior similar to that of amorphous glassy polymers in the temperature range between $-210\,°C$ and $-20\,°C$. In the presence of spherulites, on the other hand, crazes were generally restricted in length to that of single spherulites, emanating from the center and travelling along radii, perpendicular to within about 15°, to the direction of stress. Crazing along spherulite boundaries was not observed.

The environmental effects on deformation varied with the type of environment and with temperature. It was argued that gases such as N_2, Ar, O_2, and CO_2, especially near their condensation temperature, were strongly adsorbed at the craze tips, substantially reducing the surface energy of the polymer and causing a drop in yield stress for plastic flow. Hence, the hole formation involved in crazing was eased and further crazing favored. Similar observations were made by Brown et al. [53-55], who studied environmental crazing in semi-crystalline PTFE and PCTFE. Further investigations on the micromechanics of deformation of semi-crystalline PE in environmental stress cracking agents [12, 18, 56-58] have elucidated the role of spherulites, their microcrystalline block structure and their boundaries on the disintegration into independent nonuniform fibrils. Another environmental effect on crazing was reported for unstabilized PP oven aged at 90 °C [59]. Prior to tensile testing, samples aged for 400 h exhibited numerous surface crazes, which were composed of fibrils oriented parallel to the long axis of the tensile bars (melt flow direction). The high extent of deformation within these crazes was explained by relatively large internal stresses from molding in cooperation with an oxidative network degradation. Under external mechanical stress the surface crazes appeared to initiate catastrophic brittle failure of the entire sample even at room temperature and low loading rate.

Harris and Ward [60] have observed conventional crazes nearly normal to the tensile stress axis (tensile crazes), as well as what appear to be crazes along the shear direction (shear crazes), in uniaxially drawn, crystalline PETP sheets. Tensile crazes were formed parallel to the initial draw direction. The "shear crazes" which were the first of their kind to be reported, were seen in specimens under all directions of orientation. They formed always at the same angle as that of the shear bands which appeared subsequently upon yielding. An explanation for this "shear-craze"-phenomenon was offered by Brady and Yeh [61] based on their own studies of crazes and shear bands in amorphous and crystalline, isotactic PS films. First they produced a set of crazes and shear bands by stretching a film in one direction. When this film was then redrawn in a second direction, the authors observed that the stress component, which was perpendicular to the original shear direction of the original shear bands, tended to "open up" shear bands that lay along this direction. Since shear band and craze morphologies are very similar, an open shear band looks very much like a craze and could be what Harris called a "shear craze". A similar shear band-to-craze transition was recognized by Friedrich [62], who studied the correlations between shear bands and fracture in crystalline, spherulitic PP under compression at low temperatures (Fig. 5).

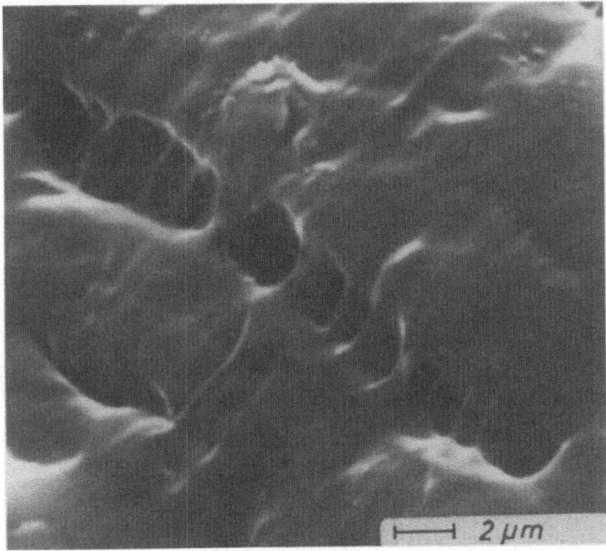

Fig. 5. Craze-like fibrillation inside normally deformed shear bands of type B in crystalline PP

Concerning the results of Brady and Yeh [61] on partially crystalline PS, the authors showed that the crystalline spherulites deformed only after substantial deformation had already occured in the interconnecting amorphous regions. Crazing and shear-banding were the primary modes of deformation. Further studies on the deformation of the lamellae within the spherulites were carried out by stretching isotactic PS films, having oriented lamellar morphology, perpendicular to the lamellar orientation. As the crazes propagated through the crystalline regions, the lamellae separated and appeared to break up into a 10 to 20 nm fibrillar morphology. It was then essentially the same craze morphology that the authors had also found for amorphous atactic PS.

Post-failure studies of the fracture surface morphology of bulk semi-crystalline polymers are more difficult than those of amorphous materials due to the more complex multiphase structure associated with semi-crystalline materials [63]. However, it was shown that some fractographic details point to the formation of a stress whitened region ahead of a notch prior to final fracture of the material. In particular, stress-whitened regions were easily visible in semi-crystalline polymers such as LDPE and HDPE [20, 64, 65]. The resulting, macroscopically apparently brittle fracture surfaces were often covered with drawn fibrils, like a brush or a dimple structure [66-68]. This indicates that the surface layer had failed after void nucleation and multiple formation of micronecks as many miniature tensile specimens. Fracture surfaces obtained by Gaube and Kausch [69] under impact loading at 20 °C from a standard-HDPE bar also clearly revealed this kind of morphology in a zone of thermally and mechanically activated crack growth starting from a knife cut. The authors gave an impression of the large plastic deformation of the cellular and void-containing fracture surface area where the walls of rosettes and ridges were drawn by up to several hundred per cent. This suggests that initial crack growth was confined to the plane strain region involving slow crack propagation through a craze as known from amorphous polymers in the mirror-like zone [70]. Fracture surfaces somewhat reminiscent of a void coalescence mechanism as a typical craze-breakdown phenomenon were also found in dry PA 6 specimens under impact [71] and fatigue loading [63].

For the following discussions, it might be convenient at this point to summarize the phenomena observed with the formation of voidy fibrillar zones (crazes) and their final breakdown in semi-crystalline polymers:

— break up of single crystals and lamellae by normal and shear displacements,
— voiding and the formation of micro-stretched bundles under normal and shear stress conditions,
— stress whitening,
— dimple formation on fracture surfaces.

Some other craze phenomena now being recognized as intrinsic crazes (crazes II) have already been discussed in Chapter 2. They will be put into context in a later section of this chapter.

2.2 Temperature Dependence of Deformation

In view of the multitude of observed deformation mechanisms it is useful at this point to examine the effects of external variables, especially that of ambient temperature, on the deformation behavior of semi-crystalline thermoplastics. At room temperature many of these polymers are above their glass transition point and owe their strength and stiffness to the crystalline phases. The first displacements start in the relatively soft amorphous layers, but the stress-strain curve is largely determined by the presence and arrangement of the crystals. Interlamellar slip has been identified as an important mechanism, but, in addition, crystalline deformation mechanisms occur at moderate strains [1, 3, 20]. The corresponding stress-strain curve shows an upper yield point associated with the initiation of a neck within the gage length. As the neck stabilizes (after the lower yield point) and the shoulders of the neck move along the gage section during the drawing operation, the drawn portion shows a tendency to become completely whitened. This stress whitening of polymers has convincingly been attributed to formation of voids [72] and/or intrinsic crazing (see the article of Dettenmaier in this volume). Void formation appears to be associated with the local inhomogeneity of the deformation as the original spherulitic structure begins to transform towards the fibrillar structure characteristic of drawn material. Microscopic studies made by Yoon et al. [73] on stress whitening of PP revealed that three distinct morphological features are present, namely, surface crazes on lateral surfaces near the neck, microvoids of 1–5 μm diameter on the fracture surface, and cracks or cleavages parallel to the draw direction at the end of the neck and in the drawn region.

Fischer [74] found that this process is combined with a decrease in density in the necked region down to 80% of the undeformed material. Using newly developed preparation techniques for SEM and TEM investigations, the author demonstrated that the beginning of stress whitening in the necked region is associated with the formation of a large number of crazed zones, the size and orientation of which can be correlated with the spherulitic structure being inhomogeneously deformed (Fig. 6).

When the temperature is lowered, on the other hand, deformation within the amorphous phase becomes more and more restrained and the material can be expected to behave more like a glassy polymer [20]. In this case, the crystals will have less freedom to reorient, due to the reduced mobility of the amorphous regions, and there will be less opportunity for them to take up an orientation which favors further drawing under stress. It is to be expected, therefore, that initial drawing will be more pronounced, and confined to those regions which have crystallized in an appropriate direction of orientation. Some reorientation of lamellae will doubtless occur just outside of the initially drawn regions. Due to a certain stress concentration effect planes of fibrillation around the initial "drawing points" will then be formed. In fact, the deformation of PP at temperatures around $-30\,^{\circ}$C, that is just below its glass transition temperature, is quite analogous to the well-known crazing of an amorphous polymer such as atactic PS below its glass transition (Fig. 7).

2.3 Low Temperature Crazing

2.3.1 Craze Microstructure

As indicated in Figure 7, low temperature crazes can form in a semi-crystalline polymer when the material is loaded below its glass transition temperature and when a critical tensile stress is exceeded. Density and lengths of these crazes increase rapidly until the breaking stress is reached. Their visible appearance is very similar to those crazes found in glassy amorphous polymers. Thin sections taken from the bulk tensile specimens reveal transmission light microscopy that the crazes have propagated perpendicular to the external load, often starting at the surface of the sample (Fig. 8a). But it is also observed that they are completely within the sample. Their lengths are much greater than the diameter of the spherulites, especially in the fine spherulitic morphologies. The crazes run straight through the microcrystalline structure of the fine spherulites, and it appears that they ignore any differences in the local lamellar orientation. Additional experiments with a partially coarse spherulitic morphology gave the impression that crazes initiate and propagate predominantly within the surrounding fine spherulitic matrix. The crazes appear to be stopped when they approach a zone of a coarse spherulite structure in which the local orientation of the radially arranged lamellae differs highly from the direction of the propagating craze (Fig. 8b). On the other hand, in a completely coarse spherulitic structure of PP 1120 with extremely high molecular weight crazes develop and grow also along secant planes of the spherulites (Fig. 8c). Spherulite boundaries could not be found to be favorable sites for craze initiation or propagation under this low temperature condition.

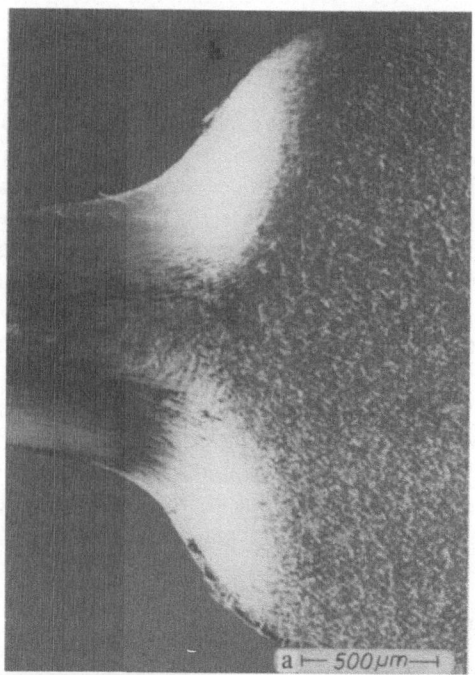

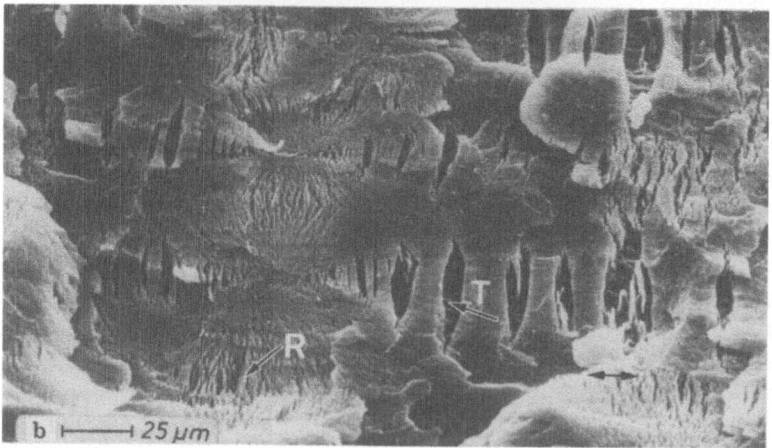

Fig. 6a. Quasi-homogeneous deformation in the necked region of fine spherulitic PP 1120 at room temperature; Fig. **6b—e.** Inhomogeneity of spherulitic deformation (arrow = loading direction): **b** radially (R) and tangentially (T) arranged craze bundles in spherulites (SEM); **c** fibrillar network in tangentially oriented crazes (low number, highly opened, SEM); **d** fibrillar fine structure in a dense bundle of numerous, radially oriented crazes (SEM); **e** TEM-micrograph of the coarse fibrillar structure in a tangentially arranged craze. (Thanks are due to P. Eyerer and G. Fischer, Stuttgart, who presented figures 6b)—e) at the DFG-Meeting on "Flow and Deformation Behavior of Polymers", Bad Honnef, October 1982)

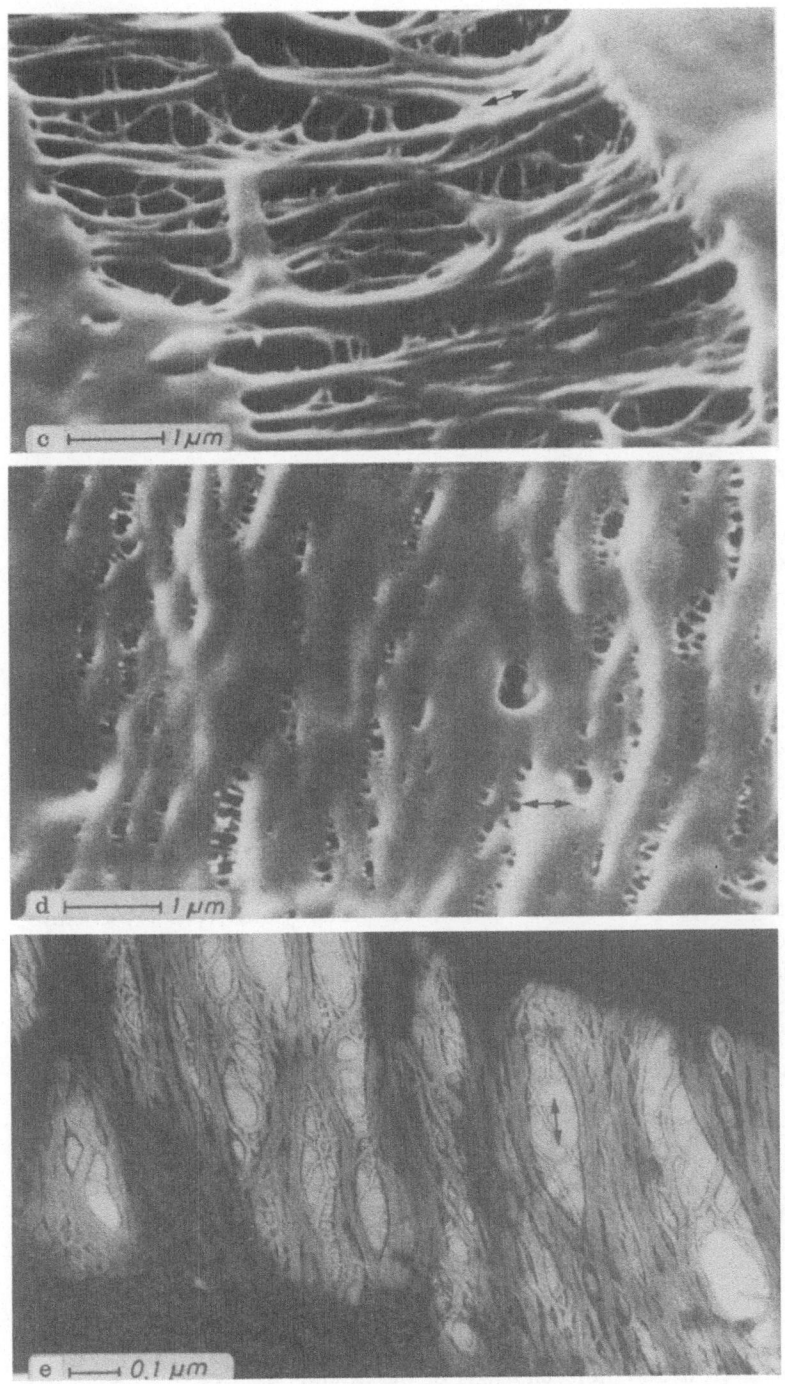

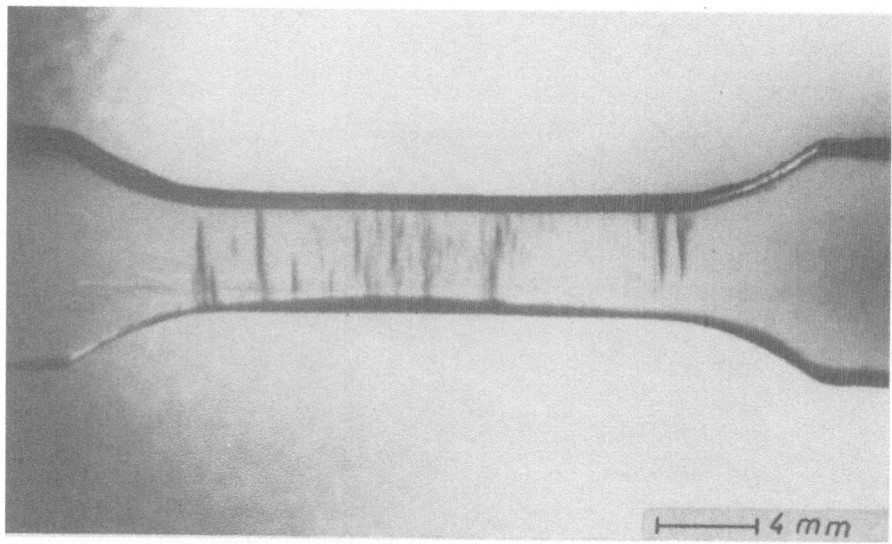

Fig. 7. Low temperature crazes in a tensile specimen of PP

The internal structure of these low temperature crazes can be detected by elastically opening a crazed bulk sample by bending and by taking a carbon-gold replica from the crazed area. Further viewing of the replica under high tilting angles (45–55°) in the transmission electron microscope indicates that the structure inside the crazes is built up of highly stretched fibrils with an average diameter of 15 to 40 nm (Fig. 9). The thickness of the crazes is in the range from 1 to 3 μm, the volume density within a craze varies between 35 to 55 %. As a matter of fact, all the data of these low temperature crazes in PP reveal essentially the same dimensions of craze morphology as known for amorphous polymers.

The observations give rise to the assumption that the process of low temperature craze formation in semi-crystalline polymers is very similar to the three-step craze nucleation mechanism described by Kausch [75] for glassy amorphous polymers. The only modification which has to be considered is the partially ordered microstructure. Here, some recent investigations by Petermann, Gleiter and Hornbogen [76] on the fracture behavior of small PE single crystals as well as the results of Brady and Yeh [23] on isotactic PS-films and some electronmicroscopic studies on drawn PP films [8] may be helpful for a further interpretation of the mechanisms. The following sequence of events can be expected to take place during craze formation in bulk semi-crystalline thermoplastics (Fig. 10):

I. First of all, at sites of enhanced stress concentration (i.e. at preexisting voids in the microstructure or at sites where amorphous layers between lamellae are oriented nearly under 45° to the applied load) a local sliding occurs between individual lemllar ribbons having only a low degree of entanglement in their common amorphous interlayer. At his stage, strain is accomodated almost entirely by the interlamellar amorphous regions. During this process the crystals change their orientation and the shear stress component decreases while the stress in direction of the molecular axis of the crystals increases.

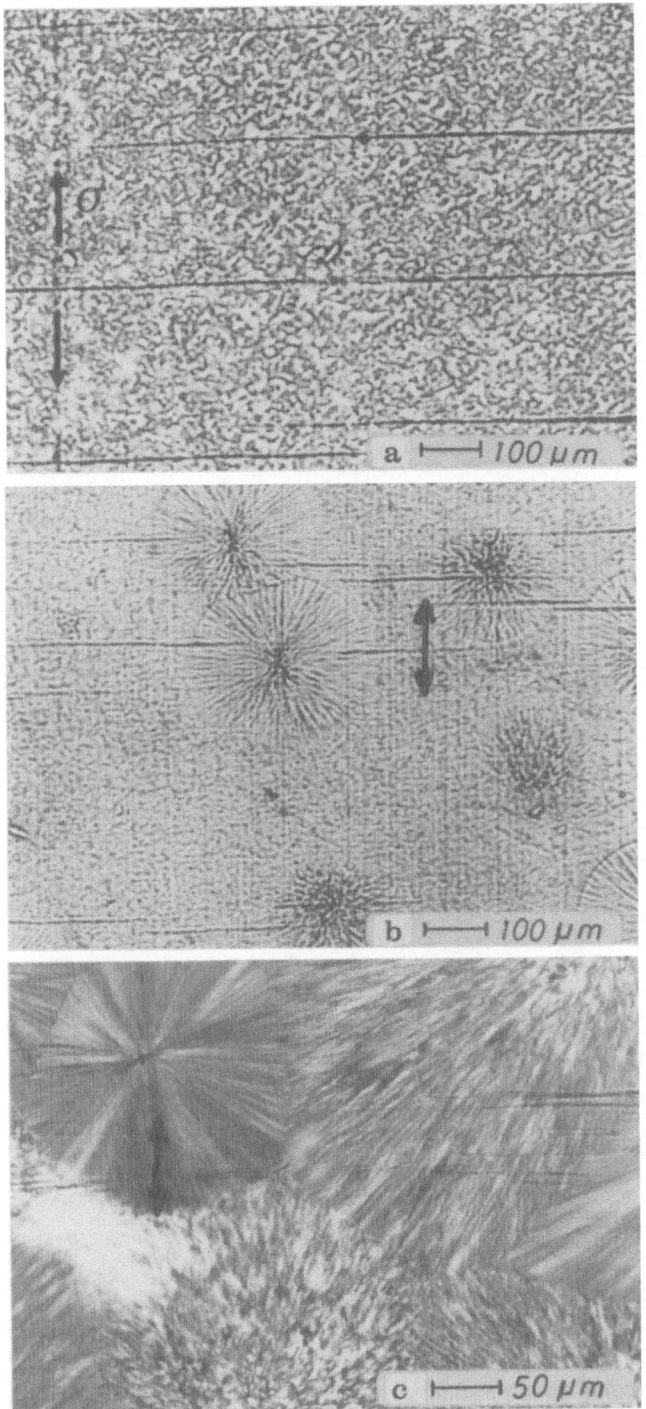

Fig. 8a—c. Thin sections of PP-tensile bars containing low temperature crazes in different morphologies: **a** fine spherulitic structure (arrow = stress direction); **b** partially coarse spherulitic structure; **c** completely coarse spherulutic structure

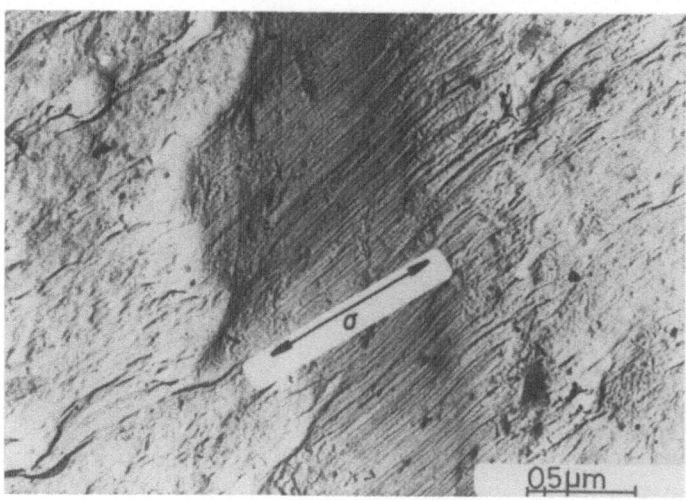

Fig. 9. Carbon-gold replica of the fibrillar structure of a low temperature craze in PP (due to the high tilting angle, the orientation of the fibrils is not perpendicular to the craze edges)

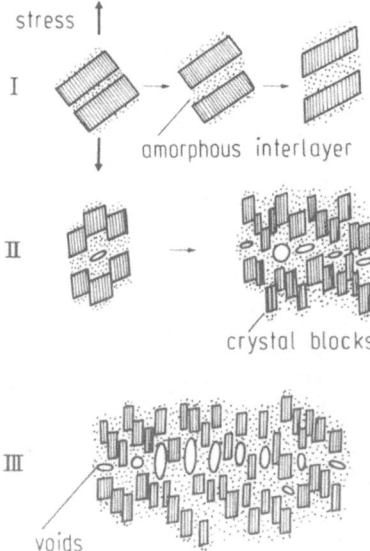

Fig. 10. Schematic steps of crazes formation in a semi-crystalline polymer structure

II. In a secondary step, once the local tensile stress reaches a critical value (equal to the stress at which macroscopic yielding can take place), individual blocks, about 10 to 30 nm in size [23], are pulled out of the crystal ribbons. Due to this local yielding process, submicroscopic defects having an ellipsoidal shape are created between the lamellae [72]. Typical dimensions are 10 to 50 nm for the large and 2 to 6 nm for the short half axis; thus, these microvoids are still, by an order of magnitude, smaller than micocopically visible crazes [77, 78]. However, the defects are able to cause an increase in stress in their lateral neighborhood, consequently involving adjacent lamellar ribbons in the local deformation process. Thus, the probability for further creation of voids as well as for formation of fibrils between these voids is increased. In the as-formed condition, the fibrils consist of partly extended tie molecules with an occasional block of still folded crystal between the strings.

III. The third step of direct longitudinal transmission of strain onto connected crystalline blocks leads to a perfect stretching of these fibrils. Because of the alignment of the molecules the fibers in this condition should possess a strength about 1 to 2 orders of magnitude higher than the yield stress of randomly distributed folded polycrystals. As the fibrils are able to stabilize the enhanced microvoid volume between them, a lateral coalescence of these voids finally provides a local deformation zone in the shape of a craze as known from amorphous polymers.

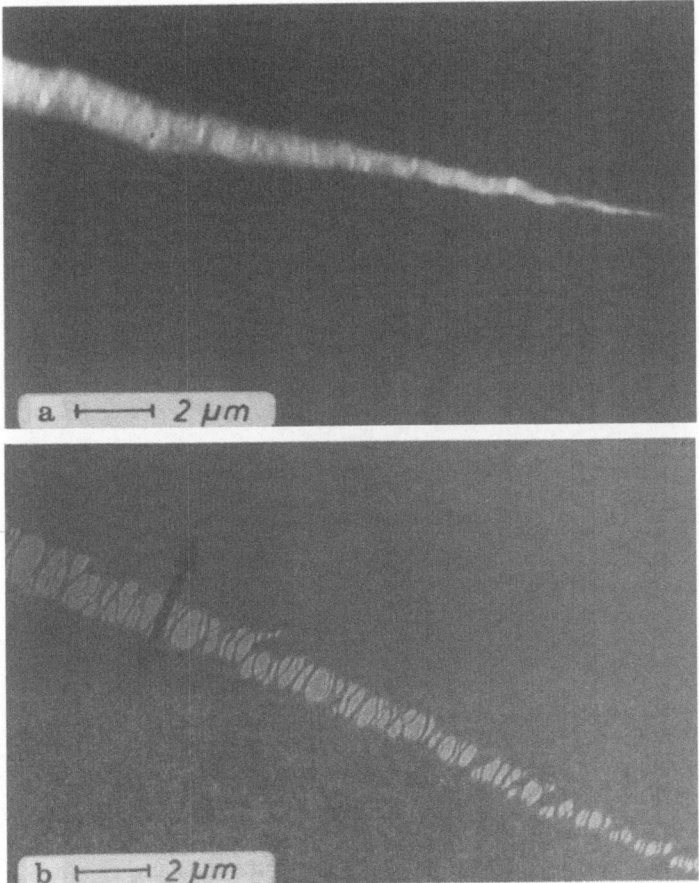

Fig. 11a and b. Micrographs of craze tips in a solvent-produced, spherulitic film of PP 1120 before **a** and after long time irradiation in the TEM **b**

From all similarities of craze initiation in amorphous and semi-crystalline polymers it can be concluded that the micromechanisms of craze growth, which have been intensively studied in amorphous polymers (see the articles by Kramer and Dettenmaier in this volume) should also be closely related. In amorphous polymers crazes grow in thickness by drawing further matrix material. Only in some cases crazes grow by an increase in the orientation and elongation of the fibrils [70]. This leads to a change in the micromorphology of the craze along its length, with coarser, thicker fibrils being found in the tapered section near the craze tip [79-81]. The morphologies of craze tip regions in thin cast films of spherulitic PP, shown in Figure 11, are found to be rather similar to those observed in amorphous PS-films [82, 83].

2.3.2 Molecular Aspects of Craze Strength and Fracture

Figure 12 represents all steps of craze formation in crystalline polymers in a single model. It is based on Hornbogen's model [84] for a crack tip in a polymer crystal, under the utilization of individual block drawings by Schultz [1] for the fine scale nature of plastic deformation in semicrystalline thermoplastics. The classification into four regions A to D (after [85]) helps to describe and understand the influence of molecular parameters on craze strength and craze breakdown.

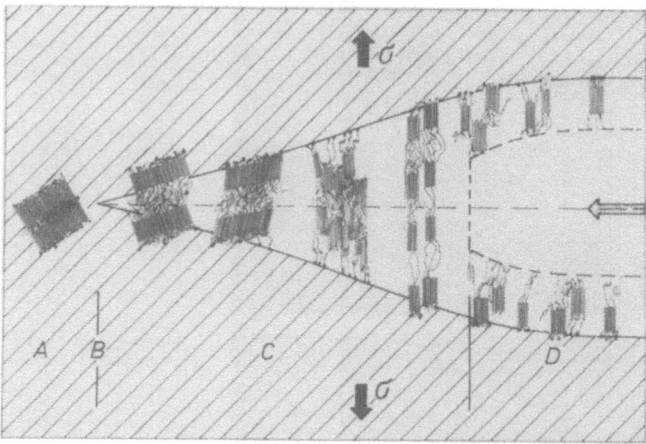

Fig. 12. Model of craze initiation and propagation in a semi-crystalline polymer: A. elastically strained region, B. region of void nucleation and fibril formation, C. craze growth due to fibril extension and continued fibrillation, D. transformation of craze into crack (after [85])

At lower temperatures craze fibrils remain in their highly stretched condition and should finally fail by bond rupture. At higher temperatures, on the other hand, a thermally activated rearrangement may occur simultaneously during the drawing process as shown for PE-single cryst.ls [28-30, 76]. In this condition the strength of the fibers may decrease because sliding parallel to the molecular axis and in the fold interfaces can take place. Nevertheless it must be noted that even in such a condition the craze matter can sustain tensile stresses which are only slightly less than the true fracture stress of a bulk, tensile specimen at room temperature. This was demon-

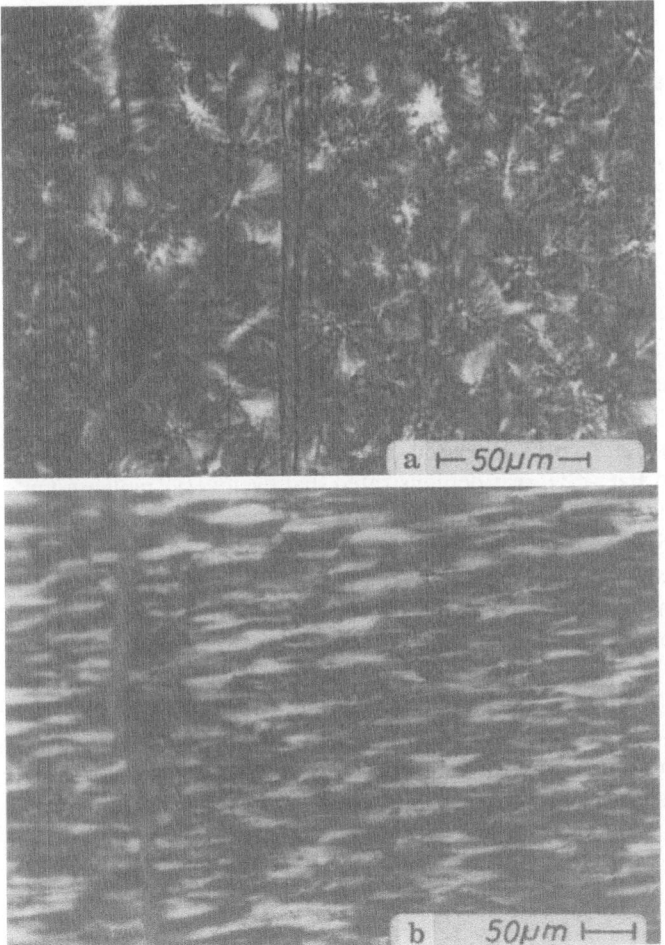

Fig. 13a and b. Polarized light micrograph of a precrazed PP-thin section **a**, which was subsequently stretched under the microscope at room temperature **b**

strated in a simple experiment with thin PP-films (about 20 μm in thickness) which were cut from a bulk specimen[86]. Before cutting, the tensile bar had been loaded at —40 °C until crazes had formed perpendicular to the stress direction. Thus, the microtomed films contained several craze-segments which partly spread across the whole width (cf. Fig. 13a). Stretching such a film at room temperature under the transmission light microscope gave evidence for the sequence of the following events:

1. Opening of the individual crazes, combined with elastic/plastic straining of the craze matter.
2. Deformation of the spherulites in the material between the crazes into a highly stretched, ellipsoidal shape (Fig. 13b; in a tensile test of a bulk PP specimen this internal deformation condition is found at nominal plastic strains of more than

200 %, the corresponding true stress, as measured in the necked region of such a specimen, is in a range of between 60 and 120 MPa).

3. Subsequent fracture along one of the craze segments which completely crossed the film.

Assuming that 60 % of the craze volume consists of voids, the true stress in the craze fibrils at break can be calculated to be 150 to 300 MPa, which is in the same order of magnitude as the true fracture stress of a bulk PP-sample after necking and strain-hardening [10].

Studies on molecular weight dependence of crazing in amorphous PS [87] have shown that in the case of high molecular weight polymer, the crazes were very numerous, fine in texture, long and very straigth. When the molecular weight was reduced, comparatively few crazes were formed; they were coarse in texture, somewhat shorter than the former, and could be jagged. At very low molecular weight the tendency for craze formation was greatly decreased, very few if any crazes formed. In a paper by Wellinghoff and Baer [88] it was demonstrated that even very small amounts of high molecular weight polymer can significantly modify the high strain-low temperature plastic flow properties of low molecular weight PS. The mixture gave rise to a highly fibrillated craze containing a few interfibrillar cross-links. The basic polymer, on the other hand, showed a tendency to cleave by crack growth; this resulted in the sparsity of the fibrillar morphology. Studies of the fracture surfaces of amorphous PETP by Foot and Ward [89] also suggested a change in failure from crazing in the high- and medium weight molecular-weight material to failure from inherent flaws in the low molecular weight material. In crystalline PETP failure was observed to arise from inherent flaws in very instance, though the authors stated that this process might be governed by craze growth from an internal flaw rather than from a surface flaw.

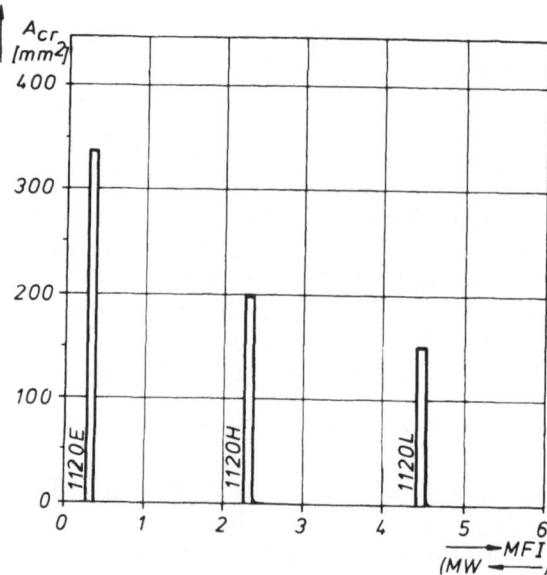

Fig. 14. Influence of MW on area deformed by crazing (A_{cr}): PP 1120 L = low MW; PP 1120 H = high MW; PP 1120 E = extremely high MW

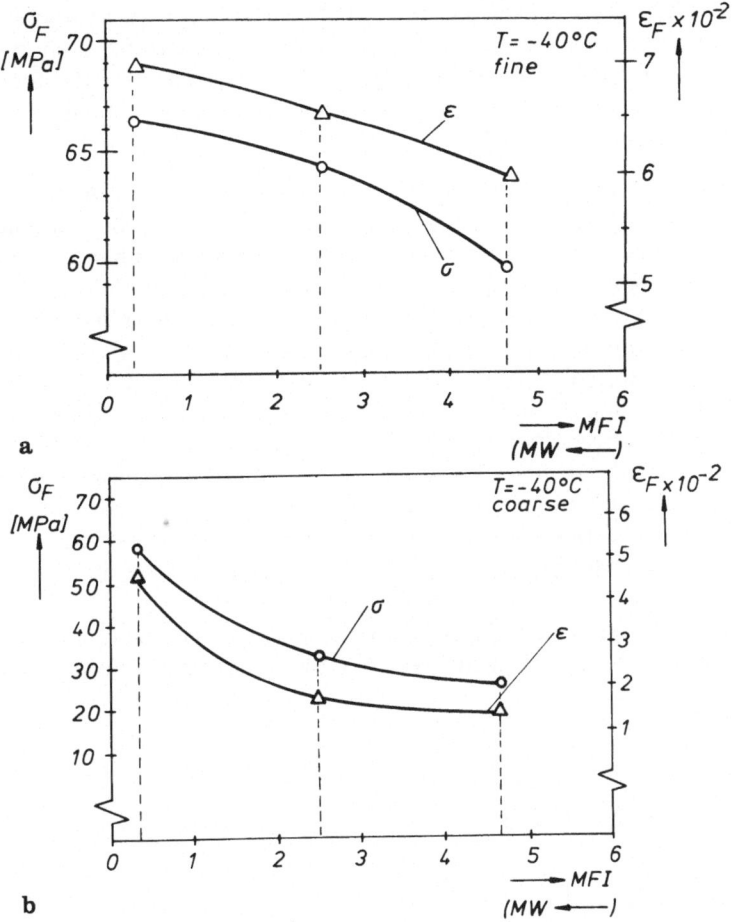

Fig. 15a and b. Diagrams demonstrating the influence of MW on fracture stress, σ_F, and fracture strain, ε_F, in **a** fine and **b** coarse spherulitic PP 1120

An interesting finding was presented by Fellers and Kee [87], namely that the applied stress at craze initiation, often called the critical crazing stress, was independent of molecular weight. From these results, Kausch [12] concluded that craze initiation is an event which depends primarily on the interaction between chain segments. On the other hand, the mean values of craze breaking stress fell with decreasing molecular weight, as well for PS [87] as for amorphous and crystalline PETP [89]. Parameters such as stress at break, strain at break and strain hardening determine plastic deformation, which, therefore, depends on molecular weight.

For the case of crystalline PP, Friedrich and Karsch [90] have observed similar results. The sum of crazed area, A_{cr}, (as calculated from the number and length of all crazes found on both sides of the gage length of broken tensile bars, and under the assumption of a quarter circular area of each craze) increased with increasing molecular weight, respectively decreasing melting flow index MFI (Fig. 14). Assuming a constant craze thickness, it must be concluded that the volume deformed by crazing increases with molecular weight. In fact, an obvious improvement of tensile fracture strain, as well as of fracture stress, with decreasing MFI was measured for both fine and coarse spherulitic PP structures at −40 °C (Fig. 15). The effects can be understood by analysing how the different stages of craze formation and craze growth are influenced by the molecular weight (cf. Fig. 12):

   I. As molecular weight increases the degree of crystallinity in the material decreases. Thus the amount of amorphous interlayers in which craze nucleation can take place first is increased. This leads to a number of crazes which enhances with sample molecular weight. This is especially true for the fine spherulitic structures which have a smaller degree of crystallinity due to the quenching treatment; the effect is less pronounced for the isothermally crystallized, coarse spherulitic morphologies. The molecular weight has very little influence on the elastic straining behavior (region A) and on the stress or strain at which crazes are initiated (point B), as already pointed out [12] for crazing in amorphous PMMA.

  II. After craze initiation, the stronger fibrils are formed and which can grow with increasing stress and time, to greater length the higher the molecular weight; consequently, region C is the more extended the higher the molecular weight.
     This fact, in turn, provides together with (I) the observed increase in fracture strain.

 III. The forces transmitted by region C grow with molecular weight due to a higher amount of topological entanglements between the crystalline blocks in the

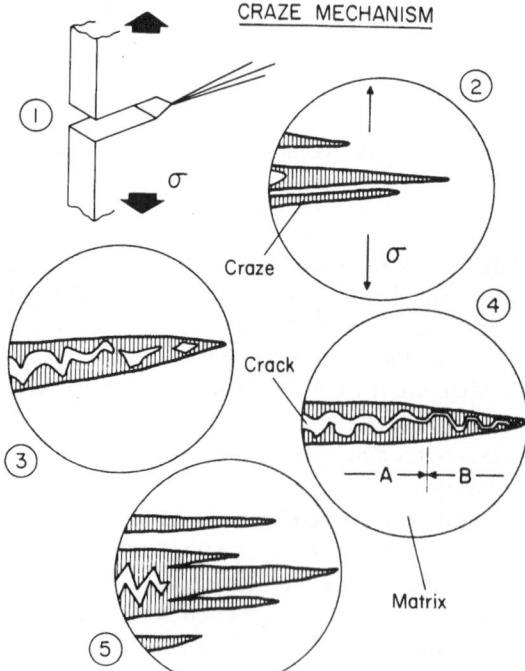

CRAZE MECHANISM

Fig. 16. Schematic of crack development in crazes: 1. stress direction and craze orientation in a CT-specimen, 2. Growth of individual crazes, 3. Nucleation of voids and their coalescence for crack formation (A), 4. Crack through craze acceleration (B), 5. Lateral coalescence of crazes with tendency to quasi-homogeneous deformation and subcritical crack progress

craze fibrils. This fact is reflected in the observed increase in stress at break with increasing molecular weight. In this condition the relevant stress is not the applied stress, but the true fracture stress in the craze fibrils [15].

Region D in the model of Figure 12 represents that part where a crack has nucleated by brakdown of the fibril structure within the craze. The fibrils fail either by continued chain reptation, including slippage and disentanglement of molecular coils, or through scission of entangled chains, frequently in the central section of the fibrils (Fig. 16). In this stage, the surface of the resulting crack shows a large number of cold stretching zones, forming a typical dimple fracture pattern [64-69]. When the crack approaches a critical size, the crack growth rate accelerates, until finally, so called "brittle, unstable fracture" of the bulk specimen takes place. During this crack acceleration, craze breakdown occurs primarily by fibril-matrix separation along the interface between crazed and uncrazed material at the upper or lower craze edge. The process results in a patchwork structure of residual craze matter on the final fracture surface. The features were reported very frequently for amorphous PS [70, 91, 92]. But there is some evidence that these fracture surface appearances occur in crystalline polymers during the transition from initial crack extension to fast fracture as well, e.g. in HDPE [68]. Also, the "island" structures observed in the second part of the popped-in crack in PA-6 specimens [71] must be very closely related to this phenomenon. Figure 17

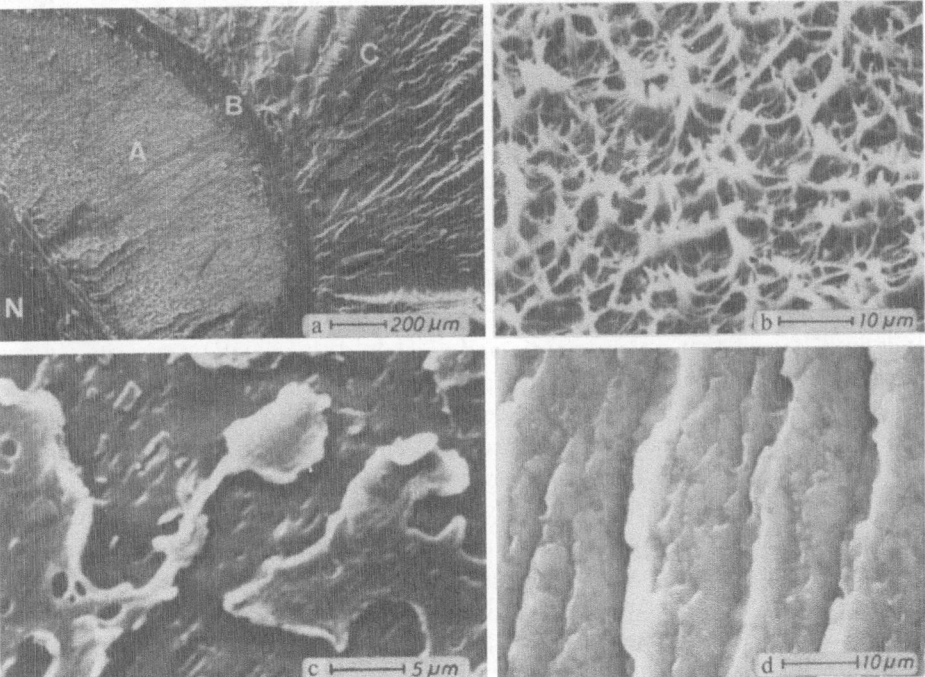

Fig. 17a—d. Details of the fracture surface on a CT-specimen of crystalline PETP: **a** general view of notch (N), subcritical crack growth through crazes (A), transitional region (B), and brittle fracture part (C); **b** dimple pattern in the central part of the broken craze (region A); **c** patchwork structure in the transition zone to crack instability (region B); **d** brittle fracture part (region C)

documents the occurance of all the three different phenomena — dimple pattern by slow crack growth, patchwork structure by crack-through-craze acceleration and brittle fracture after crack instability — on the fracture surface of a compact tension specimen of crystalline PETP [93]. Concerning tensile tests with different, semicrystalline PP specimens, the thickness of the corresponding patches on the fracture surfaces increases with molecular weight (Fig. 18). This, in turn, can be taken as indirect evidence for an increase in craze thickness with molecular weight, and strengthens the earlier conclusion that the specimen volume deformed by crazing is greater, the higher the molecular weight.

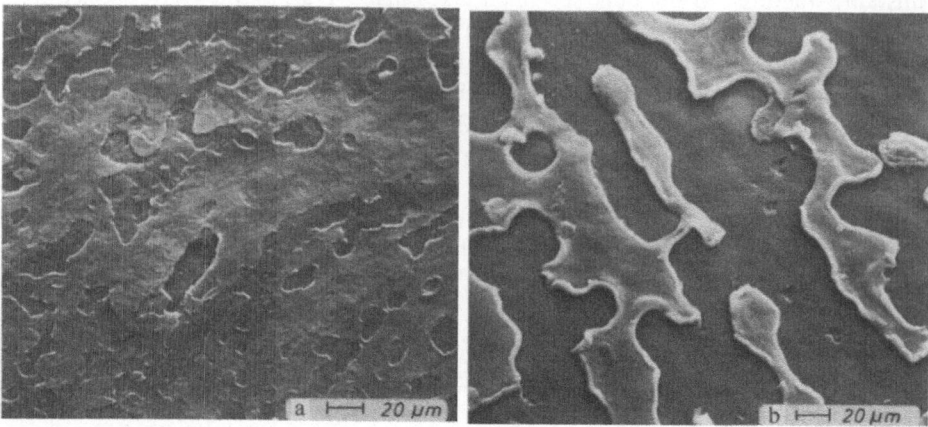

Fig. 18a and b. Patchwork pattern on the craze-fracture surface of PP with **a** ower and **b** higher MW

2.4 Craze-like Features at Room Temperature and above

2.4.1 Notch Induced Crazing

It is well known that notches, cuts, and scratches can greatly reduce the impact strength, the tensile strength, and the fatigue life of polymeric materials. Generally, notches are more detrimental to ductile polymers than to brittle ones as far as the energy to fracture is concerned. The results recently presented by Takano and Nielsen [94] indicate that expecially some very tough, semi-crystalline engineering plastics have a high notch sensitivity and may become macroscopically brittle under the influence of a sharp notch or cut. On a microscopic level, however, the polymers still demonstrate their high ductility, though this occurs in a small volume. Kausch [85] has mentioned that a macroscopically brittle fracture induced by an existing crack is preceded by the formation and/or growing of localized, stable defects. In the following, it is discussed how such defect zones in front of cracks look like and how their structure is influenced by the morphology of a crystalline polymer.

Figure 19 gives an example for the size of the plastic zone in a SEN-specimen of HDPE loaded at room temperature up to a point just before fracture occured. The zone, which is composed of a dense bundle of individual crazes, decreases in size

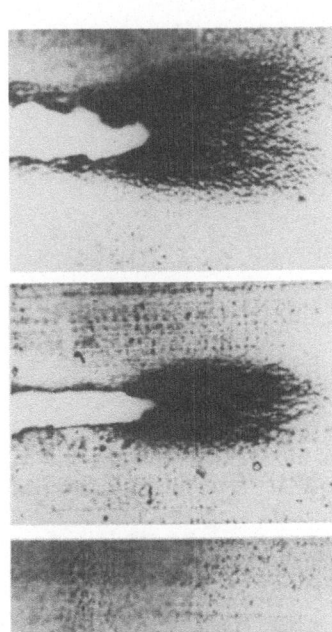

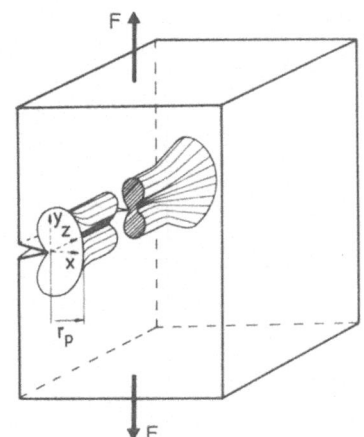

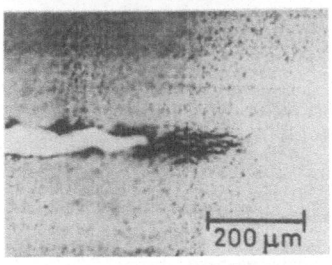

Fig. 19a. Variation of plastic zone size in a 5 mm thick SEN-specimen of HDPE: a1:0.5 mm below the surface, a2:1 mm below the surface, a3:2 mm below the surface; **b** Dog bone model of plastic zone radius, r_p, vs. SEN-specimen thickness (F = load)

from the surface into the interior of the plaque. Its shape is typical for specimens deformed under plain stress conditions. A calculation of the plastic zone radius, r_p, from equations of fracture mechanics, i.e.

$$r_p = \frac{K_{Ic}^2}{2\pi \cdot \sigma_y^2} \qquad \text{(plane stress condition, at the surface)} \qquad (1)$$

and

$$r_p = \frac{K_{Ic}^2 (1 - 2v)^2}{2\pi \cdot \sigma_y^2} \qquad \text{(plane strain condition, in the interior)} \qquad (2)$$

and by using measured data of K_{Ic}, v and σ_y, respectively (HDPE: $K_{Ic} = 1.8 \, \text{MPa} \cdot \text{m}^{1/2}$; $\sigma_y = 25$ MPa; $v = 0.4$) gives roughly the same result of r_p vs. B (specimen thickness) as is achieved by measuring r_p in variously deep, thin sections (Fig. 20).

Looking at the structure of these crack tip plastic zones in more detail, it is found that the individual crazes are less straight compared to the low temperature crazes (Fig. 21). This indicates a more pronounced influence of the crystalline microstructure on craze formation. Figure 21a and b demonstrate for fine spherulitic, highly isotactic PP the interaction between the crazes and the microstructural features. Most of the

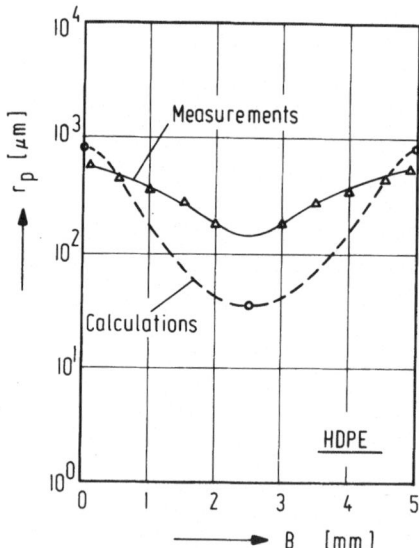

Fig. 20. Comparison of the plastic zone data obtained from Figure 19a with r_p-values calculated from mechanical tests with HDPE

crazes, which have, already, partially turned into microcracks, run along spherulite boundaries as well as along radial transspherulitic paths lying nearly perpendicular to the applied load. This is not surprising because both areas have to be considered as the weakest elements of a spherulitic structure [12, 95, 96]. The boundaries between adjacent spherulites are crossed only by short end sections of lamellae and eventually by some interspherulitic links. A little stronger are the amourphous layers between radially oriented lamellae which are fortified by a small number of interlamellar tie molecules [97, 98]. Nevertheless, at least in the fine spherulitic case, crazes which have formed in one of these regions have still enough strength to cause further crazing in their neighborhood before they break down, and thus a dense craze bundle is formed in front of the crack which can absorb much energy before final fracture occurs [10].

The internal structure of individual crazes (Figure 21c and d) can be made visible by elastically straining of the material in front of the notch using a wedge opening technique. If the craze edges are opened up to a distance of about 20 μm, the craze matter consists of 60 to 70% of voids between oriented strands having 200 to 650 nm in diameter. That means, they are about 10 times thicker than the corresponding fibril diameters of low temperature crazes formed in crystalline or amorphous polymers.

Additional effects on the structure of the plastic zone at room temperature arise from the size and distribution of spherulites in the tested polymers [99]. An extreme case is shown in Figure 22. A crack which is forced to propagate into a coarse spherulitic morphology of highly isotactic PP essentially follows the boundary zones of the original spherulites. The local direction, therefore, deviates from the normal to the applied load by up to more than 80°. The crack evidently propagates the fastest through single interspherulitic crazes (Fig. 22b) or shear zones (Fig. 22c). The fibrillar structure of these interspherulitic crazes differs in some ways from those described for the fine spherulitic morphology. Their density is only 10 to 30% of the matrix density and the diameter of the oriented strands varies between 200 and 1500 nm,

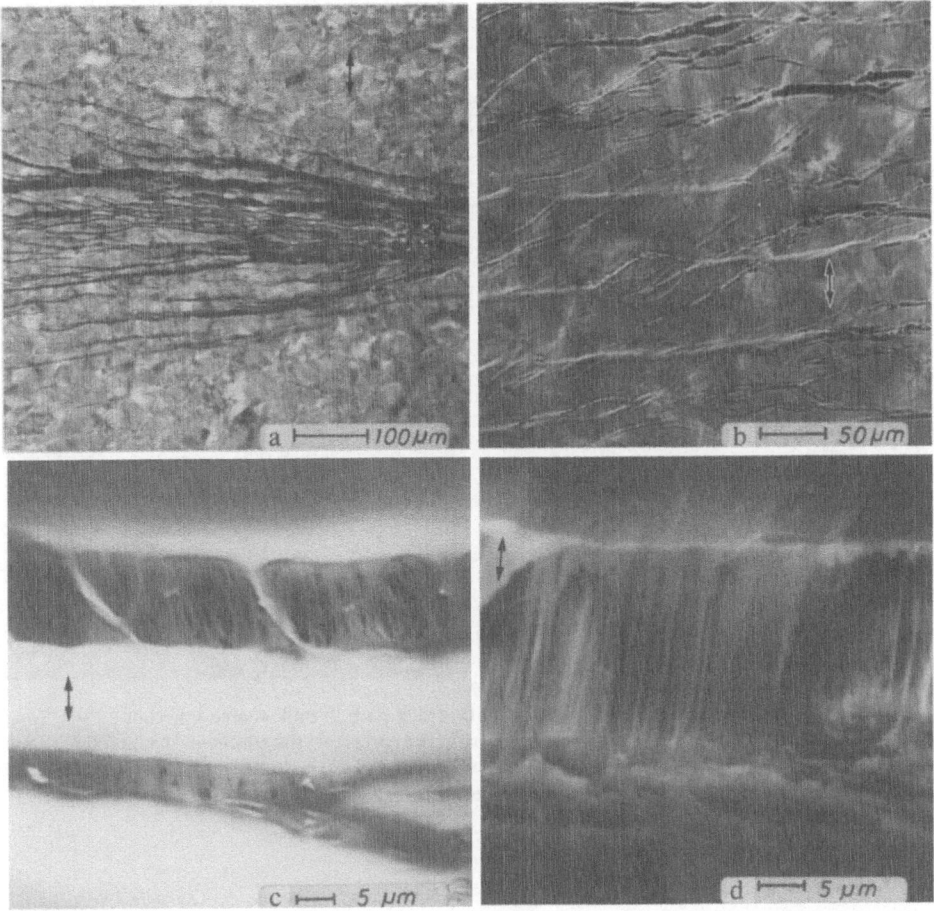

Fig. 21 a—d. Crazes in front of a crack in bulk, fine-spherulitic PP 1120 loaded at room temperature: **a** general view of a thin section in polarized transillumination; **b** higher magnification of (a) showing the preferred craze paths; **c** SEM-micrograph of the fibrillar structure in the craze bundle as seen on the surface of the bulk specimen; **d** stretched craze matter in the interior after microtoming a 2 mm thick layer from the surface of a bulk PP specimen

depending on their position (near spherulite triplepoints they are thicker than in the center of a boundary plane [100]; cf. the interspherulitic fracture surfaces in Fig. 23).

The weakness of the boundaries derives from the very slow crystallization process necessary to obtain such a morphology. Additives, impurities, low molecular weight polymer and non-crystallizable chains are partly pushed away from the spherulite growth fronts [101]. This process finally provides in the fully crystallized samples impurity-rich spherulite boundaries which are much weaker i.e. more brittle [102] than the highly crystalline, and thus harder, interior zones of the spherulites. The higher degree of crystallinity of the coarse spherulites also leads to a larger contraction of their volume and to an increased tendency for void formation along the boundaries. As a consequence, the path of least resistance follows the grain boundaries; the lowest values of fracture toughness are measured in coarsly spherulitic PP [101−103].

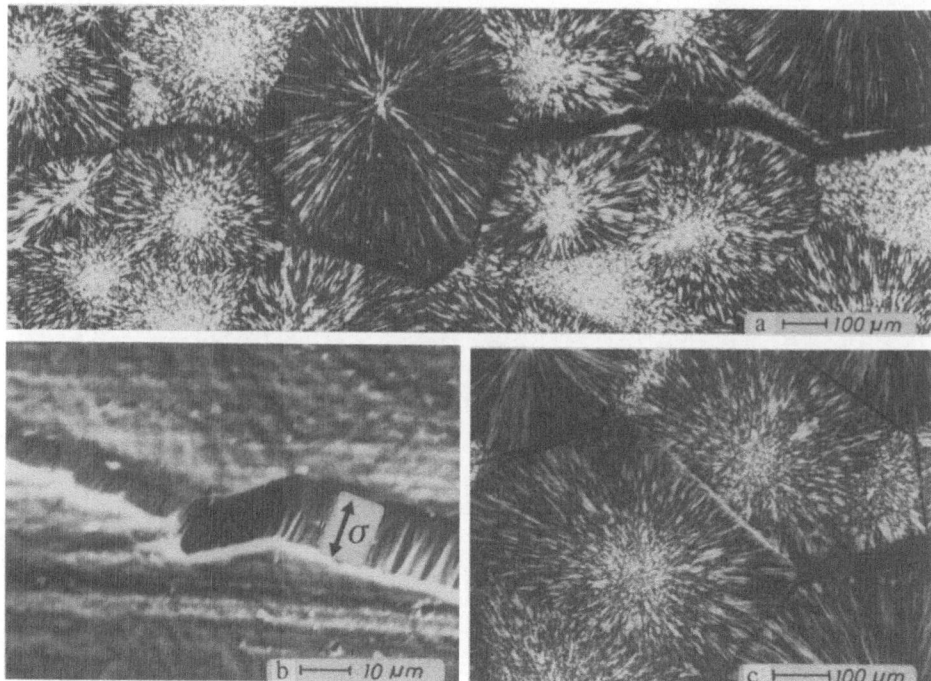

Fig. 22a—c. Microtome section of an intersperulitic crack path in bulk coarse spherulitic PP 1120 **a**. Figure **b** indicates as an SEM-micrograph taken from the surface of the specimen the interspherulitic craze formation prior to the cracking process. **c** shows a site of shear along a spherulite boundary oriented under an angle of about 60° to the horizontal crack direction

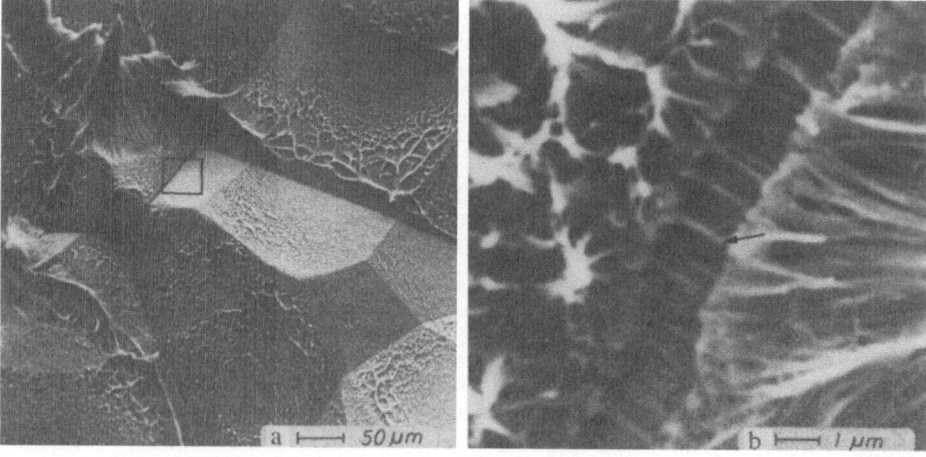

Fig. 23a. SEM-micrograph of an interspherulitic fracture surface of coarse spherulitic PP 1120 at room temperature; **b** Stretched polymer substance in (arrow, interspherulitic craze) and on the boundary surfaces

The observations can be confirmed by scanning electron micrographs of the fracture surfaces. The fine spherulitic material shows a smooth, stress whitened zone, already discussed with Figure 17 for crystalline PETP. Its formation appears to be associated with a coalescence of densely packed crazes near the fracture plane finally leading to a quasihomogeneous deformation in this region. In the coarse spherulitic material, on the other hand, plastic deformation only occurs in the boundary zones, whereas the spherulites themselves stay nearly undeformed. This results in a dimple-like pattern of ruptured interspherulitic craze matter on the polyhedron-shaped spherulites. Such fracture surface structure was also found by other investigators [104, 105] who named it "coronet" type fracture (Fig. 23).

An improvement of the fracture behavior can basically be reached by an increase in the molecular weight of the polymer. Keith and Padden [97, 98] earlier reported on PE that the interfibrillar as well as the interspherulitic link density is highly improved with increasing molecular weight. Thus crack propagation is hindered in both regions, resulting in better fracture toughness values of both the fine spherulitic and, especially, the coarse spherulitic specimens (Fig. 24). The effect mentioned last can be seen in Figure 25. It shows the notable plastic deformation of the spherulite boundary substance of the high molecular weight PP as opposed to that of the low molecular weight material. Other improvements of the fracture toughness, for example by an increase in the atactic content of PP, are discussed in earlier papers by Friedrich et al. [10, 33, 106].

Also, it is well established that incorporation of a rubber phase, particularly in glassy polymers, will greatly increase the toughness (see Chapter 6). The principle toughening mechanisms are believed to involve shear yielding and crazing, interaction between shear yielding and crazing, and the blunting, diversion and multiplication of the growing crack itself [82]. Recent investigations have shown that some of these

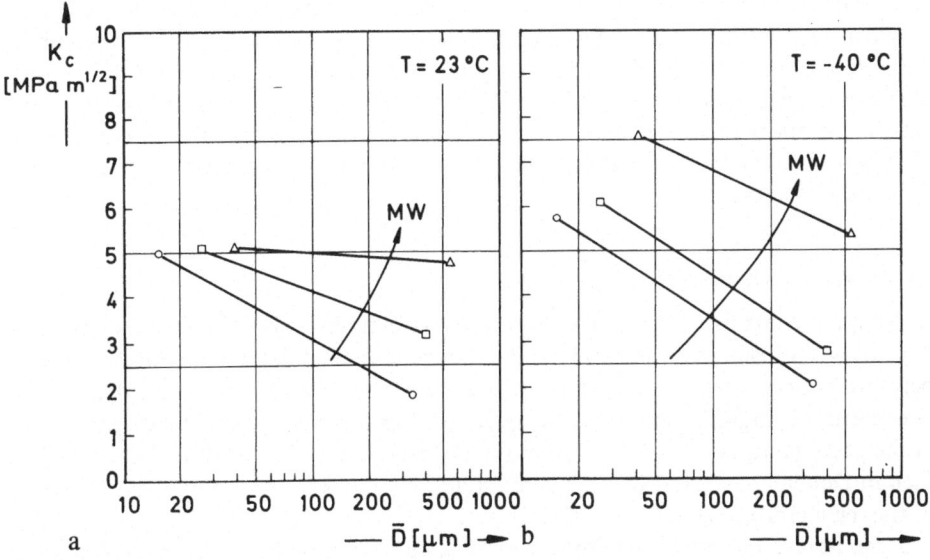

Fig. 24a and b. Fracture toughness K_c as function of both, molecular weight MW and morphology (spherulite diameter D) of PP 1120 at room temperature **a** and T = —40 °C **b**

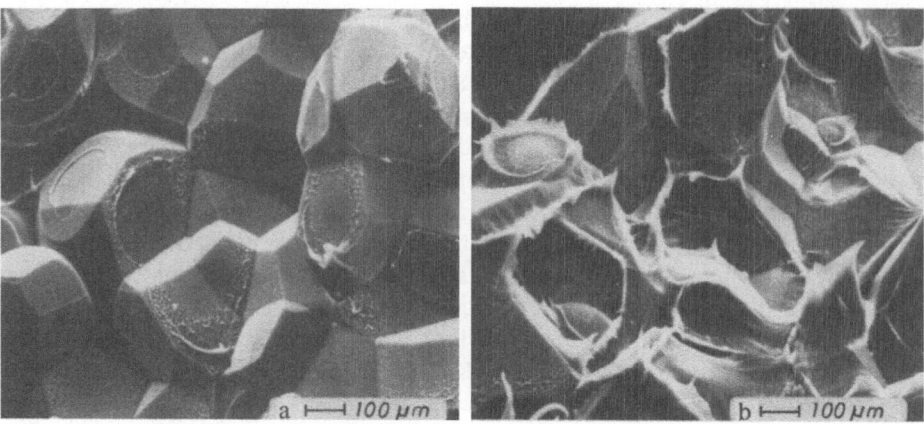

Fig. 25a and b. Interspherulitic fracture surfaces of coarse spherulitic PP 1120 with **a** lower and **b** higher molecular weight

mechanisms can also be effective in semi-crystalline PETP [107]. CT-specimens of the basic polymer containing only a few toughening particles break in a relatively brittle manner at room temperature after formation of some individual crazes in front of a notch (cf. the corresponding fracture surface in Fig. 17). A material highly modified with rubber-like particles, on the other hand, yields a macroscopially homogeneous deformation in the initial crack base providing a strong increase in fracture toughness. TEM-studies of local deformation mechanisms involved in the plastic zone region reveal that the rubber-like particles serve as stress concentraters to initiate a multiplicity of small crazes, mainly equatorial around the particles (Fig. 26a). The structure in the tapered section near the craze tip is characterized by a polymer network including a dense accumulation of microvoids with 10 to 50 nm in dimension (Fig. 26b). In the extended part of the craze, longitudinal voids with fibrils of about 15 nm in diameter are observed, bridging a craze thickness of 150 to 200 nm. If the rubber particles are sufficiently close together, the interaction of their stress fields increases the craze density, and hence increases the material's ultimate fracture energy.

2.4.2 Craze II-Obervations

In Chapter 2 intrinsic crazing has been discussed extensively. The numerous crazes appearing for instance in amorphous PC drawn to high stresses and strains in a temperature region close to the glass transition temperature, T_g, were called crazes II in order to distinguish them from the ordinary type of craze (craze I) treated in most of the craze literature [108-109]. The results showed that under the drawing conditions used in Dettenmaier's and Kausch's investigations stress whitening may be attributed to this new craze phenomenon [108-110].

Some important similarities to crazes II have also been described by Pakula and Fischer [111,112] who studied instability phenomena associated with self-oscillations during cold-drawing of partially crystalline PETP, PP and PA. The authors observed

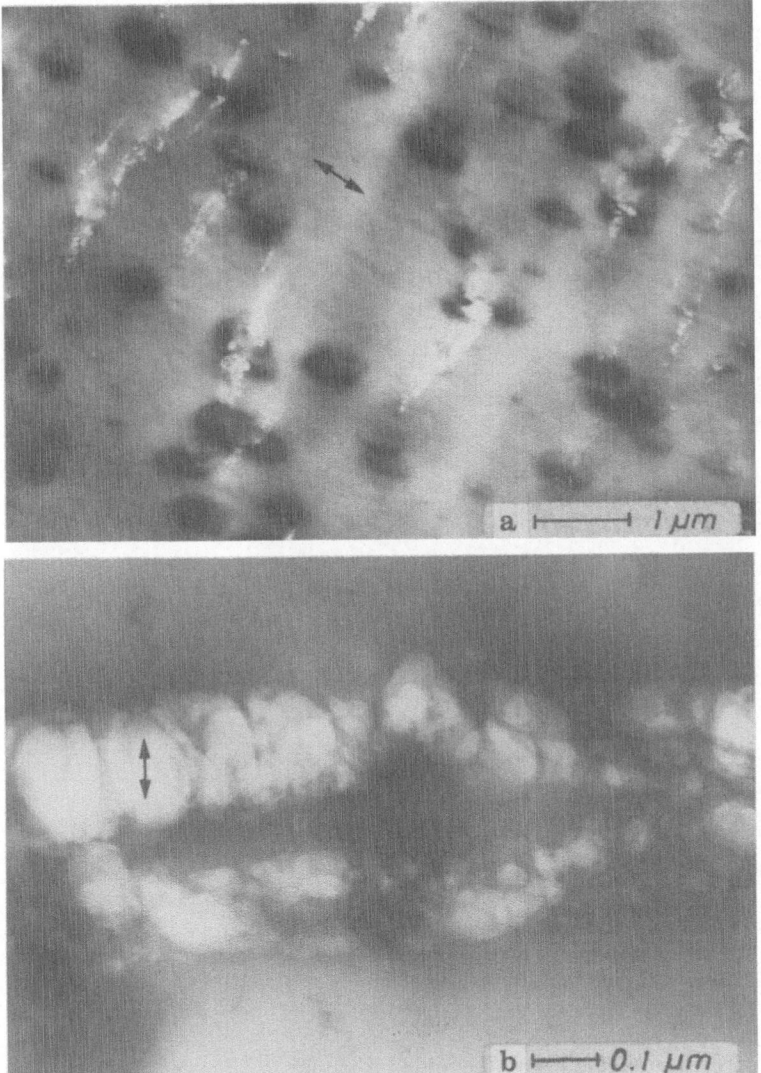

Fig. 26a and b. TEM-micrographs of ultrathin sections cut parallel to the stress-whitened surface of particle modified PETP. Dark portions in **a** represent the rubber-like particles which appear to encourage craze initiation in the matrix polymer. **b** represents the voidy and fibrillar microstructure in one of these crazes (arrow = stress direction)

at high stresses craze-like structures which they called macrocrazes. Both phenomena, crazes II in PC as well as the macrocrazes in the crystalline polymers are basically intrinsic phenomena, i.e. the local compliance fluctuations are sufficient to serve as craze nuclei at high stresses. On this basis the large number of these craze-like features, as compared to crazes I, can easily be understood [108].

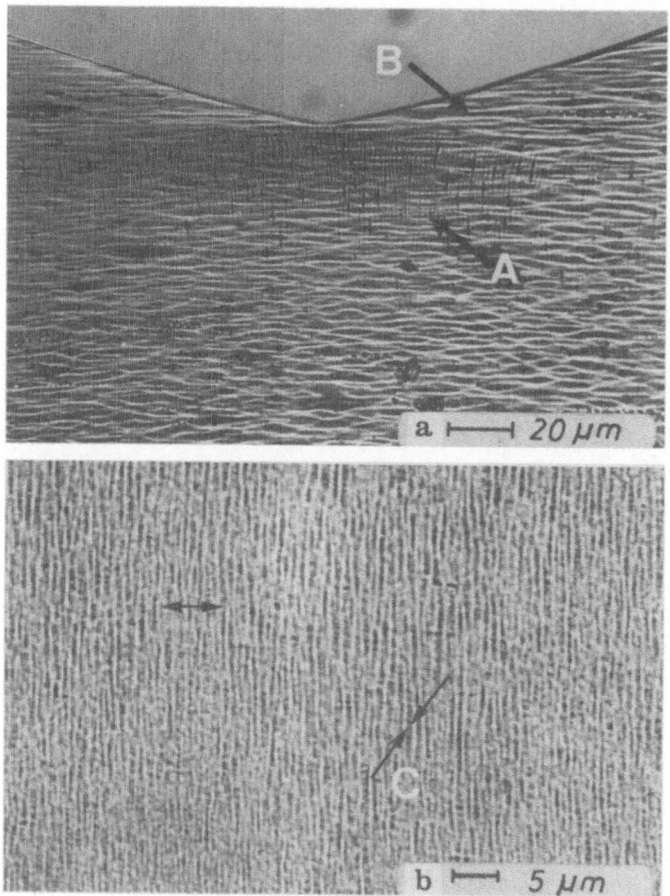

Fig. 27a. Craze-II-like features in the highly extended region of a precracked PP foil (double arrow = stress direction); **b** Higher magnification of region A using bright field transmitted light (Courtesy G. Kress, Ludwigshafen)

Craze II-like features have also been observed in Bochum [113]. Figure 27a (arrow A) demonstrates the formation of a dense bundle in highly stretched isotactic PP. The occurance of this bundle is produced in the following sequence:

a) Room temperature deformation of the spherulites in front of a knife cut perpendicular to the applied load. During this process the spherulitic structure is transformed into a fibrillar structure combined with a blunting of the initial crack tip (arrow B).

b) Subsequent initiation of a bundle of small "deformation lines" running perpendicular to the acting load in the region of highest stress concentrations (arrow A). It is followed by a separation process leading to slow crack growth of the blunted crack tip.

Higher magnification of this region under stress gives evidence for the craze-like character of these "deformation lines". The gaps are bridged by individual fibrils having a diameter between 250 and 350 nm (Fig. 27b, arrow C). This is in the same order of magnitude as the fibrils measured for crazes II in PC.

A third observation obtained from Figure 27 is the disappearance of these fissures as soon as the film is locally separated (Fig. 27a, arrow B). It must be assumed, therefore, that the "deformation lines" close upon stress relieve just as in hard-elastic materials. A characteristic feature of some of those highly crystalline, highly oriented polymers is the fact that they can be extended by 50–100%, this extension being practically completely and immediately reversible [114].

On the other hand, a number of publications which describe the preparation, morphology and properties of hard elastic materials, for example [115, 116], have also shown that stable microporous materials can be prepared from these elastic fibers or films by a variety of process conditions, by stabilizing the extended hard elastic structure. Transmission electron micrographs show that in such crazes, the material consists of ridges of very uniform lamellar regions (Fig. 28, [117]), interconnected by fibrillar links [114]. Such craze-like structures were also shown for PA-fibers [40] and for hot-drawn PP filaments after plastic deformation at room temperature [118–120]. In the latter case, plasma or chromic acid etching techniques were used to make the craze structure visible for the SEM.

Quite recently, Baer et al. [121–123], have stated that the interlamellar fibril-bridged gaps observed in "hard elastic" crystalline polymers are structurally similar to crazes occuring in glassy polymers. In other words the borderline between the structure of crazes in hard elastic and in glassy amorphous polymers is not sharp.

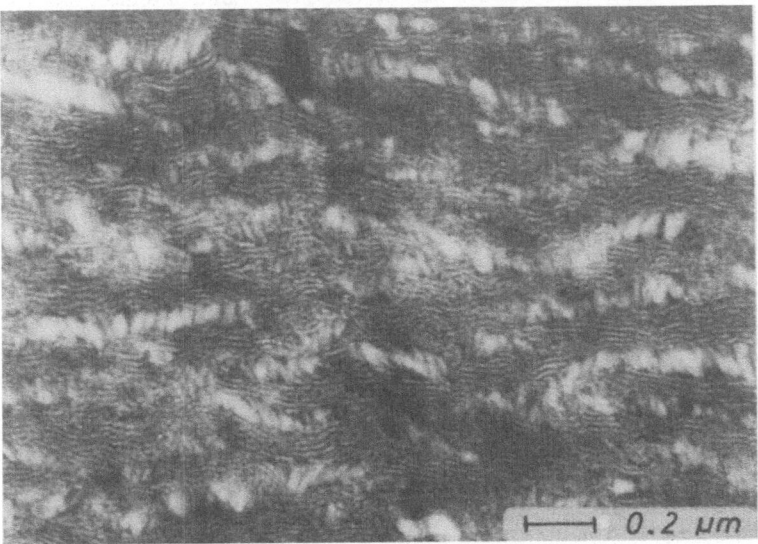

Fig. 28. TEM micrograph of craze deformation in a "hard elastic" PP-foil (Courtesy G. Kanig, Ludwigshafen). At some of the craze tips it can be recognized that some stacks of lamellae are destroyed by locally high tensile stress subsequently being transformed into fibrils

3 Shear Bands and Fracture in Crystalline Polymers

3.1 Shear Band Phenomena

Unlike inorganic glasses most glassy polymers undergo appreciable plastic deformation at room temperature and moderate strain rates before fracture occurs. In most cases this deformation develops inhomogeneously, i.e. only local regions in the material are plastically deformed. It has been pointed out in the introduction that two deformation modes are possible: shear yielding and normal stress yielding (crazing) [1].

As opposed to the literature on crazing comparatively few comprehensive studies exist which consider the formation of shear bands. Crazing and shear-banding are not mutually exclusive. Thus it is well known that glassy amorphous polymers deform by localized shear when crazing is suppressed. In that case intense shear bands and/or diffuse shear zones appear [124]. Some investigators have reported about the exact loading conditions leading to such deformation zones [125, 126], on the growth of shear bands [127], their dependence on ambient temperature [128, 129] and loading rate [130, 131], and on the morphology and molecular structure of the shear zones [61]. Wu and Li [132, 133] showed that two slip processes during the compression of bulk atactic PS are characteristic. Individual coarse shear bands appear in high speed deformation which are also observed at low temperatures. Fine slip bands arranged in a broad diffuse shear zone are found in low speed deformation and/or at higher temperatures. The authors mentioned that brittle fracture initiates in the coarse bands once they had propagated across the specimen, while the diffuse shear zone causes ductile fracture at large strains. This result was confirmed by Friedrich and Schäfer [134, 135] who mainly studied the crack initiation and the fracture mechanisms involved with shear banding. The importance of the shear band intersections for crack initiation was recently pointed out by Chau and Li [136], too.

As already mentioned there are only very few studies dealing with the formation of shear bands in semi-crystalline materials. Some recent papers on oriented [137, 138] and non-oriented semi-crystalline polymers [62, 139, 140] have indicated that under comparable conditions shear band formation can occur. Evidently the deformation process is influenced by an enhance of number of microstructural parameters, as for instance the molecular structure, the degree of orientation, the degree of crystallinity and the morphology. In the following part shear band formation under compression will be discussed in more detail.

3.2 Structure of Shear Zones under Various Conditions

In analogy to the well known observation with amorphous PS, two sets of coarse shear bands (named A and B) are also found in crystalline polymers when the deformation is performed at temperatures well below their glass transition temperature (Fig. 29a). The bands were seen to propagate under angles between 35 and 42° to the compressive axis, angles which are very similar to those observed in glassy polymers [127] or amorphous metals [141]. Their propagation paths follow spherulitic boundaries or cut through the crystalline arrangements inside a spherulite. This indicates that sliding can take place in various planes of individual crystal lamellae.

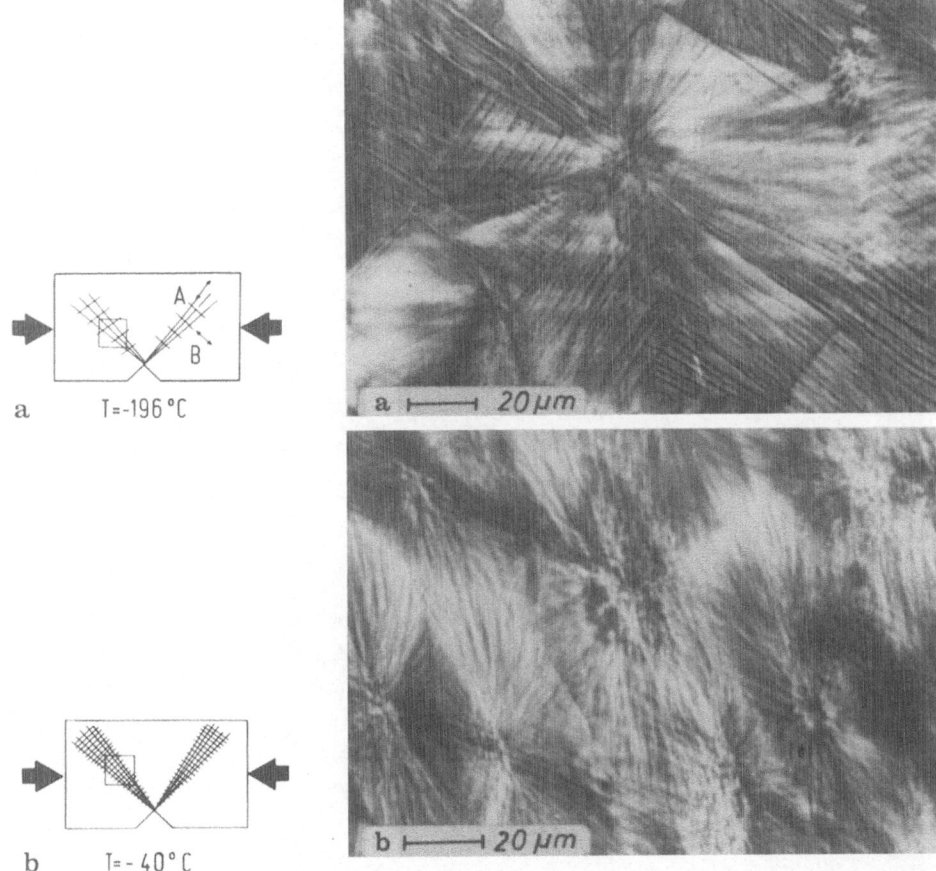

Fig. 29 a and b. Structure of the shear zone at various temperatures in PP: **a** T = −196 °C, two sets of discrete shear band A and B, **b** T = −40 °C, diffuse shear zone with quasi-homogeneous deformation of spherulites

The occurance of these processes has also been directly observed by Argon [142] in crystals of spherulitic HDPE.

At first the case where coarse shear bands are formed, will be considered . Higher magnifications obtained from thin sample sections under polarized light indicate that in crystalline polymers the shorter bands (B) are usually formed first. They are subsequently intersected by the longer shear bands (A) which have initiated from the specimen notch (Fig. 30a). Further types of intersection occur at spherulite boundaries (SB) and radiating fibrils (F) which are oriented more or less perpendicular to the shear band package (Fig. 30b). In addition it can be observed, especially in coarse spherulitic specimens under conditions which favor homogeneous shear deformation, that local shear is concentrated at spherulite boundaries running parallel to the applied shear stress (Fig. 30c).

The band thickness d ($\simeq$ 1 µm) can easily be measured and also the amount of displacement x at intersections of shear bands with each other or with certain morpho-

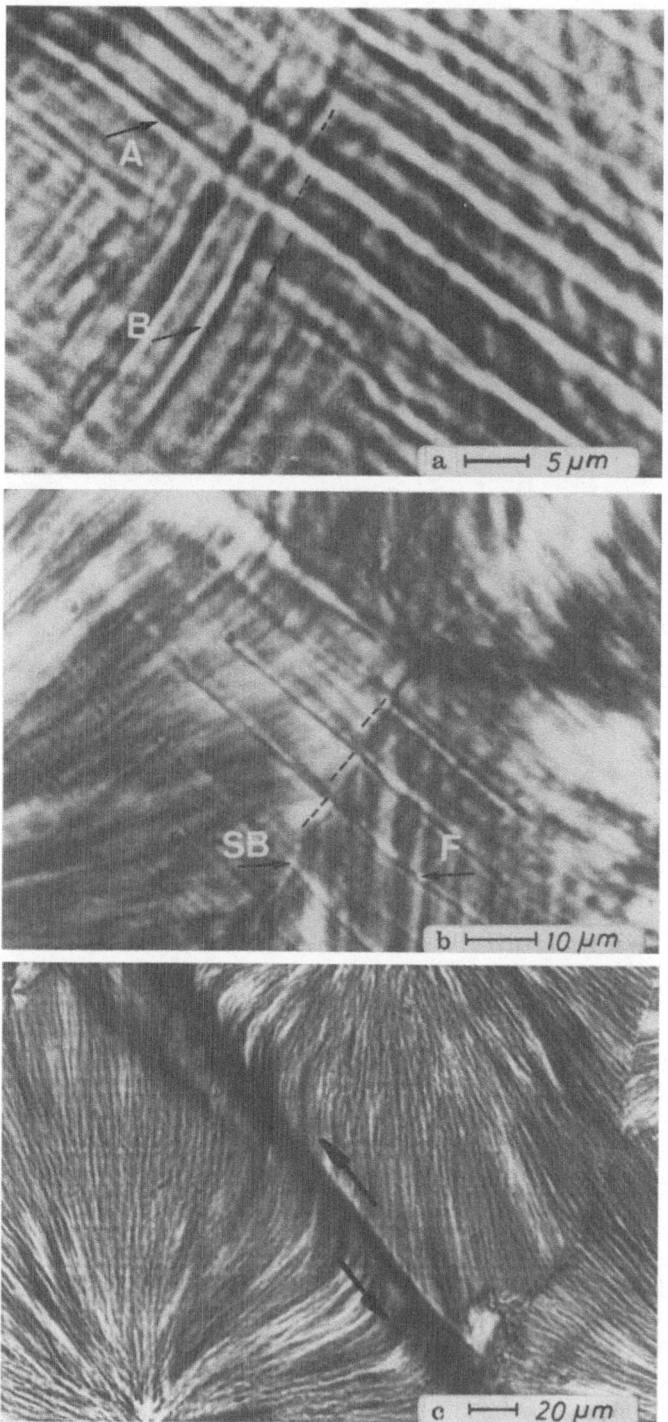

Fig. 30a—c. Interaction between shear bands of type A with bands of type B **a**, and with spherulite boundaries perpendicular **b** and parallel **c** to shear band package A in PP 1320

logical elements such as spherulite boundaries. Thus the degree of local plastic deformation inside a band, ε_i, can be calculated as $\varepsilon_i = x/d$ ($\varepsilon_i \approx 160\%$ in Fig. 30b). Its value is one or two magnitudes higher than the external strain at fracture, ε_F. The exact ratio depends on the testing temperature, similar to the results obtained with atactic PS [135]. The higher the temperature, the lower is the ratio $\varepsilon_i/\varepsilon_F$. The density of coarse shear bands, however, increases with increasing temperature.

Finally, when the testing temperature is near the glass transition temperature, a broad shear zone instead of discrete coarse shear bands develops at an angle of 45° to the direction of the external stress (Fig. 29b). Inside this zone the spherulites are homogeneously deformed in direction to the main shear stress. It is expected that

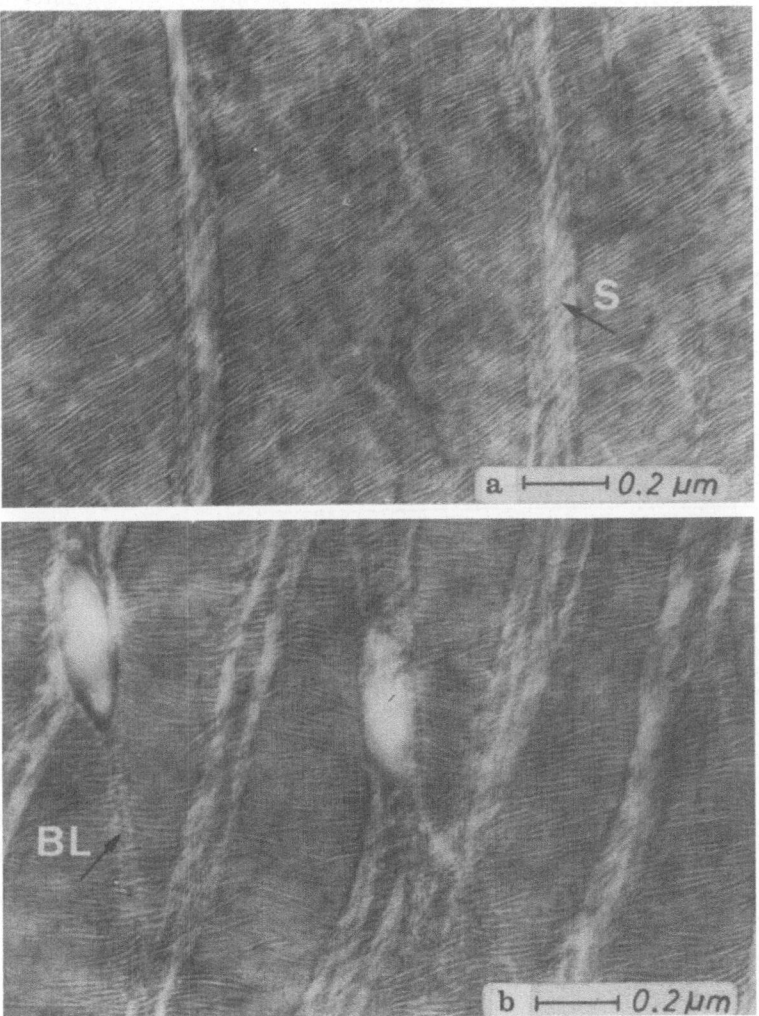

Fig. 31 a. Shear bands (S) and "bright lines" (BL) in the compressive region of a HDPE-bend specimen. The axis of the compressive stress is perpendicular to the direction of the lamellae; **b** Intersections of shear bands with "bright lines" acting as sites for crack initiation (Courtesy W. Rose, Erlangen)

this zone consists of many fine shear bands, as was observed by Wu and Li [132] in the diffuse shear zone of atactic PS.

As a matter of fact, Rose and Méurer [143-146] demonstrated on many excellent TEM-micrographs that the macroscopically homogeneous deformation of crystalline HDPE is very seldom a truly continuous deformation. Much more frequently the deformation is concentrated in particular specimen areas (discontinuous deformation zones). Typical examples, obtained from highly oriented regions of HDPE under compressive load, are discrete, fine shear bands of less than 0.1 μm in width (Fig. 31 a). From the lateral displacement of decomposed lamellae within the deformation zone it follows that the local plastic strain inside a shear band, ε_i, varies from 100 to 200 %. The shear bands point in a range of 35° to the applied stress axis. In addition, the authors observed another set of shear zones (so called "bright lines") running parellel to the compressive axis, but there is no explanation of their occurance. However, an interaction of these deformation zones with the original shear bands leads to points of intersections at which the polymeric material is highly stretched in two dimensions. Thus, at these sites voids and crack are preferably initiated either during the deformation process or, perhaps during the cutting of the ultra-thin sections for the TEM-investigations (Fig. 31 b).

The conditions under which a particular deformation mode (coarse shear banding or deformation in a diffuse shear zone) predominantes seem to depend mainly on the ambient temperature, T, as compared to the glass transition temperature, T_g, of the material. This hypothesis can be deduced from a diagram of shear modulus G vs. ratio T/T_g for the tested polymers (Fig. 32). The amorphous PS as well as the semi-crystalline polymers PP and PB-1 exhibit a tendency to formation of coarse shear bands when the ratio of T/T_g is distinctly smaller than 0.75. There exists a

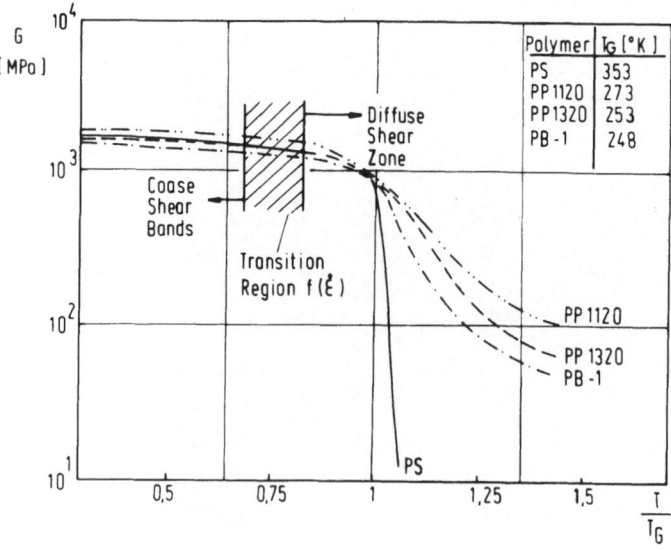

Fig. 32. Shear modulus, G, vs. ratio of undercooling, T/T_g (temperature in Kelvin), for various polymers. The modes of deformation which can predominate under compression are indicated

transition in the neighborhood of $T/T_g = 0.75$, where both phenomena, coarse shear bands and/or a homogeneous shear zone, can occur depending on strain rate $\dot{\varepsilon}$. But at elevated temperatures, i.e. near $T/T_g = 1$ or above, the formation of a diffuse shear zone becomes more and more probable under shear stress conditions.

3.3 Band Initiation Stress and Strain at Break

Figure 33 indicates the influence of the type of material on the band initiation stress σ_{BI} and on the ball thrust hardness H_K (DIN 56456) for various testing temperatures.

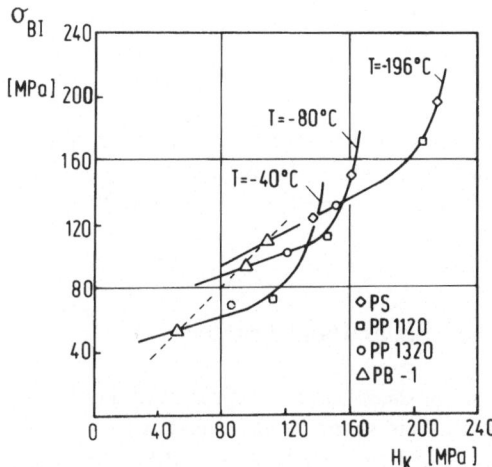

Fig. 33. Correlation between shear band initiation stress, σ_{BI}, and ball thrust hardness, H_K, for various polymers at different temperatures

For all materials (amorphous PS, semi-crystalline PP and PB-1 with fine spherulitic morphology) both properties increase with decreasing temperature. The dotted line in Figure 33 refers to PB-1. There exists an increase of σ_{BI} and H_K in the sequence PB-1 $\rightarrow$ PP 1320 $\rightarrow$ PP 1120 $\rightarrow$ PS. Both effects seem to be related to differences in the material, probably to the flexibility of molecular chains and/or whole morphological regions. Flexibility shall be understood as the capacity of individual molecular segments or crystalline blocks to undergo localized slipping and slip-tilting during plastic deformation. Evidently this property is more pronounced when the shear modulus G is low (PB-1 and PP 1320) as compared to the stiffer materials PP 1120 and PS. In the latter amorphous polymer slipping of molecular segments is strongly hindered by the benzene rings along the chains. In the other polymers the flexibility is mainly influenced by the degree of crystallinity. The higher amount of crystallinity in PP 1120 ($X_c = 60\%$, determined by density measurements) as compared to PP 1320 ($X_c = 43\%$) and PB-1 ($X_c = 40\%$) provides more stiffness and less ductility. This tendency becomes also clear in a comparison of σ_{BI} for fine and coarse spherulitic morphologies of the three polymers (Fig. 34). Due to the slower crystallization process the coarse spherulitic structures possess a higher degree of crystallinity and thus higher values of σ_{BI}. At the same time the fracture strain ε_F decreases with crystallinity.

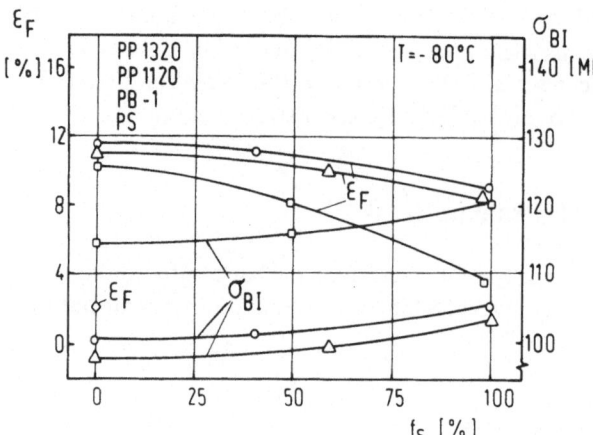

Fig. 34. Influence of morphology (f_s = content of coarse spherulites) on band initiation stress, σ_{BI}, and strain at fracture, ε_F

This seems to be due to the fact that, once a shear band has formed in a highly crystalline structure, it is easier to concentrate further deformation in this band than to form new ones in the highly stiff environment. Thus fewer bands are formed and fracture occurs at lower values of ε_F.

3.4 Mechanisms of Crack Initiation and Shear Band Fracture

The crack initiation by interaction of individual shear bands is represented schematically in Figure 35a. The stretched polymer fibrils in shear band formed at first will be further stretched when sheared by another slip band. At these sites fibril breakage occurs preferentially, and thus intersections of the shear bands act as nuclei for crack formation. Hull has pointed out the similarity between shear bands in glassy polymers and in metals in this respect [147]. After such a crack initiation further crack growth will continue along the second shear band (type A) due to enhanced shear stress concentration at the crack tips. An example for such a stage is given in Figure 35b showing crack initiation in a band of type A formed in a compressive specimen of highly atactic PP at −196 °C. Crack initiation can simultaneously occur in several bands parallel to each other. To interconnect these cracks in order to complete by rupture the specimen, craze-opening and cracking also takes place along shear bands of type B. This can lead to a catastrophic onset of fracture perpendicular to the initially activated fracture plane, and thus to a splintering of the specimen.

In addition to the described role of shear bands in crack initiation, the influence of spherulitic boundaries on crack propagation has to be considered, especially in coarse spherulitic, highly isotactic PP. Cracks which may have been formed at a particular point of a shear band according to the mechanism described above can leave their shear plane with further growth in order to move into an adjacent spherulite boundary (Fig. 36, Point I). This is especially probable when the boundaries are cut by the shear plane under a small angle. The selection of the particular crack path is determined by the partial fracture mechanical properties of the shear band and the spherulite boundary, respectively, as well as by the local geometrical conditions [103].

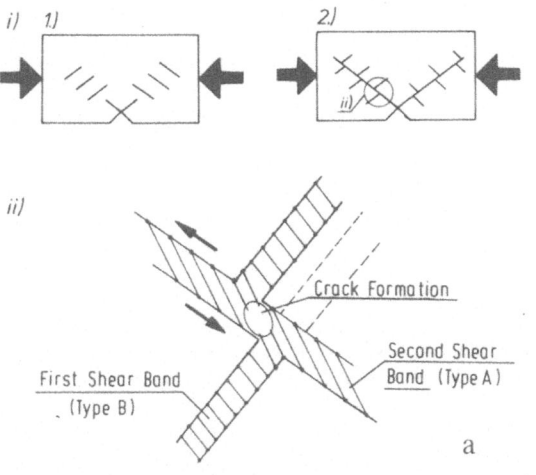

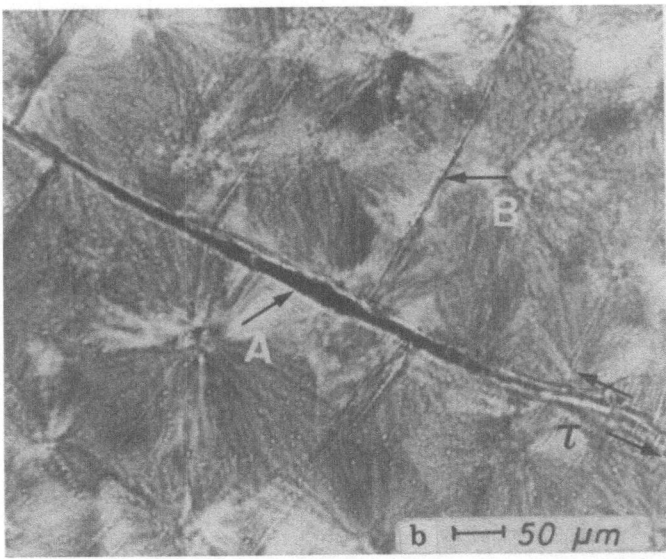

Fig. 35a. Mechanism of crack formation at intersections of shear bands: (i) sequence of shear band formation, (ii) chain scission at the intersections (schematic); **b** Shear crack formation (arrow) in a shear band of type A, leading to a 50 μm wide displacement of type B — shear bands on both sides of the crack

As the deviation of the actual crack path from the original plane becomes too large, further crack propagation starts to follow one of the parallel oriented shear bands (Fig. 36, Point II). Jumps in the shear plane can also be induced in this morphology in the direction of the short bands of type B, especially when these bands run through a boundary of the same orientation (Fig. 36, Point III). In this way, sites of local hinderance to sliding can occur. As a consequence, secondary crack formation in neighboring spherulite boundaries is initiated, leading to localized interspherulitic fracture of the specimen.

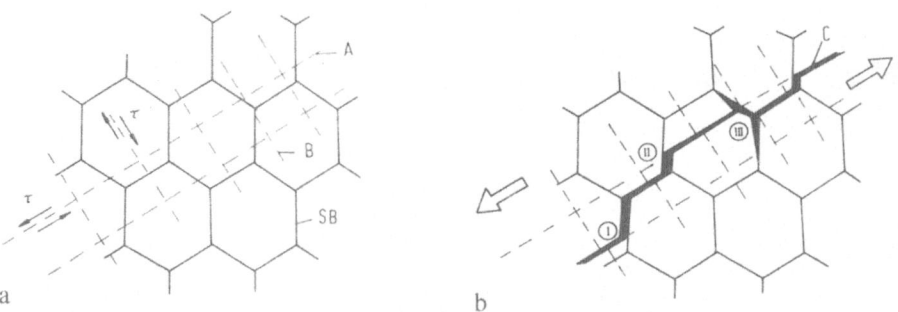

Fig. 36a and b. Crack initiation as a result of interactions between shear bands (A, B) and coarse spherulite boundaries (SB): **a** before crack initiation (τ = shear stress in the bands); **b** after crack initiation (C) ↘

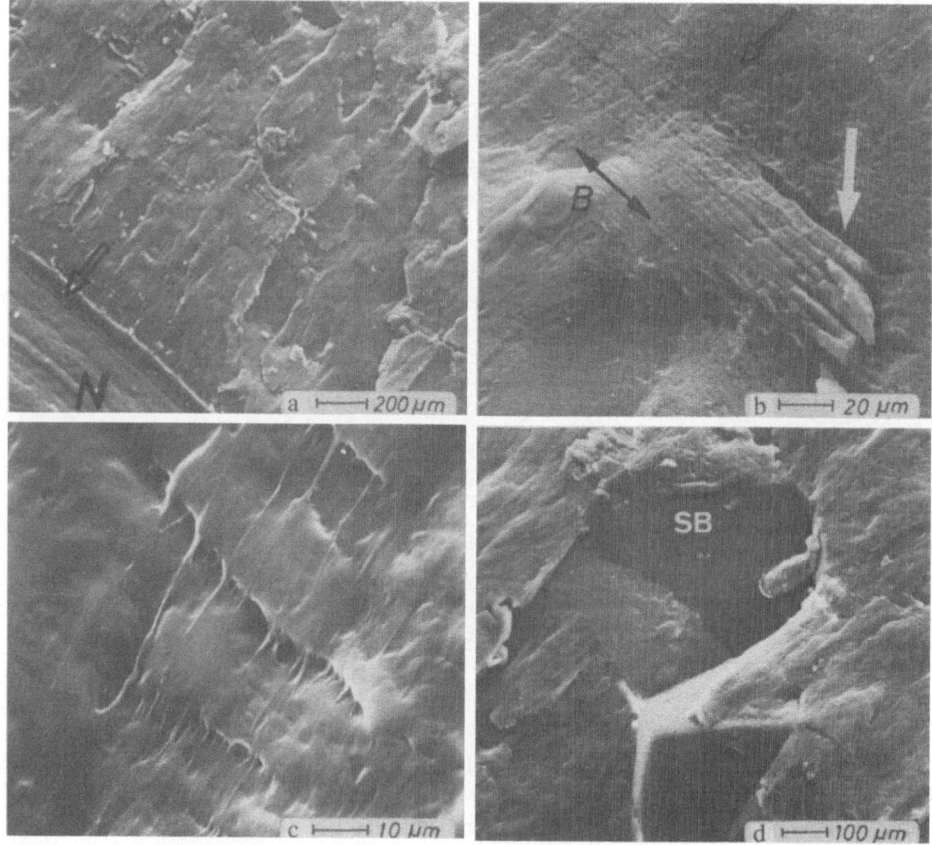

Fig. 37a—d. Structure of shear fracture surface: **a** SEM-micrograph of a shear fracture in PP 1120 (T = −80 °C); **b** secondary crack formation (white arrow) in one of the shear bands of type B in fine spherulitic PB-1 (T = −196 °C); **c** traces of shear bands "B" containing fibrillated polymer substance on a shear fracture plane in fine spherulitic PP 1120 (T = −196 °C); **d** preferred shear fracture along spherulite boundaries (SB) of coarse spherulitic PP 1120 (T = −80 °C)

Those parts of the fracture surface in which fracture has developed inside of shear bands have similar appearance in the different polymer materials. Figures 37a–d give some impressions of the different features involved in the shear fracture process. The surface developing from the initially machined notch (N) into the direction of one of the shear band packages is marked by relatively smooth plateaus connected by some fracture surface steps (Fig. 37a). The individual plateaus are interrupted by traces of intersected sets of type B shear bands, being oriented nearly perpendicular to the shear fracture plane and the sliding direction. In some cases these bands are opened up showing inside a craze-like structure (cf. Fig. 5) or real cracks (Fig. 37b). Higher magnifications of the smooth areas between the traces show a fibrillar pattern of separated shear band matter (Fig. 37c). This leads to the conclusion that high amounts of plastic deformation must have occured in the shear fracture plane. Also, it appears that the fiber ends have been heated up to their melting temperature during the shear fracture process. Williams and Mayer [139] arrived at similar conclusions for comparable observations on shear fracture surfaces in HDPE and PC.

Further fracture surface features, especially distinct for the highly isotactic, coarse spherulitic PP, are polyhedron shaped regions at which the shear crack has left the shear fracture plane in order to stay in a spherulitic boundary region oriented at a flat angle to the sliding direction (Fig. 37d). These sites often constitute secondary crack nuclei in boundaries perpendicular to the shear fracture plane.

3.5 Calculation of a Mode II-Fracture Toughness

Finally it is attempted to calculate the fracture toughness of the various polymers and morphologies breaking under mode II-conditions. Following earlier calculations of the mode I-fracture toughness, K_{Ic}, from tensile tests by Hahn and Rosenfield [148], the mechanical data obtained in the compression test can be used to determine a mode II-fracture toughness K_{IIc}. It can be regarded as a measure for the resistance of each material and morphology against crack propagation and final fracture under shear stress conditions [149].

The following equation will be used for calculation:

$$K_{IIc} = (\tau_{BI} \cdot \varepsilon_F \cdot G \cdot n^2 \cdot L)^{1/2} \tag{3}$$

with
K_{IIc} — mode II-fracture toughness
τ_{BI} — shear stress for band initiation
$\quad (\tau_{BI} = 1/2\sigma_{BI} \cdot \sin 2\varphi; \varphi \simeq 45°)$
ε_F — strain at fracture
G — shear modulus
n — work hardening exponent
L — empirical constant with the dimension of a length ($L \simeq 1$ m)

Using the data listed in Table 1 K_{IIc}-values are obtained which are a little higher than the values measured for these polymers under mode I-conditions. The factor n^2 was estimated as $5 \cdot 10^{-4}$, that means the work hardening exponent n is near zero.

Table 1. Mechanical shear properties of various polymers at T = −80 °C

Polymer	Morphology	Mechanical Data at T = −80 °C			
		Shear modulus G [MPa]	Strain at fracture ε_F	Band initiation stress σ_{BI} [MPa]	Calculated mode II-fracture toughness K_{IIc}^* [MPa · m$^{1/2}$]
Polystyrene PS	amorphous	1600	0.022	155	3.7
Polypropylene PP 1120	fine spherulitic	1600	0.105	120	7.1
Polypropylene PP 1120	coarse spherulitic	1900	0.035	122	4.5
Polypropylene PP 1320	fine spherulitic	1400	0.115	100	6.4
Polypropylene PP 1320	coarse spherulitic	1750	0.090	105	6.5
Polybutene-1 PB-1	fine spherulitic	1200	0.110	95	5.6
Polybutene-1 PB-1	coarse spherulitic	1400	0.095	95	5.7

This assumption is based on the fact that in the shear bands separation takes place by viscous flow of the material. A temperature rise in the bands seems to be very likely, considering the criteria of Argon [150] for unstable deformation, which deals with the occurrence of localized shear instability from strain softening or adiabatic heating. His adiabatic term θ is high; this means that adiabatic mechanisms are favored when the material shows a high plastic resistance and low thermal conductivity and when high deformation rates and low testing temperatures are applied.

Finally, the K_{IIc}-values are compared with the band initiation stress σ_{BI} (Fig. 38). Such a diagram can give informations about polymers with "optimum strength", defined after Hornbogen [151] as the property of a material with high resistance to plastic deformation (σ_y) as well as high resistance to crack propagation and fracture (K_{Ic}). Only a significant combination of both mechanical properties guarantees a certain security in a structural element. In particular, this diagram contains the following information:

1. Inspite of a high band initiation stress σ_{BI} the resistance of PS to subsequent crack growth and final fracture is relatively low. The K_{IIc}-values of the crystalline polymers are much higher. A similar connection has also been established recently for various polymers under tensile test conditions [106].

2. The fracture toughness K_{IIc} also seems to increase with increasing degree of crystallinity be it by making spherulites coarser by slower undercooling (PP 1320),

be it by employing a polymer of a higher tacticity (PP 1320 → PP 1120). However, this concept does not hold in the case of the coarse spherulitic morphology of PP 1120, due to the weakening effect at the spherulite boundaries within this polymer.

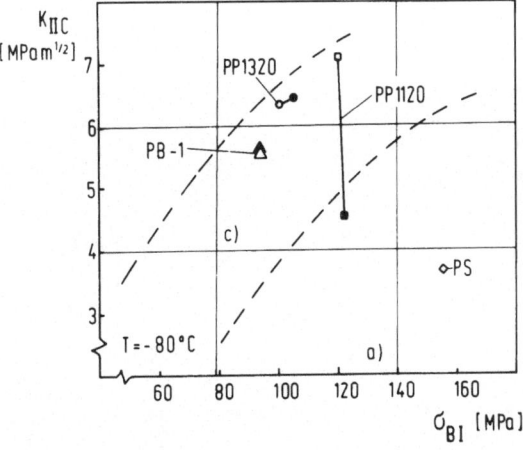

Fig. 38. Mode II-fracture toughness, K_{IIC}, vs. band initiation stress, σ_{BI}, for various polymers at T = −80 °C (open symbols: fine spherulitic, dark symbols: coarse spherulitic morphology)

4 Concluding Remarks

In view of what is known now, it has to be accepted that the formation of crazes in semi-crystalline thermoplastics is a common response to an applied tensile stress. Under plane strain conditions (favored by thick sections, sharp notches, or low temperatures) crazing provides a mechanism for absorbtion and dissipation of energy. Crazes formed at temperatures much below the glass transition temperature of the crystalline polymers have similar dimensions as those crazes known from glassy polymers, although the mechanisms of fibril formation and the molecular structure of the craze fibrils are somewhat different in the two cases. The fracture surface features resulting from the craze separation process have also very much in common with those found on amorphous polymers (Fig. 39 a–c). A modification of the crazing behavior by the morphology of the crystalline polymer must be considered for plane stress conditions, especially at temperatures above the glass transition temperature of the amorphous phase. As an extreme case, in coarse spherulitic, highly isotactic PP plastic deformation is confined to the weak spherulite boundaries, resulting in interspherulitic craze formation and subsequent boundary fracture.

Finally, stress whitening after neck formation in tensile bars of crystalline polymers under plane stress conditions may be associated with some kind of craze-II-formation, in analogy to the corresponding observations in amorphous PC.

The formation of shear bands under compression is found in crystalline polymers when loaded at temperatures lower than 0.75 T_g. Under such a condition the shear bands interact with certain morphological features such as spherulite boundaries or lamellar arrangements inside the spherulites. The band initiation stress, σ_{BI}, increases and the strain at break, ε_F, decreases with decreasing temperature and increasing stiffness of the tested polymer, i.e. increasing degree of crystallinity.

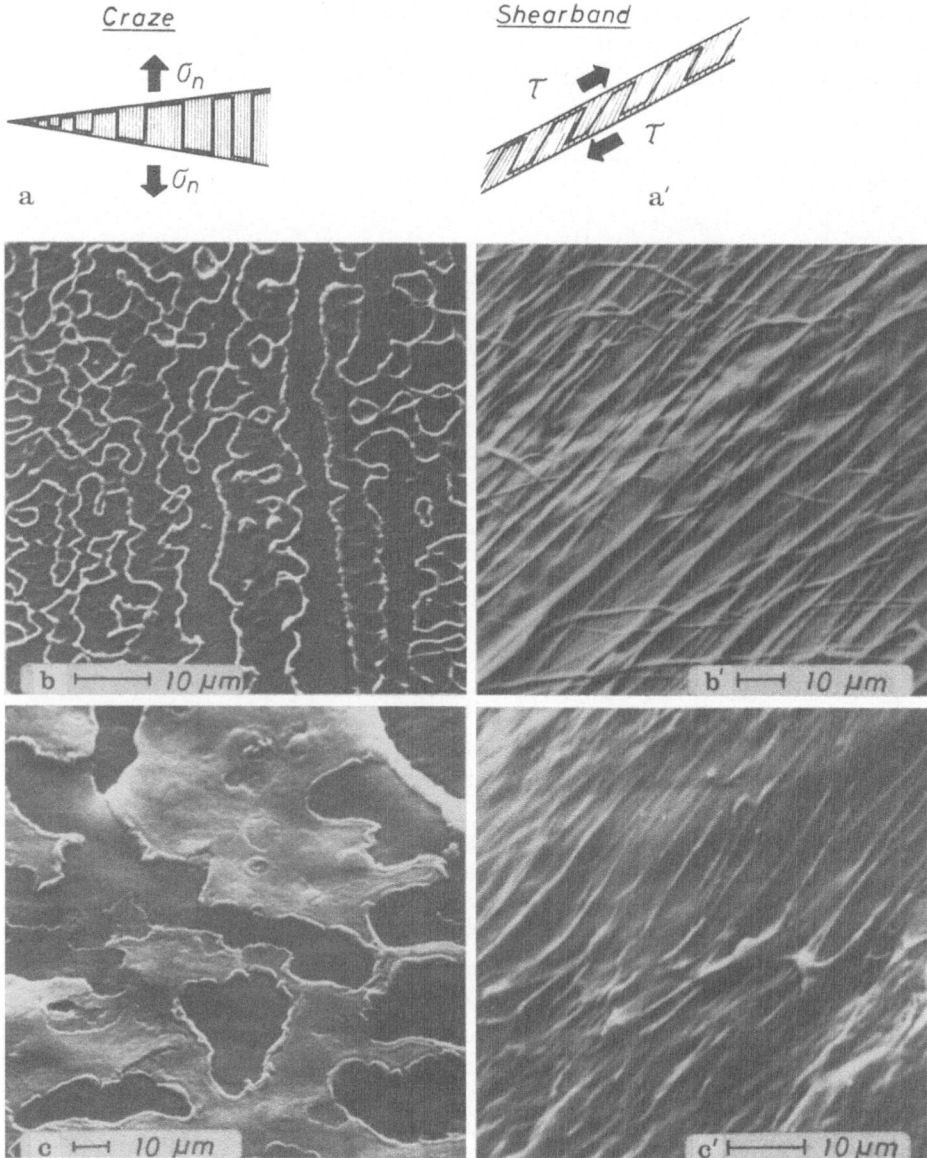

Fig. 39a. Schematic comparison of the deformation processes in crazes and shear bands. Corresponding fracture surfaces are shown in **b** for amorphous PS and in **c** for semicrystalline PP (craze) and PB-1 (shear band)

In comparing the shear fracture surfaces of amorphous and semi-crystalline polymers, it appears that the features in both cases are quite similar (Fig. 39a′–c′). This indicates that, under comparable conditions, the local stress field rather than details of the crystalline-amorphous microstructure of the polymers tested determines the operating deformation mechanism. Only secondary effects arise from the morphology of the crystalline material.

A final comparison of low temperature crazes with shear bands reveals that both deformation phenomena are related. The surface morphologies are quite similar because both modes of plastic deformation depend upon the relative displacement of domains of a size of 10 to 100 nm. However, crazing is controlled by a tensile stress and the fibrous matter contains voids. Shear banding, on the other hand, is controlled by a shear stress which encourages lateral movements without voiding. The final breakdown process may then be initiated in both cases by a random rupture at the upper or lower edge of the deformation zone (Fig. 39 a, b).

Acknowledgements: The author would like to thank Prof. H. H. Kausch for the invitation to this contribution. The valuable discussions on this topic in recent years with Prof. E. Hornbogen and Prof. H. H. Kausch are also highly esteemed. Thanks are due to Prof. G. Kanig, Ludwigshafen, Dr. G. Kress, Ludwigshafen, Dr. G. Retting, Ludwigshafen, Dr. W. Rose, Erlangen, Prof. P. Eyerer and G. Fischer, Stuttgart, and Dr. J. Petermann, Saarbrücken, for contributing valuable schematics or micrographs and to BASF A. G., Ludwigshafen, Chemische Werke Hüls, Marl-Hüls, E. I. du Pont de Nemours & Co., Wilmington, for providing various testing materials. The financial support by the Deutsche Forschungsgemeinschaft (DFG) during some of the studies carried out is gratefully acknowledged. Also, the author appreciates the help and cooperation of his Ruhr-University colleagues and students in the Institute for Materials Science, and in particular, the secretarial and photographical assistance.

5 References

1. Schultz, J. M.: Polymer Materials Science, p. 466, Englewood Cliffs, New Jersey, Prentice Hall, Inc. 1974
2. Magill, J. H.: Morphogenesis of Solid Polymer Microstructures, in: Treatise on Materials Science and Technology, Vol. 10, Part A (ed.) Schultz, J. M., p. 1, New York—San Francisco—London, Academic Press 1977
3. Bowden, P. B., Young, R. J.: J. Mater. Sci. *9*, 2034 (1974)
4. Kambour, R. P., Robertson, R. E.: The Mechanical Properties of Plastics, in: Polymer Science, Vol. 1 (ed.) Jenkins, A. D., p. 687, Amsterdam—London, North-Holland Publ. Comp. 1972
5. Andrews, E. H.: Fracture, in: Polymer Science, Vol. 1 (ed.) Jenkins, A. D., p. 608, Amsterdam—London, North-Holland Publ. Comp. 1972
6. Padden, F. J., Jr., Keith, H. D.: J. Appl. Phys. *37*, 4013 (1966)
7. Binsbergen, F. L., de Lange, B. G. M.: Polymer *9*, 23 (1968)
8. Sakaoku, K., Peterlin, A.: J. Polymer Sci. *A2, 9*, 895 (1971)
9. Keller, A.: Faraday Discussions of the Chemical Society *68*, 145 (1979)
10. Friedrich, K.: Progr. Colloid & Polymer Sci. *66*, 299 (1979)
11. Kausch, H. H.: Fracture in High Polymers. A Molecular Interpretation, in: Fracture 1977, ICF4, Vol. 1 (ed.) Taplin, D. M. R., p. 487, Waterloo, Canada, Univ. of Waterloo Press 1977
12. Kausch, H. H.: Polymer Fracture, a) p. 28, b) p. 290, c) p. 272, Berlin—Heidelberg—New York, Springer-Verlag 1978
13. Haward, R. H.: Brit. Polymer J. *10*, 65 (1978)
14. Vincent, P. I.: Plastics *29*, 79 (1964)
15. Donald, A. M., Kramer, E. J.: J. Polymer Sci., Polymer Phys., to be publ. 1982
16. Samuels, R. J.: Quantitative Structural Characterization of the Mechanical Properties of Isotactic Polypropylene, in: Plastic Deformation of Polymers (ed.) Peterlin, A., p. 241, New York, Marcel Dekker, Inc. 1971
17. Morbitzer, L.: Colloid & Polymer Sci. *259*, 832 (1981)
18. Bandyopadhyay, S., Brown, H. R.: Polym. Engrg. Sci. *20*, 720 (1980)

19. Peterlin, A., Olf, H. G.: J. Polymer Sci., Symp. No. *50*, 243 (1975)
20. Teh, J. W., White, J. R., Andrews, E. H.: Polymer *20*, 755 (1979)
21. Rabinowitz, S., Beardmore, P.: CRC Crit. Rev. Macromol. Sci. *1*, 1 (1972)
22. Kambour, R. P.: J. Polymer Sci., Macromol. Rev. *7*, 1 (1973)
23. Olf, H. G., Peterlin, A.: J. Polymer Sci., Polymer Phys. *12*, 2209 (1974)
24. Geil, P. H.: Morphological Aspects of Deformation and Fracture of Crystalline Polymers, in: Fracture Processes in Polymer Solids (ed.) Rosen, B., p. 551, New York—London—Sydney, Interscience Publ. 1964
25. Morrow, D. R., Jackson, R. H., Sauer, J. A.: Surface Morphology and Deformation Behavior of Polypropylene Single Crystals, in: Advances in Polymer Science and Engineering (ed.) Pae, K. D., Morrow, D. R., Chen, Y., p. 285, New York, Plenum Press 1972
26. Kojima, M.: J. Polymer Sci. *A2*, *5*, 597 (1967)
27. Gohil, R. M., Patel, K. C., Patel, R. D.: Colloid & Polymer Sci. *252*, 358 (1974)
28. Gleiter, H., Petermann, J.: J. Polymer Sci., Polymer Lett. *10*, 877 (1972)
29. Petermann, J., Gleiter, H.: J. Polymer Sci., Polymer Phys. *10*, 2333 (1972)
30. Petermann, J., Gleiter, H.: J. Polymer Sci., Polymer Phys. *11*, 359 (1973)
31. Siegmann, A., Peterlin, A.: J. Macromol. Sci., Phys. *B4(3)*, 557 (1970)
32. Vadimsky, R. G., Keith, H. D., Padden, F. J., Jr.: J. Polymer Sci. *A2*, *7*, 1367 (1969)
33. Schäfer, K., Friedrich, K.: Beitr. elektronenmikroskop. Direktabb. Oberfl. *12/1*, 125 (1979)
34. Sandilands, G. J., White, J. R.: J. Mater. Sci. *12*, 1496 (1977)
35. Hassak, C.: Österr. Chem.-Z., *1* (1901)
36. Herzog, A.: Kunstseide *16*, 128 (1934)
37. Cumberbirch, R. J. E., Dlugosz, J., Ford, J. E.: J. Textile Inst. *52*, T513 (1961)
38. Ford, J. E.: J. Textile Inst. *54*, T484 (1963)
39. Farrow, B., Ford, J. E.: Nature *201*, 183 (1964)
40. Prevorsek, D., Lyons, W. J.: J. Appl. Phys. *35*, 3152 (1964)
41. Daniels, B. K.: J. Appl. Polym. Sci. *15*, 3109 (1971)
42. Zhurkov, S. N. et al.: paper no. 46, in: Fracture, Proc. Int. Symp. on Fracture, Brighton, England, April 1969, p. 531, London, Chapmann and Hall 1969
43. Zhurkov, S. N., et al.: J. Polymer Sci. *A2*, *10*, 1509 (1972)
44. Williams, D. R. G.: Appl. Polymer Symp. *17*, 25 (1971)
45. Flexmann, E. A., Jr.: Polym. Engrg. Sci. *19*, 564 (1979)
46. Carlowitz, B.: Kunststoffe *67*, 449 (1977)
47. Menges, G., Alf, E.: Kunststoffe *62*, 259 (1972)
48. Menges, G.: Kunststoffe *63*, 95 (1973)
49. Menges, G., Alf, E.: J. Elastomers and Plastics *7*, 65 (1975)
50. Menges, G., Horn, B.: Kautschuk, Gummi, Kunststoffe *28*, 444 (1975)
51. Menges, G.: Kunststoffe — Plastics *8*, 15 (1977)
52. Van den Boogart, A., in: Proc. Conf. on Physical Basis of Yield and Fracture, Oxford, England, September 1966, London, Inst. Phys. and Phys. Soc. Conf. Series *1*, 167 (1966)
53. Fischer, S., Brown, N.: J. Appl. Phys. *44*, 4322 (1973)
54. Imai, Y., Brown, N.: J. Polymer Sci., Polymer Phys. *14*, 723 (1976)
55. Imai, Y., Brown, N.: Polymer *18*, 298 (1977)
56. Howard, J. B.: SPE Journal *5*, 397 (1959)
57. Bubeck, R. A., Baker, H. M., in: Proc. 5th Int. Conf. on Deformation, Yield and Fracture of Polymers, Cambridge, U.K., March 1982, London, Plastics and Rubber Inst., paper 31.1. (1982)
58. Belcher, J., Brown, H. R.: ibid, poster paper Cii (1982)
59. Wyzgoski, M. G.: J. Appl. Polym. Sci. *26*, 1689 (1981)
60. Harris, J. S., Ward, I. M.: J. Mater. Sci. *5*, 573 (1970)
61. Brady, T. E., Yeh, G. S. Y.: J. Mater. Sci. *8*, 1083 (1973)
62. Friedrich, K.: J. Mater. Sci. *15*, 258 (1980)
63. Bretz, P. E., Hertzberg, R. W., Manson, J. A.: J. Mater. Sci. *16*, 2070 (1981)
64. Fremel, T. V., et al.: Polymer Mech. *8*, 824 (1972)
65. de Charentenay, F. X., Laghouati, F., Dewas, J., in: Proc. 4th Int. Conf. Deformation, Yield and Fracture of Polymers, Cambridge, U.K., April 1979, London, Plastics and Rubber Inst., paper 6.1 (1979)
66. Hannon, M. J.: J. Appl. Polym. Sci. *18*, 3761 (1974)
67. Bragaw, C. G.: Plastics and Rubber: Materials and Applications *11*, 145 (1979)

68. Frank, R.: Der Maschinenschaden *53*, 104 (1980)
69. Gaube, E., Kausch, H. H.: Kunststoffe *63*, 391 (1973)
70. Beahan, P., Bevis, M., Hull, D.: Proc. Roy. Soc., London, *A343*, 525 (1975)
71. Russell, D. P., Beaumont, P. W. R.: J. Mater. Sci. *15*, 216 (1980)
72. Wendorff, J. H.: Progr. Colloid & Polymer Sci. *66*, 135 (1979)
73. Yoon, H. N., Pae, K. D., Sauer, J. A.: J. Polymer Sci., Polymer Phys. *14*, 1611 (1976)
74. Fischer, G.: Beitr. elektronenmikroskop. Direktabb. Oberfl. *14*, 233 (1981)
75. Kausch, H. H.: Angew. Makromol. Chem. 60/61, 139 (1977)
76. Gleiter, H., Hornbogen, E., Petermann, J.: Fracture and Fiber Formation of Polyethylene Crystals, in: Deformation and Fracture of High Polymers (ed.) Kausch, H. H., Hassell, J. A., Jaffee, R. I., p. 149, New York—London, Plenum Press 1972
77. Wendorff, J. H.: Angew. Makromol. Chem. *74*, 203 (1978)
78. Balkowski, M.: Mikrohohlräume in orientiertem Polypropylen, Diplomarbeit am Deutschen Kunststoffinstitut (DKI), Darmstadt, Dez. 1981
79. Kausch, H. H., Dettenmaier, M.: Pol. Bull. *3*, 565 (1980)
80. Farrar, N. R., Kramer, E. J.: Polymer *22*, 691 (1981)
81. Beahan, P., Bevis, M., Hull, D.: J. Mater. Sci. *8*, 162 (1972)
82. Manson, J. A., Sperling, L. H.: Polymer Blends and Composites, p. 103, New York, Plenum Press 1976
83. Donald, A. M., Kramer, E. J.: Phil. Mag. *A43*, 857 (1981)
84. Hornbogen, E.: Microdeformation and Toughness, Summer-Colloquium, Ruhr-University Bochum, unpubl. data 1974
85. Kausch, H. H.: Kunststoffe *66*, 538 (1976)
86. Adler, W., Friedrich, K.: unpubl. results, 1979
87. Fellers, J. F., Kee, B. F.: J. Appl. Polym. Sci. *18*, 2355 (1974)
88. Wellinghoff, S., Baer, E.: J. Macromol. Sci., Phys. *B11(3)*, 367 (1975)
89. Foot, J. S., Ward, I. M.: J. Mater. Sci. *7*, 367 (1972)
90. Friedrich, K., Karsch, U. A.: The Influence of Molecular Weight on Crazing and Fracture in Polypropylene, in: Proc. 27th Int. Symp. on Macromolecules, Vol. 2, p. 1035, Strasbourg, France, July 1981
91. Hsiao, C. C., Sauer, J. A.: J. Appl. Phys. *21*, 1071 (1950)
92. Friedrich, K.: J. Mater Sci. *12*, 640 (1977)
93. Friedrich, K.: Microstructure and Fracture of Fiber Reinforced Thermoplastic Polyethylene Terephthalate, Center for Composite Materials, University of Delaware, U.S.A., Report No. CCM-80-17, 1980
94. Takano, M., Nielsen, L. E.: J. Appl. Polym. Sci. *20*, 2193 (1976)
95. Stuart, H. A.: Dechema Monographien *39*, 99 (1961)
96. Matsuoka, S.: Polym. Eng. Sci. *7*, 142 (1965)
97. Keith, H. D., Padden, F. J., Jr., Vadimsky, R. G.: J. Appl. Phys. *42*, 4585 (1971)
98. Keith, H. D., Padden, F. J., Jr., Vadimsky, R. G.: J. Polymer Sci. *A2*, 267 (1966)
99. Friedrich, K.: Progr. Colloid & Polymer Sci. *64*, 103 (1978)
100. Friedrich, K.: Über den Einfluß der Morphologie auf Festigkeit und Bruchvorgänge in iso-taktischem Polypropylen, PhD-Thesis, Institute for Materials Science, Ruhr University Bochum, July 1978
101. Calvert, P. D., Ryan, T. G.: Polymer *19*, 611 (1978)
102. Friedrich, K., Karsch, U. A.: J. Mater. Sci. *16*, 2167 (1981)
103. Hornbogen, E., Friedrich, K.: J. Mater. Sci. *15*, 2175 (1980)
104. Way, J. L., Atkinson, J. R., Nutting, J.: J. Mater. Sci. *9*, 293 (1974)
105. Reinshagen, J. H., Dunlap, R. W.: J. Appl. Polym. Sci. *20*, 9 (1976)
106. Friedrich, K.: Kunststoffe — German Plastics *69*, 796 (1979)
107. Wittkamp, I., Bowe, K. H., Friedrich, K.: Beitr. elektronenmikroskop. Direktabb. Oberfl. *15*, 245 (1982)
108. Dettenmaier, M., Kausch, H. H.: Pol. Bull. *3*, 571 (1980)
109. Dettenmaier, M., Kausch, H. H.: Colloid & Polymer Sci. *259*, 937 (1981)
110. Kausch, H. H.: Molecular Mechanisms in Polymer Fracture, in: IUPAC Macromolecules (ed.) Benoit, H., Rempp, P., p. 211, Oxford—New York, Pergamon Press 1982
111. Pakula, T., Fischer, E. W.: Europhysics Conf. Abstracts *4A*, 39 (1980)
112. Pakula, T., Fischer, E. W.: J. Polymer Sci., Polymer Phys. *19*, 1705 (1981)

274   K. Friedrich

113. Kress, G.: Einfluß von Kühlbedingungen und anorganischen Zusätzen auf die mechanischen
     Eigenschaften von Polypropylen-Schlauchfolien, PhD-Thesis, Institute for Materials Science,
     Ruhr-University Bochum, October 1980
114. Noether, H. D., Hay, I. L.: J. Appl. Cryst. *11*, 546 (1978)
115. Noether, H. D.: private comm. 1982
116. Park, I. K., Noether, H. D.: Colloid & Polymer Sci. *253*, 824 (1975)
117. Kanig, G.: J. Cryst. Growth *48*, 303 (1980)
118. Garton, A. et al.: J. Mater. Sci. *13*, 2205 (1978)
119. Garton, A., Carlsson, D. J., Wiles, D. M.: Text. Res. J. *2*, 115 (1978)
120. Gronle, H., Schwarz, G.: Beitr. elektronenmikroskop. Direktabb. Oberfl. *12/1*, 137 (1979)
121. Miles, M. J., Baer, E.: J. Mater. Sci. *14*, 1254 (1979)
122. Moet, A., Palley, I., Baer, E.: J. Appl. Phys. *51*, 5175 (1980)
123. Baer, E., Moet, A., Walton, K., in: Proc. 5th Int. Conf. on Deformation, Yield and Fracture
     of Polymers, Cambridge, U.K., March 1982, London, Plastics and Rubber Inst. poster paper
     Ci (1982)
124. Whitney, W.: J. Appl. Phys. *34*, 3633 (1963)
125. Argon, A. S.: Phil. Mag. *28*, 839 (1973)
126. Bowden, P. B., Raha, S.: Phil. Mag. *29*, 149 (1974)
127. Kramer, E. J.: J. Polymer Sci., Polym. Phys. *13*, 509 (1975)
128. Haward, R. N., Murphy, B. M., White, E. F. T.: J. Polymer Sci. *A2*, 801 (1971)
129. Bowden, P. B., Raha, S.: Phil. Mag. *22*, 463 (1970)
130. Argon, A. S. et al.: J. Appl. Phys. *39*, 1899 (1968)
131. Kramer, E. J.: J. Macromol. Sci.-Phys. *1*, 191 (1974)
132. Wu, J. B. C., Li, J. C. M.: J. Mater. Sci. *11*, 434 (1976)
133. Li, J. C. M., Wu, J. B. C.: ibid *11*, 445 (1976)
134. Friedrich, K., Schäfer, K.: J. Mater. Sci. *14*, 480 (1979)
135. Friedrich, K., Schäfer, K.: Progr. Colloid & Polymer Sci. *66*, 329 (1979)
136. Chau, C. C., Li, J. C. M.: J. Mater. Sci. *14*, 2172 (1979)
137. Duckett, R. A., Zihlif, A. M.: J. Mater. Sci. *9*, 171 (1974)
138. Attenburrow, G. E., Bassett, D. C., in: Proc. 4th Int. Conf. on Deformation, Yield and Fracture
     of Polymers, Cambridge, U.K., April 1979, London, Plastics and Rubber Inst., paper 13.1
     (1979)
139. Williams, J. D., III, Mayer, G.: ibid, paper 11.1 (1979)
140. Friedrich, K.: Colloid & Polymer Sci. *259*, 190 (1981)
141. Hillenbrand, G., Hornbogen, E., Köster, U., in: Proc. 4th Int. Conf. on Rapidly Quenched
     Metals, 1391, Sendai, Japan 1981
142. Argon, A. S.: priv. comm. 1979
143. Rose, W., Meurer, Ch.: Transmissionselektronenmikroskopische Gefügeuntersuchungen an
     teilkristallinen Polymeren, in: Verformung und Bruch (ed.) Stüwe, H. P., p. 219, Wien, Verlag
     der Österreichischen Akademie der Wissenschaften 1981
144. Rose, W., Meurer, Ch., in: Proc. 5th Int. Conf. on Deformation, Yield and Fracture of Polymers,
     Cambridge, U.K., March 1982, London, Plastics and Rubber Inst., paper 3.1 (1982)
145. Rose, W., Meurer, Ch.: J. Mater. Sci. *16*, 883 (1981)
146. Rose, W.: priv. comm. 1982
147. Hull, D.: Sci. Progr. *57*, 495 (1969)
148. Hahn, G. T., Rosenfield, A. R.: ASTM-STP *432*, 5 (1968)
149. Friedrich, K.: Shear Bands and Fracture in Crystalline Polymers, in: Advances in Fracture
     Research, ICF5, Vol. 2, (ed.) Francois, D., p. 773, Oxford—New York—Toronto—Sydney—
     Paris—Frankfurt, Pergamon Press 1981
150. Argon, A. S.: The Inhomogenity of Plastic Deformation, p. 161, Metals Park, Ohio, ASM 1973
151. Hornbogen, E.: Z. Metallkde. *68*, 455 (1977)

Received November 1, 1982
H. H. Kausch (editor)

Crazing in Block Copolymers and Blends

A. S. Argon, R. E. Cohen, O. S. Gebizlioglu and C. E. Schwier
Massachusetts Institute of Technology, Cambridge, MA 02139, USA

Advances in Polymer Science 52/53
© Springer-Verlag Berlin Heidelberg 1983

List of Symbols

A   Area, B/kT
B   Activation energy at zero stress
C   Proportionality constant
D_i Scale factor for velocity; $i = 1$, interface convolution; $i = 2$, repeated cavitation
E   Young's modulus in general. With subscripts C, M, PB, PS, P: moduli of composite, matrix phase, polybutadiene, polystyrene, and particle, respectively
ΔG Activation energy
H   Height of channel in which interface convolutes; craze tip opening displacement
K   Bulk modulus in general. With subscript C, M, PB, PS, P: bulk moduli of composite, matrix phase, polybutadiene, polystyrene, and particle, respectively
K_I Mode one stress intensity factor
K_{IC} Critical stress intensity factor for propagation
N   Surface density of crazable sites
Q   Functional dependence of negative pressure on deviatric stress for cavity expansion
R   Radius of particulate heterogeneity
T   Absolute temperature
V   Volume
W   Toughness, total work to fracture
Y   Yield stress in general. With subscript cr, craze-yield stress
$Y(\lambda_n)$ Tensile yield stress of polymer after uniaxial extension to an extension ratio of λ_n
$\hat{Y}$ Athermal tensile yield stress of a glassy polymer
a   Craze half length
b   Craze half thickness at craze center
c   Volume fraction
f $f(x) = u(x) + k(x)$, final half thickness of craze cavity under the craze traction σ_c
g $g(x)$, unconstrained half thickness of craze lentil after undergoing expansion
k $k(x)$, primordial half thickness of craze cavity
n   Total number of crazes, strain rate exponent in non-linear viscosity
n_0 Total number of internal particulate heterogeneities
q   Stress concentration factor
s   Deviatoric (shear) stress
t   Time in general. With subscripts f, in, inact: fracture, initiation, inactivation time
u $(= u_y)$ $u(x)$, displacement of half craze cavity under craze traction σ_c
v   Velocity of craze
w   Width of sample in general. With subscripts l, s in the long and short direction
x   Coordinate
y   Coordinate
z   Coordinate
Δ Mean distance between surface crazes

Δ	Length of cavitational process zone
χ	Surface free energy, surface tension
β	Aspect ratio of a craze, $= a/b$
γ	Volumetric coefficient of expansion in general. With subscripts C, M. PB, PS, P: composite, matrix, polybutadiene, polystyrene, particle, respectively
δ	Mean distance between crazes in the volume
δ	With subscripts, A, B: solubility parameter
δ	With subscripts, A, B, solubility parameter
ε	Uniaxial strain in general. With subscripts f, m, z: fracture, maximum at cavitation, and in the extension direction, respectively
λ	Wave length of convoluted interface
λ_n	Extension ratio
λ_n'	As defined in Eqn. (51)
ν	Poisson's ratio of polystyrene. With subscripts C, M, P: composite, matrix, particle, respectively
ϱ	Active craze front length per unit volume. With subscripts C, M, P: composite, matrix, particle, respectively
σ	Negative pressure specifically. With subscripts c, e, i, m, P, ST, TH, ∞: craze traction, Mises equivalent, one of three principal stresses, maximum level of craze traction where cavitation in PB begins, negative pressure in particle, negative pressure due to one of three principal stresses, negative pressure due to thermal mismatch, uniaxial applied stress at the borders
σ	With subscripts xx, yy, zz etc. for components of the local stress tensor
θ	Ratio of slope of the falling to the rising part of the traction cavitation law
Θ	Craze dilatation
τ	Time constant
ξ	Symbolic parameter describing defects. With subscripts e, i: extrinsic, intrinsic, respectively
η_0	Non-linear viscosity coefficient

1 Introduction

Although under suitable conditions all glassy polymers, indeed all glasses, can undergo large plastic flow below their respective glass transition temperatures [1], many glassy polymers exhibit brittle behavior in tension under normal service conditions and in impact. It has been recognized now for some time that this behavior is closely linked to the phenomenon of crazing in which small planar regions, oriented normal to the maximum principal tensile stress, for all intents and purposes undergo a dilatational transformation. Well before the internal complexities of a craze had been appreciated, it was recognized that when uncontrolled, crazes usually fracture and cause brittle behavior, while controlled crazing can impart to the polymer very substantial toughess. The details of the interior make-up of crazes and how this relates to the molecular structure of the polymer are discussed in the chapters of Kramer [2a] and of Dettenmaier [2b] in this book. Here we will be primarily interested in how the control of crazing by the incorporation of compliant heterogeneities can impart very substantial toughness to a normally brittle glassy polymer.

The toughening of brittle plastics by the incorporation of compliant heterogeneities has been known for several decades and has been practiced industrially on a large scale. The history of its development and a summary of the generally accepted principles of such toughening is discussed in some detail by Bucknall [3]. Past industrial practice and some previous research has established that several factors are important in the achievement of acceptable toughness. Among these have been mentioned the size of the heterogeneities, their internal structure, the need to graft the rubbery heterogeneities to the glassy matrix to achieve good adhesion and a sufficiently high molecular weight in the glassy phase to form craze matter with high mechanical integrity. Since most commercial product is obtained by processing that tends to combine many key steps into one operation, it has usually not been possible to ascertain the validity of the many assertions on the importance of one or more of these factors by controlled experiments. This has now become recognized and has resulted in several recent studies on commercial material that have provided some useful insight into the role of some of the factors listed above. On the whole, however, the systematic study of the various ingredients of toughness requires preparation of microstructures by unconventional means and an approach that pays attention to a multiplicity of detail. A few such investigations are now in progress [4-6], utilizing block copolymers and their blends. Here we will present a summary of our investigation of the sources of toughening in both pure di-block copolymers and blends of di-blocks with additional homopolymer to coarsen microstructure, and to form particulate heterogeneities.

2 Principles of Toughening of Glassy Polymers

2.1 Distortional Plasticity, Dilatational Plasticity

Under suitable conditions such as straining in shear, under compression, or under large superimposed pressures, all glassy polymers can undergo large plastic distortions without stress whitening, crazing or cracking. Certain glassy polymers, such as some polyimides [7] and polycarbonate on some occasions can undergo large strain plastic flow also in tension. When such distortional plasticity prevails, the polymer can exhibit, very attractive toughness. As already mentioned above, however, most glassy polymers craze under a state of stress that has at least one important tensile component. In spite of this, when crazing is properly controlled, it can result in substantial levels of toughness. In many polymers, both forms of toughening by distortional and craze (dilatational) plasticity can coexist [2b, 3]. The interplay between these two modes of plasticity and their interaction is often complex and has its sources in the molecular architecture of the polymer, the state of aging of its glassy phase, and the microstructural morphology. This has been investigated, sometimes imaginatively [8], from the point of view of molecular detail on the basis of established correlations. It is, however, premature to claim fundamental and causal understanding of the conditions that govern the particular mixture of distortional and dilatational plasticity in a given polymer under a given state of stress. On a more restricted scale, however, the mechanism of distortional plasticity, its kinetics, and the conditions that govern how and why it localizes into shear bands is now quite well understood [1, 9–11]. Similarly, the conditions of initiation [12], growth [13, 14, 2, 15] and breakdown [16, 17] of crazes are also becoming well understood in polymers that exhibit crazing. There is, however, insufficient understanding on why certain polymers craze while others don't. We will provide no further insight here into these important questions, but will concentrate our attention exclusively on those polymers in which toughness results almost entirely from stable crazing. Furthermore, we will be interested in how dilatational plasticity can be enhanced through the incorporation of specially constructed microstructural heterogeneities either in the form of quasi-uniform hetero phases, characteristic of copolymers, or in the form of composite particles dispersed in a usually brittle glassy polymer phase. Although we will discuss primarily the altered behavior of polystyrene, most of the conclusions derived should be applicable also to other systems provided that in such systems crazing is prevalent, and that it occurs at levels usually insufficient to initiate distortional plastic flow. We recognize and emphasize that the attainment of similar microstructures in other polymer systems may require altered methods of processing, different from those that we will present here.

2.2 Toughness Due to Crazing

2.2.1 Ingredients of Toughness

Before concentrating on specific factors and mechanisms that influence crazing, it is instructive to take account of some general principles on how inelastic strain and toughness result from crazing. A similar but more restricted discussion has been given earlier by Brown [18].

For this purpose we view a craze as a lenticular region with its principal plane normal to the tensile axis z. Although actual shapes of isolated crazes differ significantly from regular shapes as we will discuss below, a craze can be considered as a very eccentric oblate spheroid with a radius a in its principal plane and a half thickness b parallel to the z axis. Furthermore, since the main microstructural feature of a craze is extended fibrils with a relatively constant extension ratio λ_n, we can view the crazing process as an extensional transformation in the z direction inside the craze, converting a much thinner oblate spheroid of half thickness b/λ_n into the current half thickness b, without any significant change of dimension inside the plane of the craze. This produces a dilatation of $\Theta_{cr} = \lambda_n - 1$ of the craze interior, in comparison with the initial solid polymer. Much evidence points into the direction that mature crazes that grow at constant velocity under a constant applied stress attain a terminal thickness $2b_f$ upon which they grow only laterally in the principal plane (see Sect. 2.2.3 below). When such a constant growth rate condition is reached, the volume of a craze can be idealized simply as $2\pi a^2 b_f$ in its expanded form and $2\pi a^2 b_f/\lambda_n$ in its unexpanded, "primordial" form.

Consider now a polymer sample of current volume V responding by craze plasticity to an imposed current elongational strain rate $\dot{\varepsilon}_z$, developing a tensile (dilatational) flow resistance Y_{cr} until a final fracture strain ε_f is achieved as shown in Fig. 1. The specific toughness W or the total deformational energy absorbed per unit volume for this polymer is the area under the stress strain curve, or

$$W = Y_{cr}\varepsilon_f . \tag{1}$$

The strain to fracture, however, is the product of the imposed strain rate and the time to fracture t_f, i.e.

$$\varepsilon_f = \dot{\varepsilon}_z t_f . \tag{2}$$

If the strain derived is entirely due to craze matter production, there will be no significant change in the cross-sectional area A of the part during such (dilational) flow. Then, by definition, the imposed tensile strain rate $\dot{\varepsilon}$ must be

$$\dot{\varepsilon}_z = \dot{\varepsilon} = \frac{1}{l}\frac{dl}{dt} = \frac{1}{l}\frac{1}{A}\frac{dV}{dt} = \frac{1}{V}\frac{dV}{dt} = \frac{d\Theta}{dt} , \tag{3}$$

where Θ is the macro dilatation of the polymer as might be measured by means of a dilatometer with a working fluid that is prevented from penetration into the craze cavities. Thus, the polymer responds to the imposed tensile strain rate by undergoing

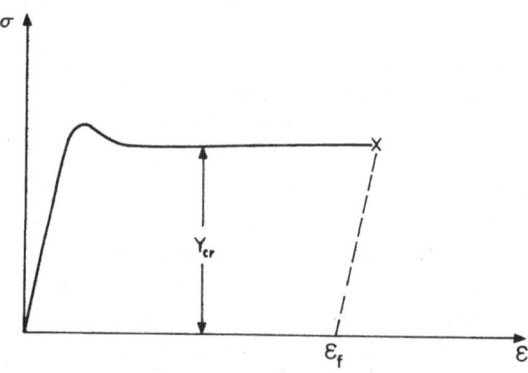

Fig. 1. Typical stress-strain curve of a polymer exhibiting (dilatational) craze plasticity

an equal rate of dilatation by crazing. On the other hand, for the terminal craze growth conditions described above it is possible to write readily

$$\frac{d\Theta}{dt} = \Theta_{cr}(2b_f) \frac{1}{V} \frac{d\alpha}{dt} \qquad (4\,a)$$

$$= (2\Theta_{cr}b_f) \left(\frac{2\pi an}{V}\right) \frac{da}{dt}, \qquad (4\,b)$$

where $2b_f$ is the craze thickness, $\Theta_{cr} = (\lambda_n - 1)$ is the average dilatation inside a craze, α the total craze area perpendicular to the direction of the maximum principal tensile stress, n the total number of crazes of average radius a in the current volume of the sample $V (= V_0 + 2n\pi a^2 b_f)$, where V_0 is the initial volume of the sample. In Eq. (4b) the terms in the first brackets that we will abbreviate with C represent the properties of mature crazes. The second set of terms in brackets which we will abbreviate with

$$\varrho \equiv \frac{2\pi an}{V} \qquad (5)$$

is the total active craze front length per unit volume. The last term in Eq. (4b) is the craze front velocity v. Re-assembling these groups of terms in Eq. (3) gives for the tensile strain rate

$$\dot{\varepsilon} = C\varrho v. \qquad (6)$$

The mechanistic parallel of this equation to the well known equation in crystal plasticity involving dislocation motion has already been pointed out by Brown [18]. We now continue and examine in detail the nature of the three terms in Eq. (6).

The craze properties represented by C will not be constant. The craze dilatation depends on the extension ratio λ_n of the fibrils which will be governed by the steady state craze traction σ_c transmitted across the craze through the solution of the equation

$$Y(\lambda_n)/\lambda_n = \sigma_c, \qquad (7)$$

where $Y(\lambda_n)$ is the true uniaxial tensile plastic resistance of the solid polymer that depends on λ_n by virtue of orientation hardening. Although $Y(\lambda_n)$ cannot usually be measured directly in macro experiments due to intervening fracture initiated by crazing, it should be readily determinable from first principles as the sum of a dissipative (von Mises) plastic resistance and an entropic resistance [19, 20]. In most instances, the extension ratio that has been measured in tufts of dry crazes is of the order of $\lambda_n \approx 4$ [21] while in small regions near the symmetry plane, (the mid-rib) considerably larger extension ratios reaching up to 8 have sometimes been found (see chapter 1). Thus, on this basis, we pick the craze dilatation to be $\Theta_{cr} = 3$. The terminal thickness $2b_f$ of a mature craze, where aging terminates the drawing process, varies considerably. While the crazes in homo-polymers tend to be thicker, in the range of 0.5 μm, those in high impact polymers with compliant heterogeneities tend to be thinner, in the range of 0.1 μm. There is as yet no complete understanding how or why craze thickening levels off at these dimensions. On the above basis, however, we evaluate the value C in Eq. (6) to be about 6×10^{-7} m in polymers with heterogeneities, and about 3×10^{-6} m in homo-polymers such as polystyrene (PS).

The quantity ϱ in Eq. (6) represents the total length of craze front per unit volume where new craze matter is being generated in planar form and added to the already existing craze matter. It results from a kinetic balance between initiation and inactivation of craze front length, and as such is expected to be influenced strongly by both the density of source sites on free surfaces and internal interfaces, where stress concentration exists, and by the processes affecting termination of craze growth such as blunting, mutual arrest, traverse through the entire cross-section etc. Both initiation and inactivation depend strongly on the local stress and temperature as we will discuss below.

The craze front velocity v can be governed by one of two distinct mechanisms of craze matter production. As Argon and Salama [13, 22] have discussed in detail, under the usual levels of service stresses or stresses under which most experiments are carried out, craze matter in single phase homopolymer is produced by the convolution of the free surface of the solid polymer at the craze tip. This occurs by a fundamental interface instability present in the flow or deformation of all inelastic media when a concave, meniscus-like surface of the medium is being advanced locally by a suction gradient. This is the preferred mechanism of craze advance in homopolymers. In block copolymers with uniform distributions of compliant phases of a very small size, and often weaker interfaces than either of the two phases in bulk, craze advance can also occur by cavitation at such interfaces to produce craze matter as has been discussed by Argon et al. [15]. Both of these mechanisms of craze advance lead to very similar dependences of the craze front velocity on applied stress and temperature that is of the basic form

$$v = D_i \exp\left(-\frac{\Delta G(\sigma_\infty/\hat{Y})}{kT}\right), \tag{8}$$

where D_i (i = 1, 2) are pre-exponential factors for the two craze extension mechanisms, dependent on some physical constants and dimensions of the microstructure, σ_∞ is the applied stress, and $\hat{Y}$ is a modulus parameter that has the meaning of the ultimate athermal plastic resistance of the polymer. We will discuss these processes in more detail and quantitatively below. There is some evidence that in pure homopolymer, an analogous mechanism of craze advance by cavitation in the solid polymer just ahead of the craze can also occur at very high stress levels producing very high craze velocities that occur in fracture.

Finally, when fracture occurs, it initiates and propagates through a craze. As we will discuss in more detail below, such initiation occurs from defects in the craze matter such as entrapped dust particles, or ultimately from large natural irregularities occurring in the craze matter [16, 17], see also chapter 3. As Eq. (2) indicates, the first formed craze matter is subjected to the full craze traction σ_c, that for the case of cessation of thickening must increase to the level of the applied stress σ_∞, for the entire duration t_f of straining, while the craze matter that is added in later stages is subjected to the same stress for proportionally shorter periods. As in all time dependent fracture processes known in solids, we expect the time to fracture of a stressed portion of craze matter to be inversely and nonlinearly dependent on the stress carried by the craze that might be typically of the form

$$t_f = t_f(\sigma_\infty, T, \xi_e, \xi_i) = A(T, \xi_e, \xi_i)/\sigma_\infty^f \tag{9}$$

where ξ_e, ξ_i stand symbolically for the description of extrinsic and intrinsic defects in the craze matter from which terminal fracture initiates, and where f is a phenomenological stress exponent. During craze plasticity, $\sigma_\infty = Y_{cr}$, the craze flow stress. We expect that when craze sources are plentiful, resulting in a high active craze front length ϱ, the craze velocity that needs to be maintained to match the imposed strain rate can be proportionally lower resulting in a lower craze flow stress and an increased craze fracture time t_f. The contrary will hold when craze sources are few, requiring high velocities and high craze flow stresses to be maintained. There is very little information available currently on the important dependence of the craze fracture time t_f on the applied stress σ_∞ implied by Eq. (9).

Assembling all detailed ingredients into Eq. (1) for the toughness of the polymer resulting from crazing, we obtain

$$W = Y_{cr}\varepsilon_f = CA\varrho(Y_{cr}) v(Y_{cr}, T)/Y_{cr}^{f-1} . \tag{10}$$

In Eq. (10) that gives the toughness in dilatational plasticity the factors C and A are dependent on craze microstructure and will not vary significantly. The stress and temperature dependence of the craze velocity while quite determinate in the interface convolution process of craze matter production will also be quite sensitive to microstructural detail of phase distribution in block copolymers. The applied stress $\sigma_\infty = Y_{cr}$ producing craze yielding is integrally locked in with ϱ and v. This makes $\varrho(Y_{cr})$ one of the most important factors accessible to the materials scientist in governing toughness, as has been recognized intuitively by most investigators. The desirable response of a polymer is a high degree of toughness, independent of the rate of deformation $\dot\varepsilon$ which normally can vary from 10^{-4} sec^{-1} of a slow tension experiment to $5 \cdot 10^2$ sec^{-1} typical in a situation of impact. Since the dependence of all other terms in Eq. (10) on stress are either known or determinable as in the case of the expression for the time to fracture, it should in principle be possible to design a set of heterogeneities that can initiate craze front at the required rate with the desired stress dependence, i.e.

$$\varrho(\sigma_\infty) = \frac{W\sigma_\infty^{f-1}}{ACD_i} \exp\left(\frac{\Delta G(\sigma_\infty/\hat{Y})}{kT}\right) \tag{11}$$

for the level value of the toughness W, with the meaning of the various terms as defined earlier, and the form of $\Delta G(\sigma_\infty/\hat{Y})$ to be outlined below.

The present discussion indicates that there is a balance that needs to be attained between the various factors that influence the toughness. As mentioned above, although the total active craze front length per unit volume is the most important parameter available to the investigator, previously recognized by many, the actual kinetics of the growth response of a craze and the integrity of craze matter under stress must also be controlled if frustration is to be avoided in the attainment of high toughness. Thus, as in metallurgy, it is difficult to achieve both a high craze yield stress and a large strain to fracture simultaneously under any given condition of straining. On the other hand, achieving higher toughness with larger strain rates in a given microstructure should be readily possible if the stress dependence of ϱ and v are stronger than that of t_f.

In the sections immediately following we will review briefly the state of understanding of processes governing the initiation and inactivation of craze fronts, the velocity of craze fronts, and the time dependent fracture of craze matter under stress.

2.2.2 The Active Craze Front Density

Craze front initiation and inactivation

As in other branches of materials science where stress assisted nucleation of new phases, cavities, or initiation of microcracks are of concern, the initiation of crazes in stressed glassy polymer is also subject to considerable complexities that have still not been completely resolved mechanistically. The early and largely phenomenological approaches to the problem have been reviewed by Kambour [23]. Clearly, since the state of strain in a polymer can be affected by hot stretching in any number of ways free of any crazing, criteria that are based on critical strain states are not acceptable. At best, such criteria imply elastic strain which is governed by the state of stress, hence leading to a stress description in the end. The earliest statement of an acceptable criterion for craze initiation is due to Sternstein and Ongchin [24], who parameterized their observations on craze initiation under bi-axial stress. Their description, however, could not be generalized unambiguously to be applicable in a three-dimensional state of stress. Such a generalization was proposed by Argon [25, 26] and refined further on the basis of detailed experiments on craze initiation under biaxial states of stress, carried out with Hannoosh [12]. On the basis of the experiments of these last authors where craze initiation times were studied on surfaces with controlled topography subjected to combinations of a deviatoric shear stress s, and a negative pressure σ, the isochronal craze iniation, at a time t under stress, has been stated to be

$$\frac{AQ}{(s/Y)} - \frac{3}{2}\left(\frac{\sigma}{Y}\right) = Q \ln (t/\tau), \qquad \left(\frac{\sigma}{Y} > 0\right) \tag{12}$$

where Y is the tensile plastic resistance, $A(= \Delta G_0/kT)$ is a temperature dependent constant with a magnitude of 9.54 in PS at room temperature, $Q(= 0.0133)$, the measure of interaction of the deviatoric shear stress and the negative pressure, and τ is a characteristic time constant, dependent primarily on temperature, and possibly on some elusive supra-molecular microstructural detail, has the dimensions of 6.0×10^{-8} sec at room temperature in PS. The deviatoric shear stress $s(= \sigma_e/\sqrt{3})$ where σ_e is the well known Von Mises equivalent stress, and the negative pressure σ are the local values concentrated by surface irregularities or by differences between the stiffness of the particulate phases and the matrix. The form of Eq. (12) has a mechanistic basis discussed in detail in the references cited above and represents the trend of the measurements of Argon and Hannoosh [12] quite well in the range where the negative pressure is positive and non-zero. There is no craze initiation in pure shear or when the pressure is positive. Previously initiated crazes, however, can grow under any state of stress that contains at least one principal component of tensile stress regardless of whether the total state of pressure is positive, as we will discuss further in the next section. The above criterion of craze initiation which requires a key step of plastic deformation governed by the deviatoric stress to initiate a precursor density of submicroscopic cavities that subsequently expand under negative pressure is the preferred mode of craze initiation in homogeneous glassy polymers containing no heterogeneities with weak interfaces.

In heterogeneous polymers containing either single phase or multiphase particles, the craze initiation condition is expected to be applicable to the glassy polymer outside the particles — provided that the stress is concentrated there over a large enough volume element to initiate the set of representative precursor processes [27, 28]. The exact minimum size of the critical volume element is not clear, but experiments on the craze initiating efficiency of particles suggest that it is of the order of 50–100 nm. That macroscopically determined craze initiation criteria are applicable in the vicinity of heterogeneous particles of the appropriate size, usually in the range of microns, has been demonstrated by Oxborough and Bowden [29, 30], albeit using a complex strain criterion.

In block copolymers where large differences exist between the shear moduli of the rubbery phases and the glassy matrix, significant concentrations of negative pressure can arise inside the compliant phases upon deformation to produce internal cavitation and result in the formation of a craze nucleus as has been observed in some detail by Argon et al. [15] in the deformation of the KRO-3 Resin, a di-block copolymer of PS and polybutadiene (PB) of corrugated lamellar morphology. The cavitational strength of PB spherical domains has now been well established in experiments based on the thermal stress induced shift of T_g [31]. This information makes possible an accurate statement of a craze growth model based on propagating a damage zone.

Craze fronts become inactivated when the local conditions there become too radically modified to produce continued craze growth. This can occur in a number of ways. First, the stress at the tip of a craze could drop below the level necessary for propagation by the encounter of a nearby craze, by coming too close to the shielded poles of a compliant heterogeneity, or by being arrested by heterogeneities stiffer than the glassy polymer. Second, a craze can be terminated by running across the entire cross section. Finally, if the growth of a craze is sluggish due perhaps to visco-elastic stress relaxation at its tips, general yielding of the polymer may occur that can readily blunt and consume the craze. Of these possibilities, the first two are likely to be the most prominent in polymers with craze yield stresses significantly below the distortional yield stress.

The Kinetic Balance

With the above considerations on initiation and inactivation of craze fronts we now develop relations for ϱ, the steady state active craze front length per unit volume. Here it is necessary to distinguish two separate cases: surface initiation limited crazing; and volume initiation limited crazing. The first category applies to the cases of most homopolymers and to many block copolymers with phase dimensions too small to result in stress concentrations in the requisite volume element that was mentioned above. In these instances crazes are observed to initiate from sources at free surfaces, and the specimen or part dimensions become an important consideration. The second category applies to cases of glassy polymers containing heterogeneities of a size large enough to initiate crazes on or near their highly stressed interfaces.

Surface initiation limited crazing. Consider a rectangular slab of length L, and small and large lateral dimensions w_s and w_1, with craze initiating sources distributed uniformly along the entire lateral surface at a density of N (m^{-2}). The sample is under a state of uniaxial tension[1] parallel to the long dimension L. We model the situation by considering, as in Fig. 2, that crazes initiate along the entire length w_1 and propagate across the narrow dimension w_s, and become inactivated when they reach the other side. In this geometry, mutual arrest of craze fronts is unimportant since even with mutual arrest a stepped craze covers the entire cross-section.

The rate of increase of the craze front length ϱ is then

$$\left(\frac{d\varrho}{dt}\right)^+ = \frac{w_1}{V} \frac{NA}{t_{in}} = 2\left(1 + \frac{w_1}{w_s}\right) \frac{N}{t_{in}} \tag{13}$$

where $V (= w_1 w_s L)$ is the total volume of the sample, $A = 2(w_1 + w_s) L$ the total stressed surface area, and t_{in} the initiation time of a craze from a surface source according to a specific evaluation of Eq. (12).

The inactivation rate of craze fronts by exiting on the opposite surface or by mutual arrest of two fronts propagating from opposite sides,

$$\left(\frac{d\varrho}{dt}\right)^- = \frac{(3/2) w_1}{t_{inact}} \frac{1}{w_s w_1 \Delta} \tag{14}$$

$$t_{inact} = \left(\frac{3}{4}\right) \frac{w_s}{v} \tag{15}$$

where the coefficients of Eqs. (14) and (15) are appropriate averages of the modes of exit on the opposite surface and internal mutual arrest, where Δ is the mean vertical spacing between active crazes. Noting that at any one time the active craze density can be written as

$$\varrho = \frac{w_1}{w_s w_1 \Delta}, \tag{16}$$

we can combine Eqs. (13)—(16) to write the total rate of increase of active craze front length per unit volume as

$$\frac{d\varrho}{dt} = \left(\frac{d\varrho}{dt}\right)^+ - \left(\frac{d\varrho}{dt}\right)^- = 2\left(1 + \frac{w_1}{w_s}\right) \frac{N}{t_{in}} - 2\varrho \frac{v}{w_s}. \tag{17}$$

---

[1] More general, bi-axial states of stress in the plane can be readily considered similarly. Since our aim here is to develop the connection between microstructure and uniaxial tensile toughness, such cases will not be developed.

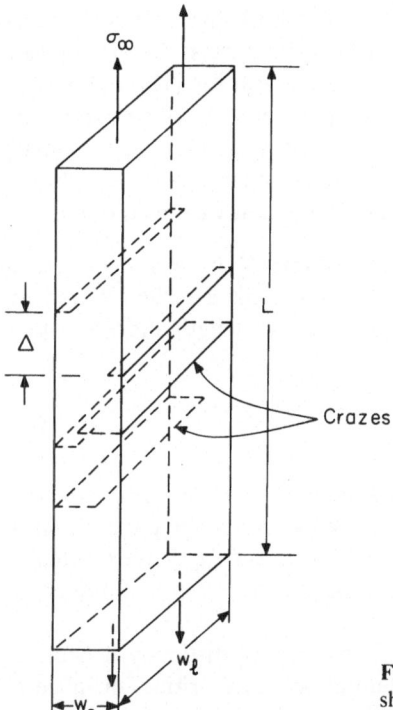

Fig. 2. Distribution of surface-initiated crazes in a thin sheet sample

For a steady state to exist, a kinetic balance is necessary between initiation and inactivation, i.e. $d\varrho/dt = 0$. This gives immediately the required steady state craze front density as

$$\varrho_{ss} = \frac{(w_s + w_l)}{v} \frac{N}{t_{in}} .\tag{18}$$

From Eq. (18) we note that the total axial craze strain rate according to Eq. (6) is simply

$$\dot{\varepsilon} = C \frac{(w_s + w_l) N}{t_{in}} .\tag{19}$$

In Eq. (19) $C \simeq 6 \times 10^{-7}$ m for heterogeneous polymers and 3×10^{-6} m for homo PS as discussed above, and where the initiation time is obtained by evaluating Eq. (12) as

$$t_{in} = \tau \exp\left[\frac{\sqrt{3}A}{q(\sigma_\infty/Y)} - \frac{1}{2Q} q\left(\frac{\sigma_\infty}{Y}\right)\right] .\tag{20}$$

In Eq. (20), q is a stress concentration factor or distribution of stress concentration factors on the surface as discussed by Argon and Hannoosh [12], and σ_∞ is the applied tensile stress. In Eq. (20), the specific substitutions $s = \sigma_\infty/\sqrt{3}$ and $\sigma = \sigma_\infty/3$ were made and it was assumed (somewhat simplistically) that both the deviatoric stress and the negative pressure are concentrated equally at the surface sites. The appropriate values for τ, A, Q, Y were cited for PS at RT in connection with Eq. (12). For a more detailed discussion of how t_{in} varies with temperature, the reader is referred to Argon and Hannoosh [12]. From Eq. (19) we see that for this limiting geometry the velocity of crazes does not directly influence the craze strain rate and therefore the toughness. This is not so in the case of volume initiation limited crazing that we now discuss.

Volume initiation limited crazing. Consider a large volume V of a polymer with internal heterogeneities of approximately spherical shape with a radius R and an initial density of n_0/V (m^{-3}). The rate of increase of craze front length under a distant stress of σ_∞ in this case is

$$\left(\frac{d\varrho}{dt}\right)^+ = \frac{2\pi R}{V} \frac{n_0}{t_{in}(\sigma_\infty)}, \tag{21}$$

where the craze initiation time t_{in} is again a strong function of the applied stress concentrated around the particle by the different elastic properties of the particle. Since any one particle can initiate more than one craze and the density n_0/V is expected to be quite high, we make no correction for the exhaustion for the potential numbers of sites for nucleation as ϱ increases.

The rate of craze front inactivation in this geometry will be distinctly dependent on mutual arrest of craze fronts when they encounter within a critical distance δ, permitting the craze fronts to sweep out a mean free path of $l/2$ in each direction from the source particle at the center, as shown in Fig. 3. Considering that the typical craze front per "cell" of $l^2\delta$ will be on the average only half as long as the periphery of the cell, we have for the active craze length per unit volume

$$\varrho = \frac{4(l/2)}{l^2\delta} = \frac{2}{l\delta}, \tag{22}$$

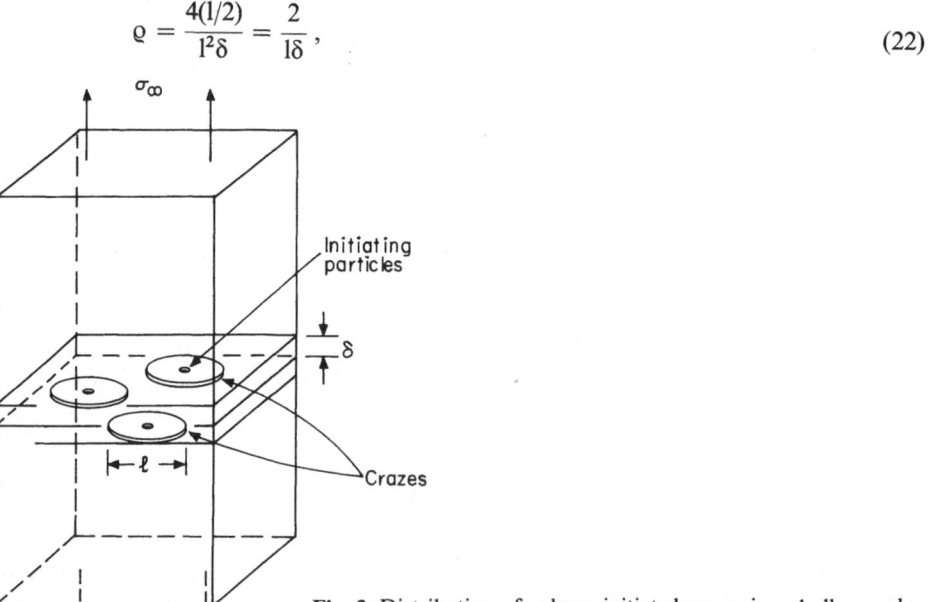

Fig. 3. Distribution of volume-initiated crazes in a bulk sample

and an inactivation rate of

$$\left(\frac{d\varrho}{dt}\right)^{-} = 2\,\frac{4l}{l^2\delta}\,\frac{1}{t_{inact}}\,, \tag{23}$$

with an inactivation time of

$$t_{inact} = \frac{1}{2v}\,. \tag{24}$$

This gives for the total rate of increase of the active craze front length per unit volume

$$\frac{d\varrho}{dt} = \left(\frac{d\varrho}{dt}\right)^{+} - \left(\frac{d\varrho}{dt}\right)^{-} = \frac{2\pi Rn_0}{Vt_{in}} - 4\varrho^2 v\delta\,. \tag{25}$$

From Eq. (25) we find the steady state ϱ_{ss} by setting this equation to zero,

$$\varrho_{ss} = \sqrt{\frac{\pi}{2}\left(\frac{n_0}{V}\right)\left(\frac{R}{\delta}\right)\frac{1}{vt_{in}}}\,. \tag{26}$$

We expect that the critical capture distance δ will not be very stress dependent. This is because the main body of the craze transmits full traction while the expected deficiency in the craze traction near the craze tip will scale with the applied stress: so will the allowable decrease in stress at the critical distance δ that stops the growth of the oncoming craze. Hence the stress dependence in ϱ_{ss} results from both the craze velocity and the craze initiation time.

In this geometry, the craze strain rate becomes

$$\dot{\varepsilon} = C\,\sqrt{\frac{\pi}{2}\left(\frac{n_0}{V}\right)\left(\frac{R}{\delta}\right)\frac{v}{t_{in}}}\,, \tag{27}$$

where the craze initiation time obeys a relation of the form given by Eq. (20) with, of course, a different factor q that results from the three-dimensional nature of stress concentration around a particle.

2.2.3 Craze Growth Velocity

Craze stresses

Studies of craze microstructure and the surrounding displacements of crazes have established that the only parts around a craze that undergo plastic deformation are concentrated into a "process zone" at the tip of the craze, and into a fringing layer all around the entire craze body. In the process zone craze matter is generated by one of the two processes discussed above, and fibrils are necked down to the final extension ratio. In the fringing layer, additions are made to craze fibrils by drawing polymer out of half space. Outside the idetifiable parts of a craze, the solid polymer remains entirely elastic while inside the craze body the fully drawn fibers carry the required craze tractions purely elastically in their orientation hardened state at the final extension ratio. While in some polymers with too low an entanglement density fibril creep can occur, this is of no interest to us in that it results in rapid craze matter break-down and fracture. Thus, for the purposes of stress analysis it is fitting to consider the craze as a dilatational transformation of the polymer. This we will do below for simplicity in a plane strain geometry.

The region of the two-dimensional craze consists of expanded fibrilar material that prior to its expansion occupied a "primordial" region of thickness $2k(x)$ where x is the coordinate in the craze plane as indicated in Fig. 5. The analysis of stresses

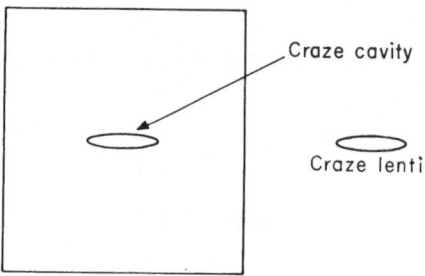

Craze cavity

Craze lentil

Fig. 4. Craze viewed as a dilatational transformation. The craze lentil removed as a thought process is to undergo an expansion transformation in the vertical direction, and is to be fitted back into the craze cavity followed by the application of a normal stress

and displacements of the craze border as well as the stresses outside the region of the craze can be best understood by a series of transformations and superposition exercises, treating the infinite solid with the craze cavity and the craze sheet separately. We proceed by cutting out a lenticular cylinder of thickness $2k(x)$ and laying aside the cut-out region for processing separately as shown in Fig. 4. The cross-sectional shape of the lentil is one of very large aspect ratio $\beta = a/b \gg 1$. Consider now a reference half space under uniform stress σ_∞. The displacements u_y are linear in y, with $u_y = 0$ along $y = 0$ everywhere. Consider also a second half space loaded along the strip $-a \leq x \leq a$ with a surface compressive stress $(\sigma_\infty - \sigma_c)$ with the rest of the half space along $x < -a$, $x > a$ being subjected to a displacement constraint of $u_y = 0$. From the theory of elasticity [32], we have for the second half space loading problem the following displacements and stresses along the plane $y = 0$.

$$u_y = \frac{2(1 - v^2)(\sigma_\infty - \sigma_c)}{E} \sqrt{a^2 - x^2} \quad \therefore \quad (-a \leq x \leq a) \tag{28}$$

$$u_y = 0 \quad (x < -a; \; x > a) \tag{29}$$

$$\sigma_{xx} = \sigma_{yy} = -(\sigma_\infty - \sigma_c) \quad (-a \leq x \leq a) \tag{30}$$

$$\sigma_{xx} = \sigma_{yy} = (\sigma_\infty - \sigma_c) \left\{ \frac{x}{\sqrt{x^2 - a^2}} - 1 \right\} \quad (x < -a; \; x > a). \tag{31}$$

Superposition of the first and second half-plane loading problems gives the desired description of a region under a distant traction of σ_∞ and a craze traction of σ_c, as

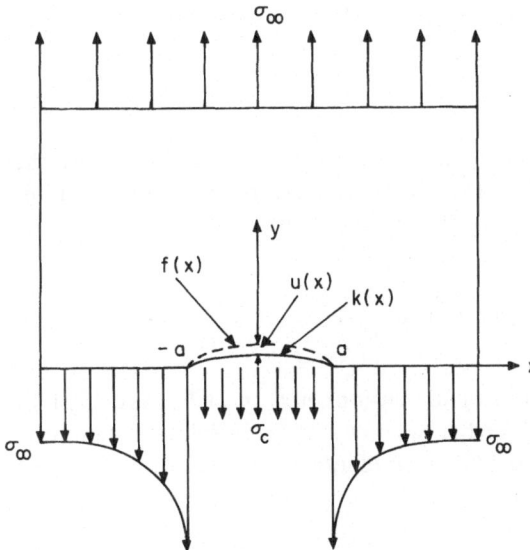

Fig. 5. Distribution of craze related tractions on and around the ellipsoidal craze cavity under stress σ_∞: $k(x)$ is the premordial craze cavity, $u(x)$ the additional displacements of the craze border upon the insertion of the expanded craze lentil and the application of the external stress resulting in the final shape of the craze border $f(x)$

shown in Fig. 5, resulting in the displacements of Eqs. (28) and (29) and the stresses that are

$$\sigma_{xx} = -(\sigma_\infty - \sigma_c) \tag{32}$$

$$\sigma_{yy} = \sigma_c \tag{33}$$

along the strip $(-a \leqq x \leqq a)$, and

$$\sigma_{xx} = (\sigma_\infty - \sigma_c) \left\{ \frac{x}{\sqrt{x^2 - a^2}} - 1 \right\} \tag{34}$$

$$\sigma_{yy} = (\sigma_\infty - \sigma_c) \frac{x}{\sqrt{x^2 - a^2}} + \sigma_c \tag{35}$$

on the outside of the strip along $(x < -a; x > a)$. Adding the initial "primordial" half thickness dimensions $k(x)$ of the craze cavity, we obtain the final displacement $f(x)$ of the craze cavity under the craze traction σ_c as

$$f(x) = u(x) + k(x) = k(x) + \frac{2(1 - v^2)(\sigma_\infty - \sigma_c)}{E} \sqrt{a^2 - x^2} \tag{36}$$

along the extent of the craze $(-a \leqq x < a)$.

Consider now in a corresponding way the craze lentil. Initially of a half thickness $k(x)$, the craze lentil undergoes a pure extensional transformation in the y direction without any change of dimension in the x (and z) directions, made possible by the craze fibrillation. This results in an expansion of the craze lentil in the y direction by the average extension ratio of the craze matter fibrils λ_n to give an unconstrained shape of the craze lentil

$$g(x) = \lambda_n k(x) \cdot \tag{37}$$

Under the action of the craze traction σ_c, the craze thickens a bit further by the elastic stretch of the craze fibrils. If E_y^c is the effective averaged craze lentil modulus in the fibril direction, the final shape h(x) of the craze lentil under full traction σ_c becomes

$$h(x) = g(x) + g(x) \frac{\sigma_c}{E_y^c} = \lambda_n k(x) \left(1 + \frac{\sigma_c}{E_y^c} \right). \tag{38}$$

The expanded and stretched craze lentil of dimensions h(x) must however fit into the craze cavity f(x) exactly. To proceed further, we now make an idealized assumption that the premordial craze matter shape k(x) is an elliptical cylinder with an aspect ratio β, i.e.

$$k(x) = \frac{1}{\beta} a \sqrt{1 - \left(\frac{x}{a} \right)^2} \cdot \tag{39}$$

Combining Eqs. (36), (38) and (39), and noting that the craze matter is very much stiffer than the craze cavity, i.e. $(\lambda_n \sigma_c / (\lambda_n - 1) E_y^c) \ll 1$, we obtain the craze traction σ_c to be

$$\sigma_c = \sigma_\infty - \frac{E(\lambda_n - 1)}{2(1 - \nu^2) \beta} \cdot \tag{40}$$

From the simple exercise given above, we conclude that if the craze lentil shape remained constant with an unchanging aspect ratio β, the craze traction σ_c would always be less than the applied stress σ_∞, and the craze tip "driving force"

$$K_I = (\sigma_\infty - \sigma_c) \sqrt{\pi a} = \frac{E(\lambda_n - 1)}{2(1 - \nu^2) \beta} \sqrt{\pi a} \tag{41}$$

would increase steadily. This would have to result in accelerating crazes since there is no known way for the extension resistance of a craze with constant morphology in a pure homopolymer to increase at the same rate.

We note further, however, that if the craze thickening were to cease after a given time when a certain craze thickness $2k_0$ were achieved at the center of the craze, the aspect ratio β would monotonically increase with a, and σ_c would monotonically approach σ_∞. This would result in an ever decreasing craze tip driving force

$$K_I = \frac{E(\lambda_n - 1) k_0}{2(1 - \nu^2)} \sqrt{\pi/a} \cdot \tag{42}$$

and a decelerating craze growth rate.

Crazes in real polymers differ in some important respects from the analysis given above. First, they are generally three-dimensional and penny shaped (or half-penny shaped, if nucleated from a free surface). This, however, results only in the incorporation of a factor $\pi/2$ in the last term of Eq. (40) and on the right sides Eqs. (41) and (42). Second, and more importantly, the available evidence indicates that craze lentils are not of ellipsoidal shape but have feathered tips, not giving rise to the very high concentrations of elastic stress that is indicated by the distributions of Eqs. (34) and (35). This results from the development of a craze-tip process zone of extent Δ within which the craze tuft extension ratio develops from the initial convolutions of the interface or from the craze tip cavities to the fully established craze matter extension ratio λ_n. This process zone acts as a cavitational plastic zone at the tip region of the craze. There is no evidence for any important distortional plastic zone outside the distinguishable region of craze matter in the solid polymer. In fact, when such distortional plasticity can develop outside the craze matter region craze growth is usually observed to stop. Third, most craze velocity measurements in pure homopolymer indicate that intact crazes do not accelerate under a constant applied stress. They either grow at constant velocity, when in vacuum or in a neutral environment, or decelerate slowly to the velocity corresponding to that in the neutral environment from an initially higher velocity when the environment is not neutral. Thus, from these differences, we conclude that under a constant applied stress the actual craze tip driving force remains constant and independent of craze length, or at most decreases slightly. This is compatible only with achievement of a craze tip traction distribution that remains independent of craze length, which indicates that the craze traction σ_c along the majority of the craze body rises asymptotically to equal the applied distant stress σ_∞, as had been anticipated first by Andrews and Bevan [33]. The traction distributions on crazes, measured in thin films [21] indicate that the sought distribution for crazes growing at constant velocities must peak at the craze tip to above the applied stress, dip under the level of the applied stress somewhere toward the end of the process zone Δ, in the craze body region, and then rise again to the level of the applied stress over the remainder of the craze body (see chapter 1). The mechanistic basis of the development of such a traction distribution resulting in a constant craze tip driving force is not clear at this writing. It is, however, clear that this must involve cessation of craze thickening beyond a certain stage — most likely as a result of a reproducible history of physical aging in the fringing layer of the craze borders where polymer tufts are drawn out of the half spaces on either side. It is necessary to observe here that nearly all the craze traction distributions discussed in the literature and recently reviewed by Bevan [34] are inadequate as they incorporate a constant craze body traction $\sigma_c < \sigma_\infty$. This, however, as explained above, must result in craze acceleration — if no unjustifiable criteria, such as assuming that Δ/a remains constant, are involved. We take note of this principal deficiency in the understanding of craze growth in our developments below, and re-cast our craze growth models in such a way that this detailed traction distribution at the craze tip becomes unnecessary. A similar approach has been employed by Verheulpen-Heymans and Bauwens [35] earlier, albeit in connection with unjustifiable assumptions for the mechanical properties of craze matter.

Growth of crazes by interface convolution

Noting that the kinetics of craze growth in pure homopolymer is different from what might be expected from a mechanism of repeated cavity nucleation at the craze tip, and that the topology of craze matter is one that involves continuously interconnected

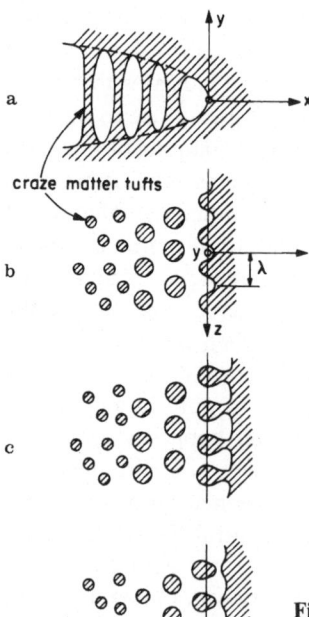

craze matter tufts

Fig. 6 a–d. The interface convolution process producing craze matter in homopolymer under stress (after Argon and Salama [13]), courtesy of Taylor and Francis)

air passages, Argon and Salama [13] proposed that craze matter is produced by the repeated convolution of the craze tip interface as depicted in Fig. 6. Apart from the fact that this process is ubiquitous as a meniscus instability in the flow of fluids, spreading and separation of fluid-like media with complex rheology, and even in the fracture of highly non-linear metallic glasses [22], it furnishes a quantitatively accurate framework to explain the kinetics of craze growth. The key to the development is a perturbation analysis by Argon and Salama [22] that has lead to the computation of the principal wave length λ of the instability that is advancing with a velocity u_x, that is obtained as

$$\lambda = 2 \sqrt{\sqrt{3} \, \frac{H\chi}{\eta_0} \left(\frac{2}{\dot{\varepsilon}_e}\right)^n}$$

(43)

$$u_x = \frac{n}{2(1 + 2n)} \sqrt{3} H \left(\frac{\dot{\varepsilon}_e}{2}\right).$$

(44)

In Eqs. (43) and (44), H is the channel height in which the meniscus convolutes (or in our case the craze tip opening displacement), χ the surface energy of the solid polymer, $\dot{\varepsilon}_e$ the equivalent inelastic deformation rate, η_0 a non-linear extensional viscosity coefficient, and n a strain rate exponent in the non-linear connection between the equivalent deformation resistance σ_e and the equivalent strain rate. The last four quantities are taken to be related by a non-linear constitutive law that is

$$\sigma_e = \frac{3}{2}\eta_0\dot{\varepsilon}_e^n .$$

(45)

Recognizing that the principal wave length of the convolution λ, must be intimately related to the opening displacement H, or vice versa, we equate λ to H and introduce Eq. (45) into Eq. (43) to obtain

$$\lambda = \frac{12\sqrt{3\pi^2}}{2^{1-n}} \left(\frac{\chi}{\sigma_e}\right). \tag{46}$$

If σ_e is taken as the tensile plastic yield strength Y of the solid polymer (= 100 MPa for PS at room temperature), χ the surface energy (= 5×10^{-2} J/m^2), and n the phenomenological plastic strain rate exponent (= 0.05 at room temperature [20]), we calculate λ to be 53 nm, i.e. exactly what is measured by transmission electron microscopy [36] or low angle X-ray diffraction [37]. The magnitude of the principal wave length of Eq. (46) can also be used in Eq. (44) to obtain an expression for the craze velocity without specific knowledge of the craze tip traction distribution. This gives

$$u_x = \frac{da}{dt} = 9\pi^2 \left(\frac{n}{(1+2n)\,2^{1-n}}\right) \frac{\chi}{\sigma_e} \dot{\varepsilon}_e. \tag{47}$$

To obtain the proper temperature and stress dependence of the craze velocity we use the specific form for $\dot{\varepsilon}_e$ that incorporates this information [20, 7]

$$\dot{\varepsilon}_e = \dot{\varepsilon}_0 \exp\left[-\frac{B}{kT}\left(1 - \left(\frac{\sigma_e}{\hat{Y}}\right)^{5/6}\right)\right] \tag{48}$$

where $\dot{\varepsilon}_0$ is a pre-exponential frequency factor (= 10^3 sec^{-1}), B a scale factor for the activation free energy for plastic flow incorporating a number of molecular parameters determined by Argon and Bessonov [7], and $\hat{Y}$ an athermal flow stress given by

$$\hat{Y} = \frac{0.133}{(1-\nu)}\mu \tag{49}$$

in which μ and ν are the shear modulus and the Poisson's ratio of the solid polymer. Thus the craze velocity becomes determinable under any condition if the appropriate effective plastic resistance of the plastically deforming solid polymer can be established. We do this by drawing from the conclusions of the preceding section. In a craze tip zone advancing with a given velocity, the equivalent plastic resistance can not differ much even though the actual traction distribution can vary due to different levels of orientation hardening. We consider the region at the end of the process zone Δ where fully developed craze tufts have been established and plastic drawing of the tufts out of the two half spaces still occurs at an average traction of σ_∞. As discussed by Argon et al. [15], in this region the root of a representative craze tuft is under a force of $\sigma_\infty A_0$ while the drawn portion of the tuft body inside the craze matter with area A is subject to the same force but has a different total deformation resistance $Y(\lambda) = Y_0(1 + Y_h(\lambda_n)/Y_0)$, i.e.

$$\sigma_\infty A_0 = AY_0(1 + Y_h(\lambda_n)/Y_0), \tag{50}$$

where Y_0 is the initial yield strength and $Y_h(\lambda_n)$ is the back stress due to the orientation hardening contribution of entropic origin. On the basis that in this drawing of a tuft, the equivalent plastic resistance over and above the back stress due to orientation hardening must be constant at all stages, we equate σ_e to the initial value Y_0 of this drawing process to obtain

$$\sigma_e \approx \sigma_\infty \lambda_n/(1 + Y_h(\lambda_n)/Y_0) = \sigma_\infty \lambda_n' \tag{51}$$

where λ_n' is defined by the second equality in Eq. (51). With this we finally state the craze velocity to be

$$\frac{da}{dt} = D_1 \exp\left[-\frac{B}{kT}\left(1 - \left(\frac{\sigma_\infty \lambda_n'}{\hat{Y}}\right)^{5/6}\right)\right] \tag{52}$$

where

$$D_1 = \frac{9\pi^2 n\chi\dot{\varepsilon}_0}{(1 + 2n)\,2^{1-n}\sigma_\infty \lambda_n'}. \tag{53}$$

Argon and Salama [13] have demonstrated that this form of the expression accounts for the measured craze growth kinetics in homogeneous glassy polymers very well, giving for λ_n' values in the range of 1.5–2.0, which is quite consistent with the definition of this quantity in Eq. (51) and the known extension ratio of $\lambda_n \simeq 4$ in craze matter and back stresses that correspond to this extension ratio.

Growth of crazes by repeated cavitation of hetero-phases

Argon et al. [15] have presented very strong evidence that crazes in the commercial block copolymer KRO-3 Resin with lamellar morphology form and grow by extensively cavitating the rubbery phases. Additional detailed studies by Schwier [38] on pure PB/PS di-block copolymers of both rod and spherical domain morphology have demonstrated directly for the former, and indirectly for the latter, that here too, craze growth involves extensive repeated cavitation of the rubbery domains (see also Argon et al. [9]). As we will discuss more specifically below in Sections 4 and 5, the cavitation results from the combined action of the large negative pressures concentrated in the rubbery domains due to differential thermal contraction between the applied stresses. As pointed out by Argon et al. [15], the growth of crazes in such hetero-phase polymers is best understood by the adaptation to crazing of a model of Andersson and Bergkvist [39] for the growth of cracks in a degrading material layer in the plane of the crack. In the context of craze growth by cavitation of an included phase, the model of Andersson and Bergkvist transforms into the geometry presented in Fig. 7a and b. The material plane of thickness δ, ahead of the plane of the craze must be brought up to a normal traction level σ_m, purely elastically, whereupon, as depicted in Fig. 7b, the cavitation strength of the rubbery phase is reached. As soon as cavitation occurs in the rubbery phase, its load carrying capacity (by virtue of the negative pressure that is supported up to that point) is completely lost, resulting in the elevation of the effective stress in the majority phase of glassy polymer to the level of plastic flow. This initiates the "degradation" of the solid by the plastic expansion

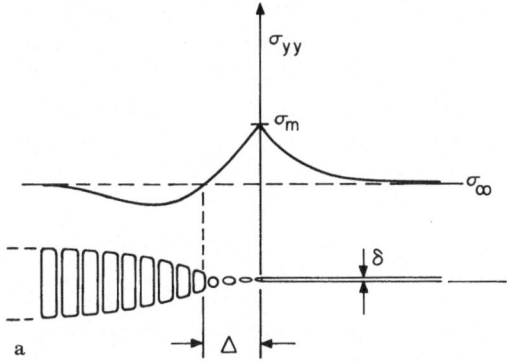

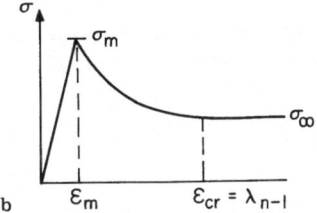

Fig. 7a and b. Sketch of a craze and its tractions for the growth process by repeated cavitation of heterophases at the craze tip. In the process zone Δ, mature craze matter tufts are established as material points enter at right and exit at left: **a** traction distribution across craze; **b** sketch of traction displacement law for craze matter production

of the now porous glassy polymer to form the craze tufts under a dropping traction down to σ_∞ as the uniaxial extension of the cavitated layer is increased from the small strain ε_m to the fully established craze strain of $(\lambda_n - 1)$. The resulting traction across the plane of the craze is sketched out in Fig. 7a, showing the "mapping" of the dropping traction profile of Fig. 7b into the craze tip process zone of extent Δ in an inverted manner. The actual traction distribution is sensitive to the specific shape of the dropping portion of the traction-displacement law and is not known. On the basis of our discussion in the section on craze stresses above, however, we expect that the traction will dip below σ_∞ in the front portion of the craze body just beyond Δ, and should again rise to the level of σ_∞ in the majority of the craze body. The dipped portion of the traction at the front part of the craze body is assumed to remain of constant extent and results in a constant craze tip driving force K_I that must be balanced by the craze tip process zone resistance K_{Ic}. While the latter can be calculated with some precision from the process history of the production of mature craze matter inside Δ, the former is not known because of the undetermined nature of the traction distribution. As discussed earlier by Argon et al. [15], the velocity of crazes in this modification of Andersson and Bergkvist's degrading material model is

$$\frac{da}{dt} = \frac{\Delta}{\varepsilon_{cr}} \dot{\varepsilon}_e(\sigma_e) \tag{54}$$

$$\Delta = \frac{K_{Ic}^2}{\pi(\sigma_m - \sigma_\infty)^2} \tag{55}$$

$$K_{Ic}^2 = \frac{E_c^2 \varepsilon_m^2 \delta \left(1 + \dfrac{1}{\theta}\right)}{(1 - \nu_c^2)} \tag{56}$$

where E_c and v_c are the average Young's modulus and Poisson's ratio of the composite polymer, ε_m the elastic strain at which cavitation in the rubbery domain begins, and θ the ratio of the descending slope to the ascending slope of the material degradation law sketched out in Fig. 7b. Considering that the descending portion of the degradation law arises from the initial drawing traction of a bar of yielding polymer undergoing negligible strain hardening, we estimate $\theta = \varepsilon_m$, and evaluate K_{Ic}^2 to be

$$K_{Ic}^2 \frac{\sigma_m E_c \delta}{(1 - v_c^2)} \, . \tag{57}$$

The initial thickness δ of the cavitating layer that appears arbitrary is actually fixed by the size of the phase domains of the rubber as we will discuss in detail in Section 4 below.

Combining Eqs. (54), (55) and (57) and using the same approach to the establishment of the equivalent plastic resistance of the deforming polymer that was introduced in connection with the mechanism of craze growth by the interface convolution process, we write the craze velocity to be

$$\frac{da}{dt} = D_2 \exp\left[-\frac{B}{kT}\left(1 - \left(\frac{\sigma_\infty \lambda_n'}{(1 - c)\,\hat{Y}_c} \right)^{5/6} \right) \right] \tag{58}$$

$$D_2 = \frac{\delta \dot{\varepsilon}_0}{(1 - v_c^2)\,(\lambda_n - 1)\,\pi(\sigma_m/E_c)\,(1 - \sigma_\infty/\sigma_m)^2} \tag{59}$$

where λ_n' has the same meaning as that in Eq. (51). The additional term $(1 - c)$ in the denominator reflects that only that volume fraction is made up of the drawing glassy polymer, while $\hat{Y}_c$ now represents the athermal plastic resistance of the composite calculated from the composite elastic constants.

2.2.4 Fracture of Craze Matter under Stress

A number of investigators [16, 17, 40-42] have studied in detail the form of breakdown of craze matter under stress in homo-polystyrene. The causes of such breakdown or macro cavity initiation by multiple and adjacent fracture of craze matter tufts has been found to result almost exclusively from particulate impurities trapped in the polymer during manufacturing [16, 17]. In the still widening portion of a craze, when such an entrapped particle reaches the craze borders and its interfaces are subjected to complex states of stress with large negative pressure components in the tuft drawing region it can be torn away from its surroundings. If the size of the particle is much larger than the mean craze matter wave length of about 50 nm, it can act as a stress concentrating imperfection on the surrounding fibrils that can enlarge the initial cavity around the particle by preferential fracturing of neighboring fibrils. Although this is probably the most important of the causes of craze matter fracture, it has been observed also that interaction of crazes with oblique surface scratches [41] and imperfections of craze matter itself, particularly in the so-called central "midrib" region [42], can initiate fractures in craze matter. In addition, many dynamic experiments under high strain rate fracture conditions [43] or when high amplitude stress waves are made to traverse through crazed polymer [44], fractures are found to initiate from the craze matter borders quite often in the absence of apparent imperfections. This suggests that the craze border is weaker [41] than either the drawn tufts or the solid polymer — being a region of partially drawn polymer with high strain gradients and large spatial variations of internal stress. Thus, although the features of the craze breakdown and initiation of the fracture process has been very well documented, there is no definitive quantitative information on the time dependent fracture of craze matter or craze interface under stress that is required for incorporation into the overall understanding of the toughening process.

3 Degrees of Freedom in the Control of Morphology

Microphase separation in narrow molecular weight distribution diblock and triblock copolymers has been described successfully in several different theoretical treatments [45-47]. In Fig. 8, the theory of Helfand [45] has been employed to generate a type of 'phase diagram' on which distinct regions of morphology are evident. Which of these five basic structures, i.e., spheres of A in a matrix of B, cylinders of A in B, alternating lamellae, cylinders of B in A and spheres of B in A, as shown in Fig. 9, is the most stable at equilibrium depends almost entirely on the copolymer composition and only very slightly on molecular weight. The boundaries between regions in Fig. 8 are nearly vertical. The characteristic sizes and spacings of the morphological features in any of the five regions of the phase diagram of Fig. 8 depend on the molecular weights of the constituent blocks of the copolymer; quantitative expressions for the molecular weight dependence have been published [45] and a considerable body of experimental evidence lends support to these power law expressions. The location of the curved boundary in Fig. 8 which separates low molecular weight homogeneous

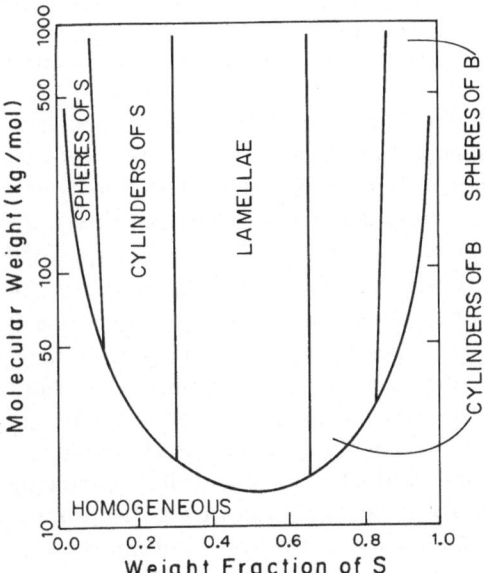

Fig. 8. Equilibrium morphologies attainable in di-block copolymers as a function of weight fraction and molecular weight of species

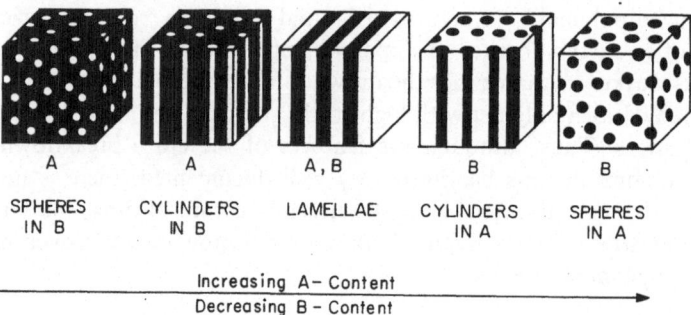

SPHERES   CYLINDERS   LAMELLAE   CYLINDERS   SPHERES
IN B   IN B   IN A   IN A

Increasing A- Content

Decreasing B - Content

Fig. 9. Equilibrium morphologies attainable in di-block copolymers as the concentration of species A increases and species B decreases

block copolymers from the microphase separated high molecular weight materials depends on the degree of incompatibility between A and B, as manifest for example [48] in the magnitude of the difference in solubility parameters, $|\delta_A - \delta_B|$. A larger solubility parameter mismatch translates into phase separated materials at lower molecular weights.

When block copolymers have broad molecular weight distributions or when blends of monodisperse block copolymers are prepared, the morphologies may be more complex and are no longer determined by considerations of copolymer composition alone. An example of this [49] is found in the two commercially available K-Resin block copolymers, both of which contain 23 wt % polybutadiene and 77 wt % poly-styrene. KRO-1 ($\bar{M}_w = 179$; $\bar{M}_n = 132$ kg/mole) exhibits a morphology best described as an interconnected network of randomly wavy cylinders of polybutadiene in a polystyrene matrix as shown in Fig. 10a, whereas KRO-3 ($\bar{M}_w = 217$; $\bar{M}_n = 106$ kg/ mole) exhibits regular but randomly disoriented domains of parallel lamellae as shown in Fig. 10b. In Fig. 11, a blend of two narrow distribution block copolymers reveals a complex mixture of two coexisting morphologies quite different from the theoretical prediction of polybutadiene cylinders for this copolymer composition of 23% polybutadiene, 77% polystyrene.

Addition of one or both of the corresponding homopolymers to a block copolymer offers further possibilities for constructing morphologies of specified form and dimensions. A first consideration in this area is the case for which the homopolymers have molecular weights which are equal to or less than the molecular weights of the corresponding block segments in the copolymer, i.e. $M_{A, homopolymer} \leqq M_{A, block}$, and similarly for B. Under such conditions it has been demonstrated [50] that, up to certain limits, the added homopolymers may become solubilized in the respective A and B domains of the block copolymer. This leads to the possibility of coarsening, or swelling, the morphological features of the block copolymers by incorporating one or both of the corresponding homopolymers into a blended sample. If the two homopolymers are added in the same proportion as the composition of the original block copolymer, the compositions of all ternary blends so produced have the same value; the unique line of constant composition, or isopleth, shown in Fig. 12 describes this process. In principle, blends along the isopleth should exhibit a single type of morphology which becomes coarser in dimension proceeding from the apex to the base of the

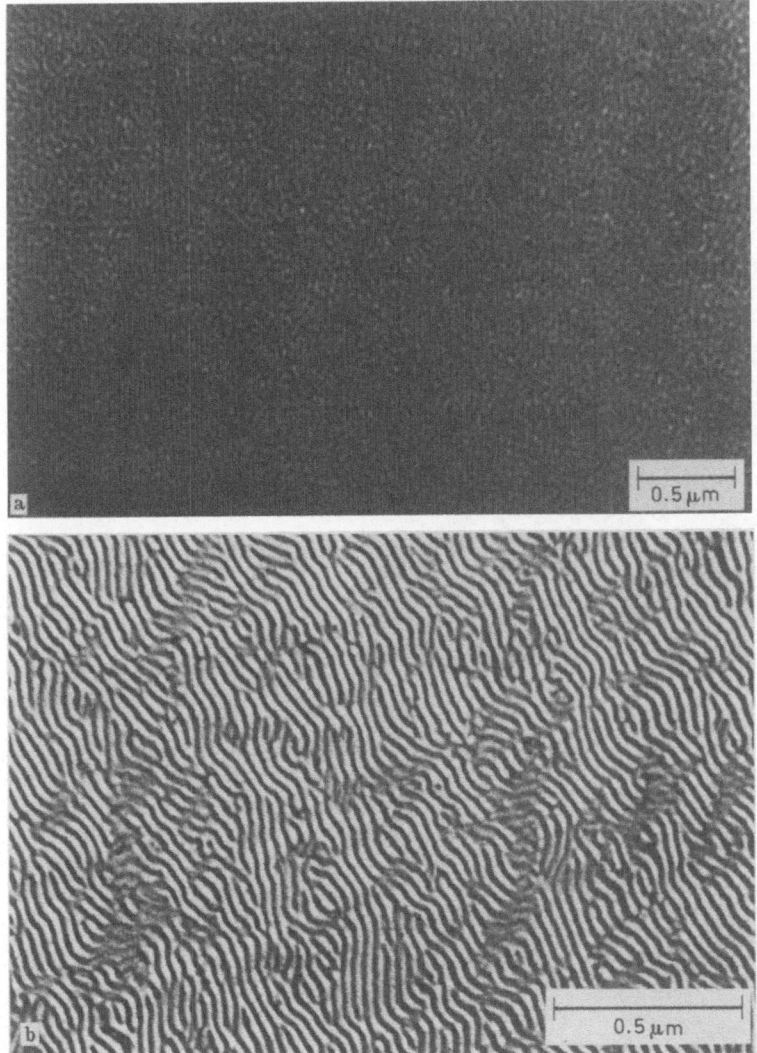

Fig. 10a. Randomly wavy and spatially interconnected rod morphology of KRO-1 Resin; **b** randomly corrugated lamellar morphology of KRO-3 Resin (after Argon et al. [15], courtesy of J. Wiley and Sons)

ternary diagram. A certain amount of supporting evidence for this behavior has appeared in the literature [50]. However, other effects such as a change in the type of morphology, broadening of the distribution of domain sizes, disruption of the long-range packing order of domains and a limit to the amount of homopolymer which can be solubilized have been observed in our work. An example of this is in Fig. 13a–c, where 13a illustrates a pure block copolymer with 23 % PB in a regular wavy rod morphology. Figures 13b and c in turn show the changing morphology at 23 % PB but where the di-block concentration is reduced to 0.5 and 0.2 respectively, illustrating

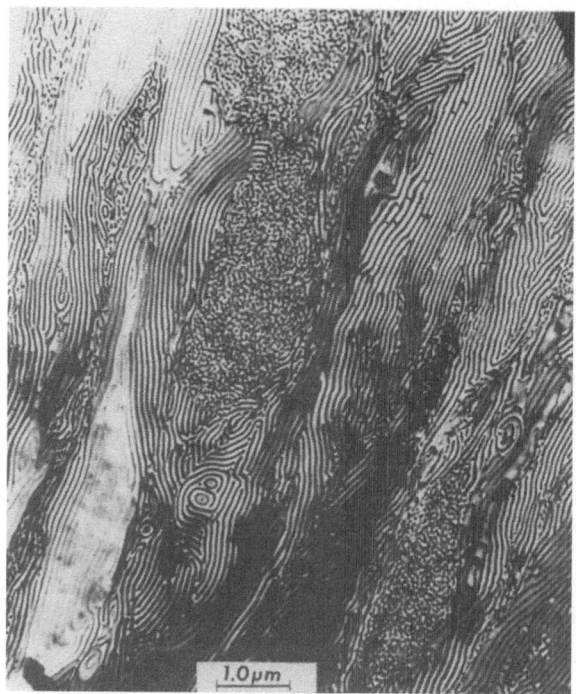

Fig. 11. Blend of two narrow distributions of di-block copolymers showing a mixture of two coexisting morphologies of wavy rods and corrugated lamellae

A/B DI-BLOCK COPOLYMER (M_A/M_B)

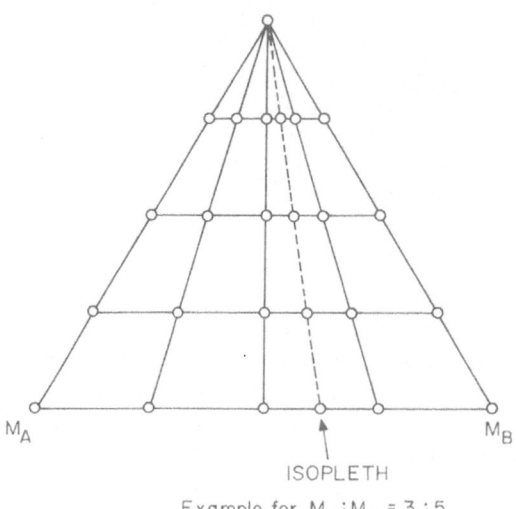

M_A M_B

ISOPLETH

Example for $M_A:M_B = 3:5$

Fig. 12. Ternary blend of an A/B di-block copolymer with additional homopolymer A and B along an isopleth to coarsen microstructure at relatively constant morphology

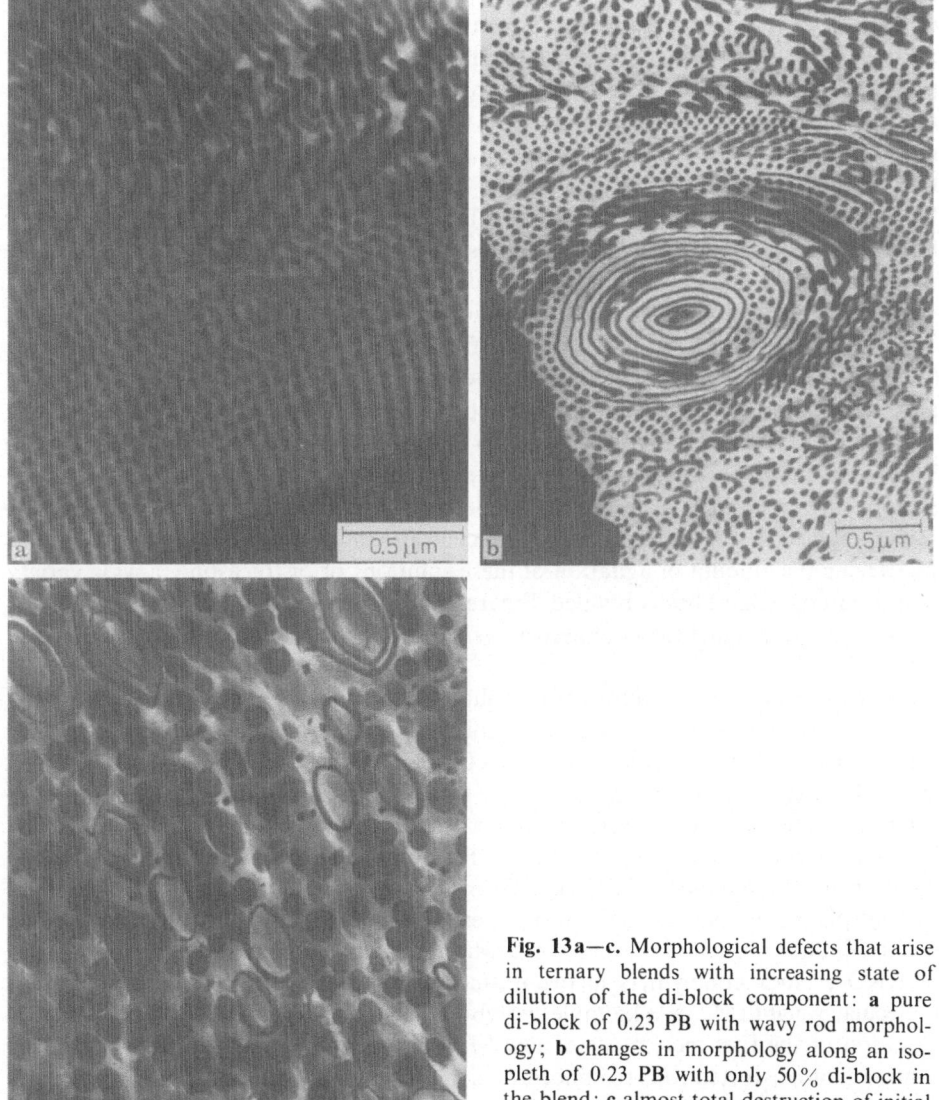

Fig. 13a—c. Morphological defects that arise in ternary blends with increasing state of dilution of the di-block component: **a** pure di-block of 0.23 PB with wavy rod morphology; **b** changes in morphology along an isopleth of 0.23 PB with only 50% di-block in the blend; **c** almost total destruction of initial morphology along the isopleth when only 20% di-block remains in the blend

the increasing morphological variation and almost complete loss of regularity of form. The solubility limit for homopolymers and block copolymers has been treated theoretically in some detail recently by Liebler [51].

When added homopolymers have molecular weights which substantially exceed the corresponding block component molecular weights, solubilization of homopolymers is not observed. Instead, the block copolymer separates from the homopolymers leading to a "three-phase" (A, B, AB) system. If only one of the homopolymers is employed, a "two phase" system is obtained: heterogeneous domains of AB embedded in the selected high molecular weight homopolymer, e.g. A. At equilibrium, the separation of the incompatible components should, in principle, be macroscopic in nature, i.e. layering of the components will occur similar to an oil/water system. In practice, microscopic particles with dimensions on the order of several micrometers of the heterogeneous block copolymer in the high molecular weight homopolymers may be desired. This type of morphology may be obtained by at least two methods. One method relies on the use of a second block copolymer to serve as an emulsifier for the first copolymer and the other method takes advantage of the relatively slow kinetics of block copolymer droplet coalescence when dispersed in a high molecular weight polymer, and perhaps solvent, medium.

Reiss [52] has demonstrated the emulsification power of a high molecular weight AB copolymer (Cop E) in the presence of a lower molecular weight AB copolymer (Cop D) and a high molecular weight homopolymer A. In this case, stable morphologies were obtained with well dispersed particles of Cop D, each containing the characteristic microphase structure of Cop D, embedded in the high molecular weight homopolymer. In Sections 4 and 5 below we present similar results obtained by solvent spin casting techniques using only a single copolymer and a high molecular weight homopolymer. Using different initial concentrations of copolymer and solvent and by varying the amount of agitation of these solutions prior to casting, a wide variety of morphologies have been obtained. Figure 25 a, b, and Fig. 30 show specific examples obtained by dispersing the commercial resins KRO-1 and KRO-3 in high molecular weight PS [53].

Another attractive possibility for morphology control is the case in which a homopolymer with high molecular weight is not solubilized by the block copolymer that forms separate heterogeneous particles but where a second homopolymer of low molecular weight is incorporated into the domain structure of the block copolymer islands. In this case, the internal microstructure of the block copolymer islands can be controllably modified: coarsened or transformed to a new phase form, by the addition of the low molecular weight homopolymer. Figure 25a and Fig. 30 show two morphologies obtained in this way, where high molecular weight polystyrene forms the continuous phase in both cases and where the embedded particles are composed of KRO-1 block copolymers incorporating, in the case of Fig. 30, additional low molecular weight ($M_w \simeq 3$ kg/mole) polybutadiene. The figures show that the incorporation of the first increments of low M_w PB produces a morphological transformation from distorted rods to concentric spherical shells while additional increments coarsen this new morphology until the solubility limit of low M_w PB is reached.

For completeness it should be pointed out that many interesting morphologies of the type described above can be obtained by polymerizing a solution of AB or ABA block copolymers in styrene monomer, analogous to the process of HIPS production in which polybutadiene homopolymer is the dissolved species. This approach has been described in detail by Echte [54].

4 Toughness of Pure Di-Block Copolymers

4.1 Morphologoical Features of Crazing

4.1.1 Commercial K-Resins

Argon et al. [15] have studied in some detail the crazing of two commercial block copolymers marketed by the Phillips Petroleum Co. under the trade names of KRO-1 and KRO-3 Resins. The chemistry, molecular weight distribution and some mechanical properties of these two block copolymers are given in Table 1. A distinguishing feature of these K-Resins is that the PB branches of the block components of the molecules are star shaped — being topologically connected together at the center. This gives the PB domains an aspect of being cross-linked. The yield stresses listed in Table 1 refer to craze yielding and should not be taken as the true yield stresses of the uncavitated material in compression. Such compressive yield stresses are considerably higher. The morphology of these two resins in fully equilibrated, annealed sheets is shown in Fig. 10a and b. They can be best characterized as an interconnected space network of randomly wavy rods of PB of 20 nm diameter surrounded by a matrix of PS for the KRO-1 Resin; and randomly wavy, quasi-parallel sheets of PB of 20 nm thickness separated by a PS matrix.

In the usual stress-strain experiments both of these block copolymers undergo craze yielding at a stress level of 21 MPa. The axial elongation of the samples is unaccompanied by any significant reduction in cross sectional dimensions all the way up to fracture. Fracture in the KRO-1 Resin is found to occur at a strain of roughly 0.09 while that of the KRO-3 Resin is as large as 1.10. The clue to the very large difference in toughness of these two block copolymers is obtained from examination of the crase microstructure.

In KRO-1, crazes were observed to nucleate predominantly on the surface and particularly from the cut edges of the thin sheet specimens. The crazes were generally straight, well defined and in most instances propagated across the entire thickness of the sheet. Transmission electron microscopy accompanied by the standard methods of accentuation of the rubbery phase by $Os O_4$ showed well defined crazes of roughly 0.2–0.5 µm thickness as shown in Fig. 14a. Figure 14b shows that the microstructure of such crazes consists of fibrils of PS with large extension ratio, with many attached islands of PB indicating that the PB phase is not significantly incorporated into the

Table 1. Some Physical and Mechanical Properties of KRO-1 and KRO-3 Resins at Room Temperature [15]

	KRO-1	KRO-3
M_w	179 kg/mole	217 kg/mole
M_n	132 kg/mole	106 kg/mole
Weight fraction PB.	0.23	0.23
Yield stress	20.95 MPa	21.0 MPa
Young's modulus	1.27 GPa	1.47 GPa
Strain to fracture	0.09	1.10
$\hat{Y}$	92.7 MPa	107.5 MPa

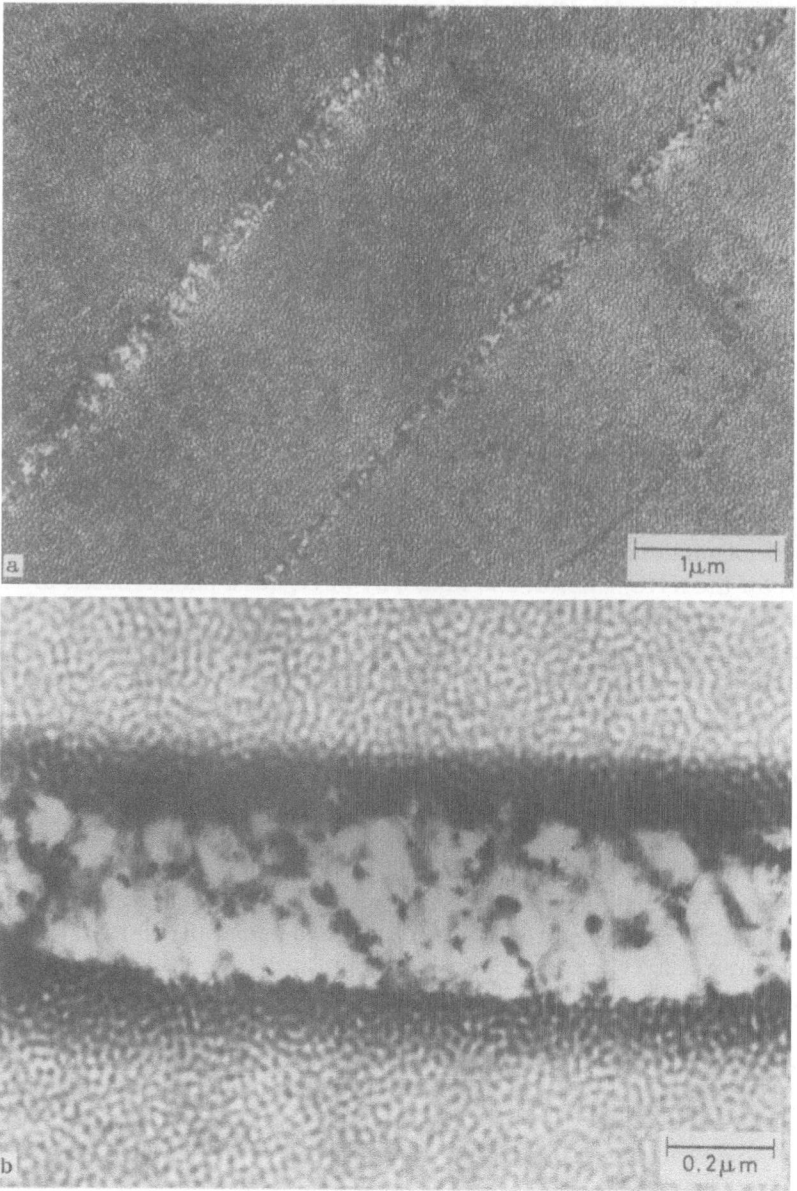

Fig. 14a and b. Crazes in KRO-1 Resin: **a** general distribution of well defined crazes in the KRO-1 microstructure; **b** detailed view of craze microstructure consisting of drawn tufts of PS incorporating some drawn PB. (From Argon et al. [15]; courtesy of J. Wiley and Sons)

craze matter fibrils and does not play an important role in the drawing of fibrils, although, as we will see below, it is vital in the crazing process as a whole.

In KRO-3, a very different situation exists. The crazes are macroscopically indistinct and wavy. Whitening of the sample is more homogeneously distributed. The micro-

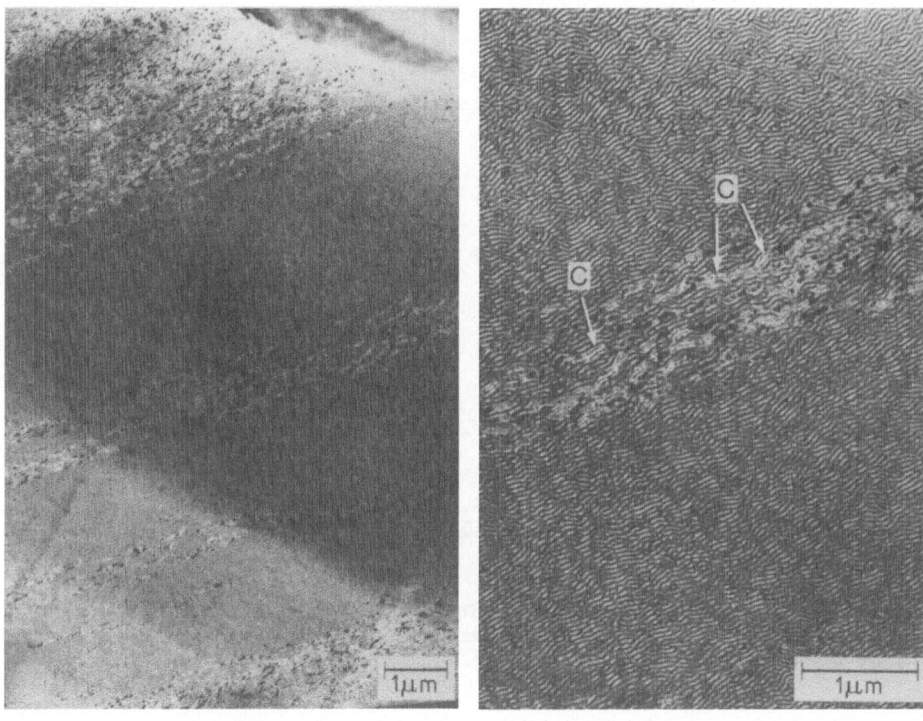

Fig. 15.   Fig. 16.

Fig. 15. Micrograph of a crazed KRO-3 Resin showing dispersed cavitation and localized cavitated regions acting as crazes. (From Argon et al. [15]; courtesy of J. Wiley and Sons)

Fig. 16. Micrograph of a craze zone in KRO-3 Resin. The dark regions inside the PB phase resulting from enhanced fixing with Os O_4 are early forms of cavitation in this phase. Note particularly the regions marked with C showing distinct splitting inside the PB phase (from Argon et al. [15]; courtesy of J. Wiley and Sons)

structure of stretched sheets shows widespread diffuse cavitation as shown in Fig. 15, which in some instances acts like a craze by association. Examination in more detail at higher magnification, as shown in Fig. 16, reveals both unambiguous quasi homogeneous cavitation in the PB lamellae as indicated in regions C of the figure, and many cases of early cavitation that manifests itself by the enhanced precipitation of Os O_4 in the form of small black zones in the PB phase, clearly visible in Fig. 16.

Craze growth in both KRO-1 and KRO-3 Resins is linear in time. The dependence of craze velocity on normalized stress at both room temperature and 253 K is presented in Fig. 17a and b for KRO-1 and KRO-3 Resins respectively. Examination shows that, although the stress dependence of craze velocity in these two different polymers is indistiguishably similar at room temperature, the craze velocities in KRO-3 Resin at 253 K are several orders of magnitude higher for the same normalized stress levels than in KRO-1 Resin. This results from the significantly higher level of thermal stresses in the corrugated lamella morphology at low temperatures.

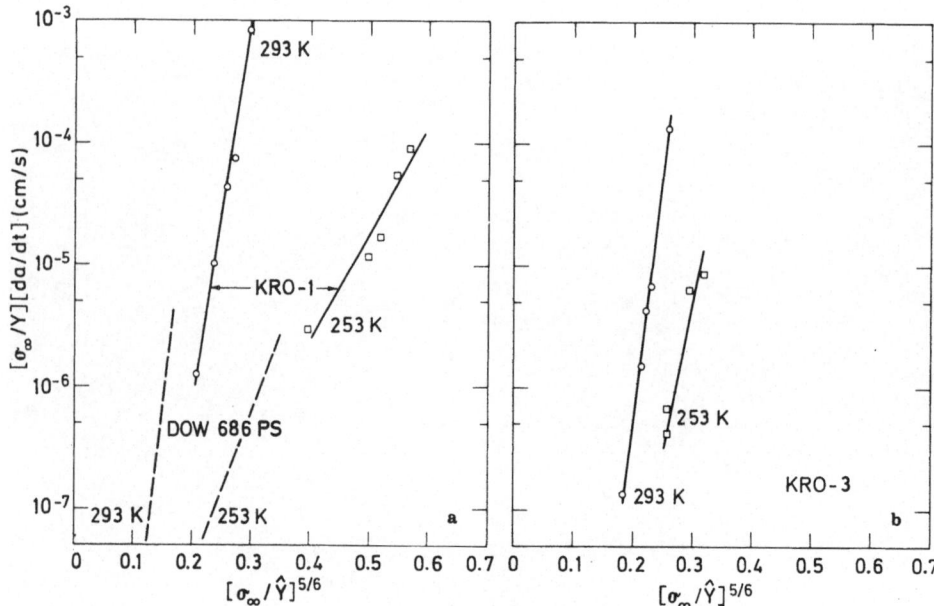

Fig. 17a and b. Dependence of craze velocity on normalized stress at 253 °K, and 293 °K: **a** for KRO-1 Resin, compared with results on a mineral-oil free PS; **b** for KRO-3 Resin (from Argon et al. [15], courtesy of J. Wiley and Sons)

4.1.2 Di-Block Copolymers with Wavy Rod Morphology

Argon et al. [6] have investigated crazing and fracture in a family of anionically poly-merized di-block copolymers of PB and PS with narrow ranges of molecular weight and molecular weight ratios of di-block components. Of these we will discuss here the behavior of a set having 23% PB in wavy rod morphology and the following molecular weights (PB/PS): (21 kg/mole)/(77 kg/mole); 45/150; 119/400; and 256/600. The morphology of the annealed samples is shown in Fig. 18. The figures show that the packing of the rods tends to be more regular for samples with low molecular weight. There was, however, no long range alignment of the rods in any of the four cases. The stress-strain response of two of the four polymers at three different strain rates is shown in Fig. 19. Whitening of the samples begins at the upper yield stress at one edge of the sample and spreads across the cross section under decreasing stress similar to the behavior of an initiating shear band [55]. Subsequent extension under a constant lower yield stress of about 19 MPa converts more of the sample into whitened zones. The total extension to fracture is directly proportional to the volume fraction that has undergone whitening, and there is no significant lateral contraction, indicating that the inelastic strain is entirely due to crazing. The curves in Fig. 19 indicate that the toughness of the polymers increase directly with increasing overall molecular weight, with a significant increase in the strain to fracture occurring with increasing strain rate. That this was not a result of merely the increase in the scale of the microstructure was confirmed by testing samples in which the morphological scale was coarsened along the isopleth as discussed in Section 3 above by blending

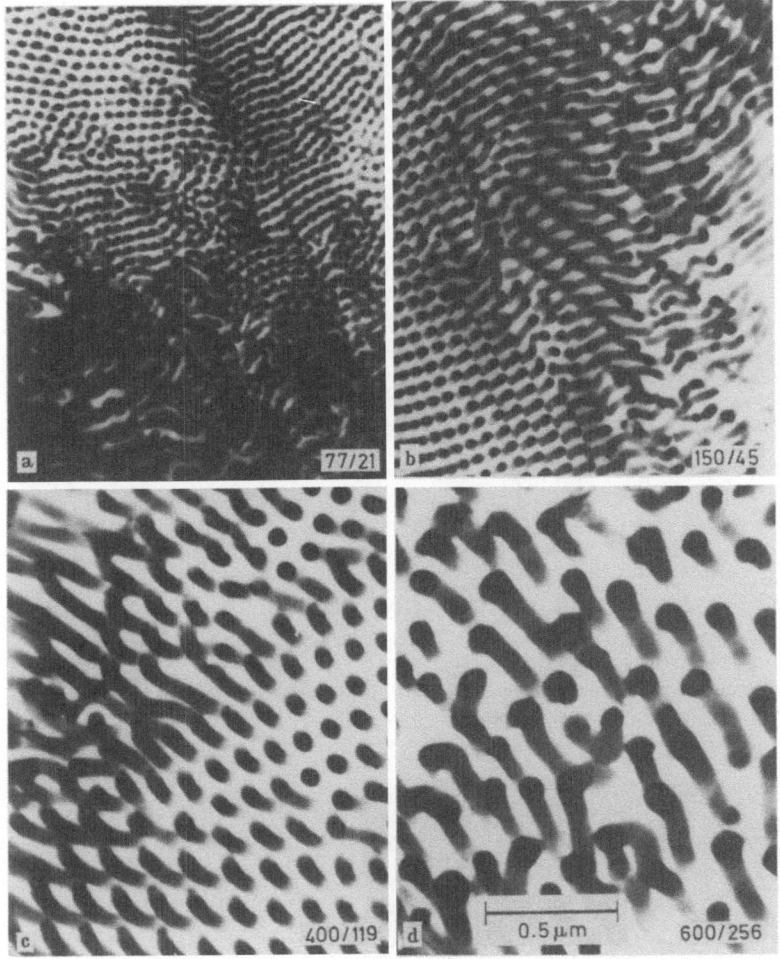

Fig. 18a—d. Morphologies of di-block copolymers with the PB phase in the form of wavy rods with PS/PB molecular weights of: **a** 77/21 kg/mole; **b** 150/45; **c** 400/119; **d** 600/256. (From Argon et al. [6]; courtesy of Plastics and Rubber Institute)

low molecular weight di-block samples with additional homopolymer of the same molecular weights as that in the di-block components. The toughness of these samples of low molecular weight but with a coarsened morphological scale was still low. That the increased toughness was due to increasing molecular weight and the attendant increase in the molecular entanglement density was confirmed further by irradiation cross-linking the low molecular weight di-blocks with 3 MeV electrons to a dose of 40 Mrad. This was found to increase the strain to fracture in the di-block PB 21/PS 77 from about 0.02 to 0.12 and in a di-block PB 48/PS 126 from about 0.02 to 0.2–0.4 respectively.

Figure 20 shows a typical craze microstructure in these strained di-block copolymers, that for PS 600/PB 256. Examination revealed that the scale of the craze micro-

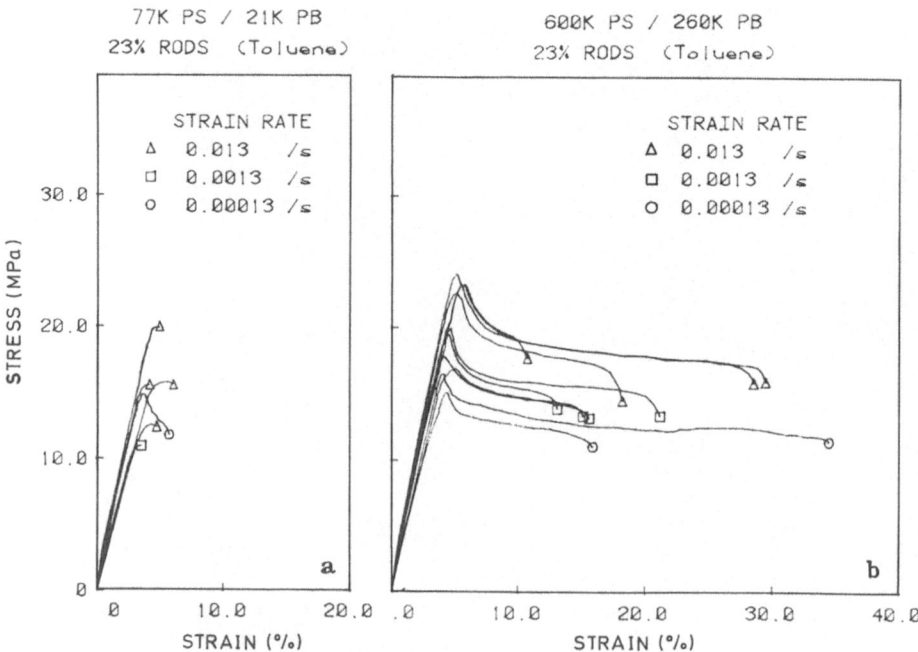

Fig. 19. Stress-strain curves to fracture of two of the four di-blocks shown in Fig. 18 (from Argon et al. [6]; courtesy of Plastics and Rubber Institute)

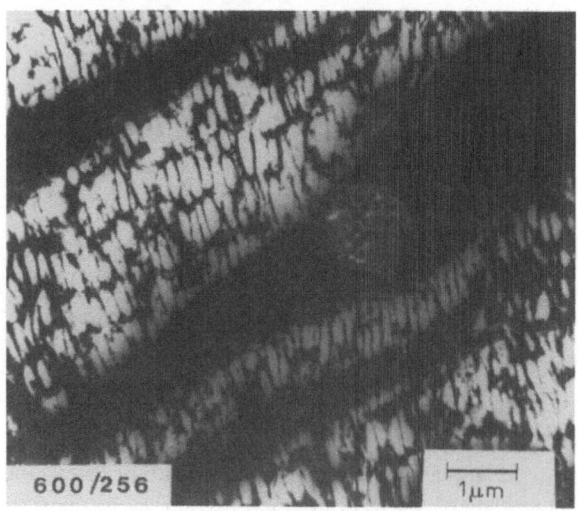

Fig. 20. Craze morphology in the di-block copolymer: PS/PB: 600/256 (from Argon et al. [6]; courtesy of Plastics and Rubber Institute)

structure and that of the phase domains is nearly identical in all cases. This is most likely the result of cavitation in the rubbery domains under the action of the applied stress accentuated by the advancing craze tip. Support for this is obtained from the apparently cellular nature of the packing of craze fibrils in the craze microstructure of Fig. 20 which shows that the material in one set of the "cell walls" of craze matter

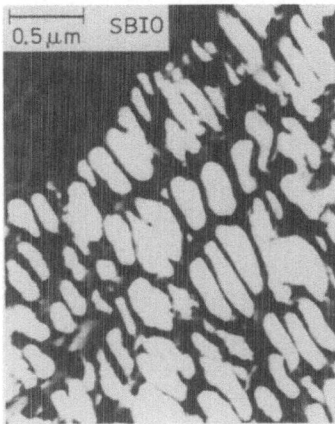
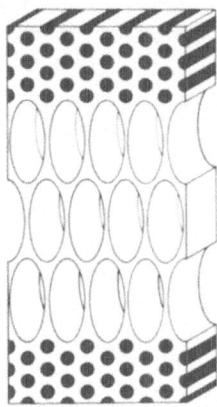

Fig. 21. Details of the cellular cavitation mode of craze matter production in the di-blocks with wavy rod morphology. Cartoon depicts the idealized form of the cavitation followed by drawing of the PS tufts

corresponds to that per one rubbery domain. This suggests a direct conversion process of the phase domains to craze matter as is sketched out in Fig. 21, in which the rubbery domain cavitates, and this is followed by the plastic extension of the PS that is left behind, once the load carrying contribution, by negative pressure, of the rubbery domains is lost. That this must be so for quantitative reasons will be discussed below.

4.1.3 Di-block Copolymers with Spherical Domain Morphology

The crazing behavior of a series of heterogeneous polymers with spherical PB domain morphology obtained from narrow distribution molecular weight and molecular weight ratio di-block copolymers of PB/PS with additional blending of both homo PB and homo PS of lower molecular weight than that of the block components to coarsen the morphological scale was investigated in detail by Schwier [38]. When in the form of pure di-block, without additional blending, the stress strain behavior of material of such spherical morphology is found to be very similar to that of the material with wavy rod morphology discussed in the preceding section. The principal difference in behavior of the two morphologies is a 16 % increase in the average modulus and craze yield stress of the material with spherical morphology at constant composition that can be attributed directly to the change in morphology itself. In the spherical morphology material, the plastic strains to fracture show the same increase with increasing overall molecular weight as in the material with wavy rod morphology, but even on a more rapid rate — going from negligible levels at 100 kg/mole to about 0.5–0.6 for 550–600 kg/mole at comparable volume fractions of PB. Figures 22a and b are two examples of the stress strain behavior of the di-block copolymers with spherical morphology.

Figure 23 shows the structure of a typical craze in a sample of PB 230/PS 200 di-block component with additional homo PS blended to bring the volume fraction down to 11 %. The micrograph shows clearly that the craze has propagated by cavitating the rubbery domains capturing many cases of the scalloped borders of cavitated particles along the craze boundary. The thickness of the craze is also seen to be of the order of the rubbery particle size with not much evidence indicating thickening

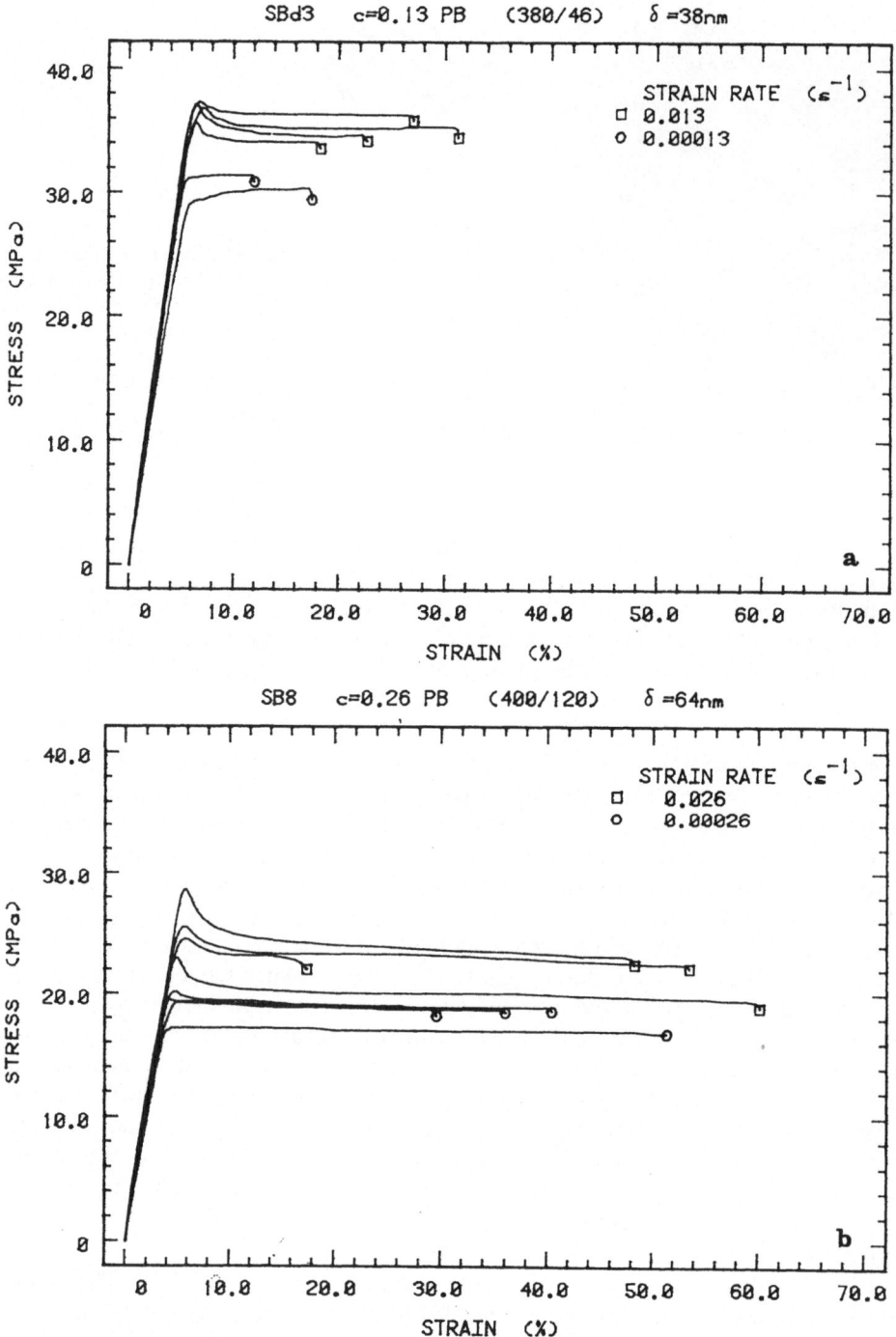

Fig. 22a and b. Stress-strain curves to fracture of two di-block copolymers: **a** SBd3 with c = 0.13 PB, and δ = 38 nm; **b** SB8 with c = 0.26 PB and δ = 64 nm

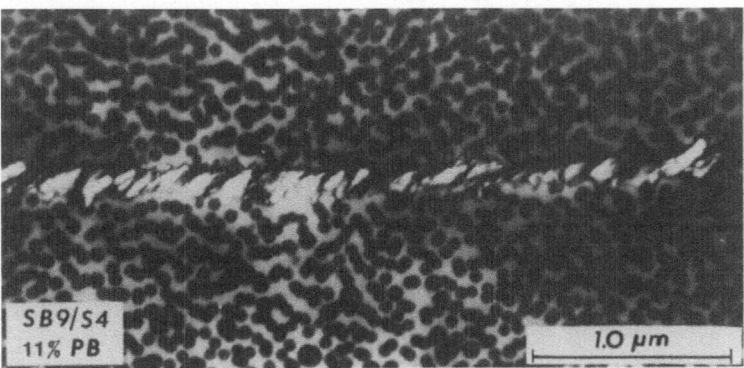

Fig. 23. Micrograph of a craze in a di-block with spherical morphology. Note the very narrow nature of the craze involving only about a layer of one spherical domain

of the craze. Thus, crazes in material with spherical morphology tend to be very thin and quite well defined.

Figure 24a–c shows the dependence of the velocity of crazes in material with spherical morphology on the applied tensile stress. The figures show results for blends made from three different di-blocks having molecular weights and weight ratios indicated in Table 2. In nearly all cases, additional low molecular weight PS has been blended into these materials to lower the volume fraction of the PB but maintain the particle size constant. Examination of the figures shows that the craze velocity increases directly with increasing volume fraction of PB at constant stress as is expected from the form of Eq. (58) for craze growth by the repeated cavitation process.

Table 2. Di-Block Blends with Spherical Morphology

Designation	Casting Solvent	vol. fract. c	PB mol. WT kg/mole	PS mol. Wt. kg/mole	δ nm	Λ nm
SB7/S4a	MEK/THF	0.058	59	284	40	103
SB7/S4b	Tol.	0.058	59	284	50	129
SB8/S3a	MEK/THF	0.08	120	250	40	93
SB8/S3b	Tol.	0.08	120	250	62	144
SB7a	Tol.	0.11	59	560	44	92
SB7b	MEK/THF	0.11	59	560	36	75
SB9/S4c	MEK/THF	0.11	230	200	73	152
SBd3	Tol.	0.13	46	380	38	
SB8/S3c	MEK/THF	0.15	120	300	65	122
SB9/S4a	MEK/THF	0.18	230	230	70	124
SB8	MEK/THF	0.26	120	400	64	100
SB9/S4b	MEK/THF	0.26	230	290	74	116

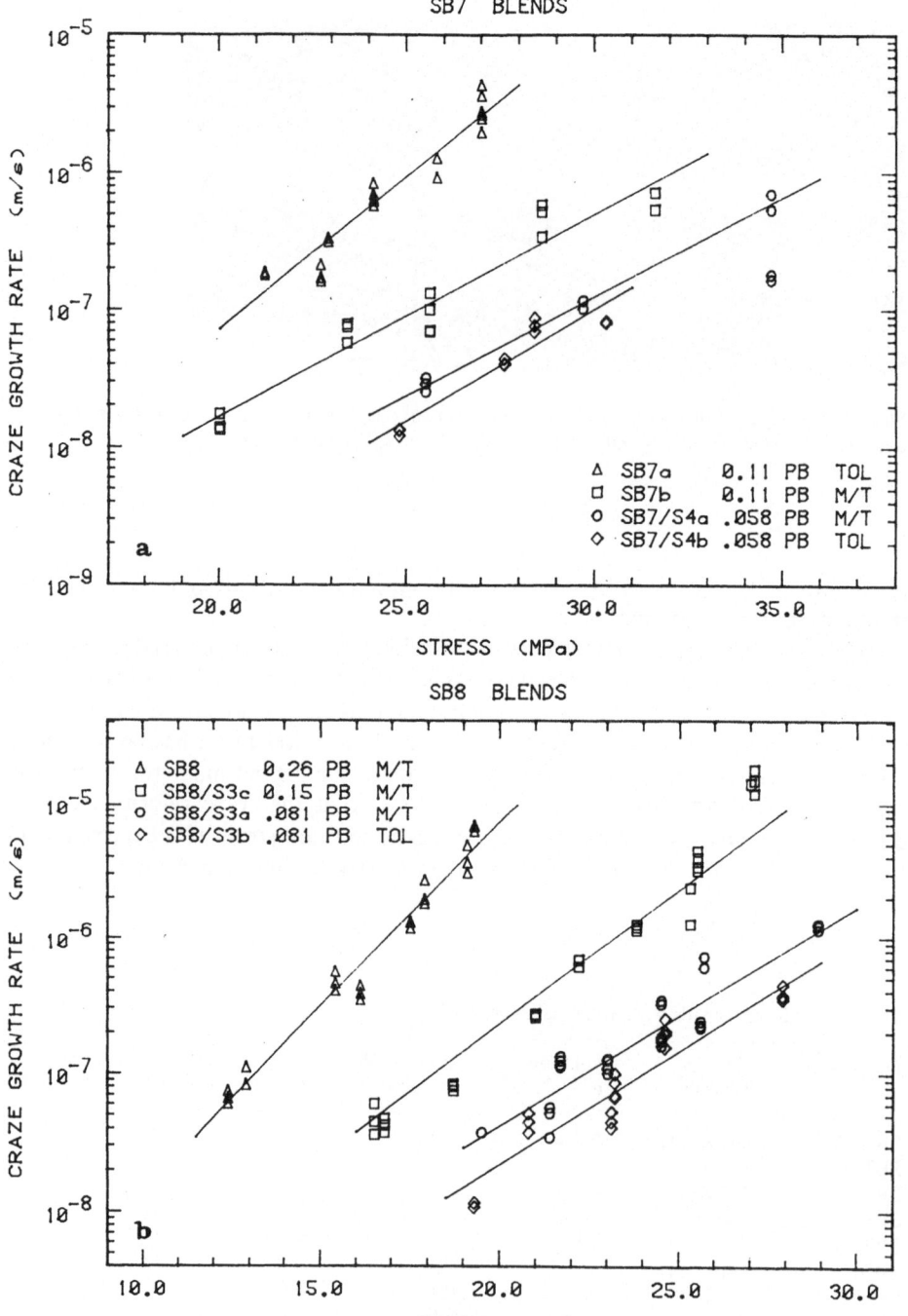

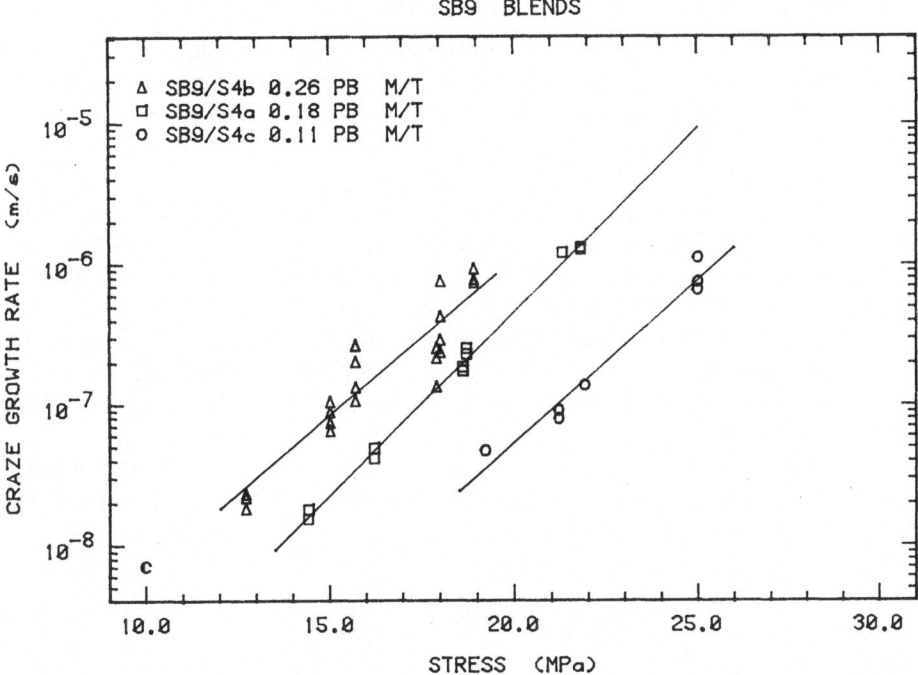

Fig. 24a—c. Craze velocities as a function of stress in blends of di-blocks with spherical morphology. **a** SB7 blends; **b** SB8 blends; **c** SB9 blends. The molecular details of these blends are given in Table 2

4.2 Kinetics of Craze Growth in Di-Block Copolymers

4.2.1 Thermal Stresses in PB

In experiments on the dynamic mechanical properties of di-block copolymers with spherical morphology, Bates et al. [31] have noted that the T_g of PB in the spherical domains could be substantially suppressed below the normal level of $-80\ °C$ in unconstrained PB. This suppression is quantitatively explainable as has been done earlier by Bohn [56] to be the result of the negative pressures induced in the PB due to the substantial thermal mismatch with PS. Bates et al., however, have observed that through changing the molecular weight of the PB in the particles, the thermally induced negative pressure in the PB could be increased only up to 78 MPa as the T_g was reduced to $-98\ °C$. From this, the very important conclusion is reached that the thermal stresses of PB are substantial and that the ideal cavitational strength of PB at $-98\ °C$ is 78 MPa. Since the ideal cavitational strength of a fluid must be directly proportional to its bulk modulus, the ideal cavitational strength of PB at room temperature can be estimated from the known bulk modulus and its temperature dependence [57] to be about 60 MPa. As we will demonstrate below, this is of key importance in explaining the crazing behavior of heterogeneous polymers containing PB.

4.2.2 Crazing in Block Copolymers with Spherical Domains

We will limit our discussion here only to the films containing spherical particles of PB, as these are of all morphologies the ones most readily amenable to analysis. Materials with rod morphology or lamellar morphology exhibit qualitatively very similar behavior but quantitatively require more detailed analysis that is, however, unfortunately less exact.

In all block copolymers investigated, the principal morphological wave length of the rubbery domains, i.e. sizes or mean spacings, is too short to concentrate the stress in a large enough volume element to nucleate crazes in the majority phase of PS. Furthermore, with the exception of perhaps the pure KRO-3 Resin with lamellar morphology, the combination of thermal stresses and the levels of applied stress are also insufficient to cavitate the rubbery domains without any further assist from more macro stress concentrations. Hence the available evidence indicates that in these polymers, crazes initiate entirely from surface stress concentrations.

Once crazes have developed, however, their growth is governed by the repeated cavitation of the rubbery domains, and the developments of Section 2.3 apply. The velocity of the craze is given by

$$\frac{da}{dt} = D_2 \exp\left\{-\frac{B}{kT}\left[1 - \left(\frac{\sigma_\infty \lambda_n'}{(1-c)\,Y_c}\right)^{5/6}\right]\right\} \tag{60}$$

$$D_2 = \frac{\delta\dot{\varepsilon}_0 E_c}{(\lambda_n - 1)\,(1 - v_c^2)\,\pi\sigma_m\,(1 - \sigma_\infty/\sigma_m)^2}\,, \tag{61}$$

where the maximum level of the craze traction develops to cavitate the PB domains. This occurs when the various components of the craze tip stresses induce enough negative pressure in the PB domains to raise them from the level of the already substantial thermal stresses to the required stress for cavitation. Thus, if the critical level of craze tip stress intensity to propagate it with the required velocity is K_{Ic}, the following local stresses should be present at the point of peak traction,

$$\sigma_{rr} = \frac{K_{Ic}}{\sqrt{\pi\Delta}} \tag{62}$$

$$\sigma_{\theta\theta} = \frac{K_{Ic}}{\sqrt{\pi\Delta}} + \sigma_\infty \tag{63}$$

$$\sigma_{zz} = v_c(\sigma_{rr} + (\sigma_{\theta\theta} - \sigma_\infty)) = 2v_c\frac{K_{Ic}}{\sqrt{\pi\Delta}} \tag{64}$$

where σ_∞ is the distant tensile stress and Δ is the cavitational process zone size given by Eqs. (55) and (56). Since at the point of propagation of the craze front $\sigma_{\theta\theta} = \sigma_m$, (or from Eq. (55)), we have

$$\frac{K_{Ic}}{\sqrt{\pi\Delta}} = (\sigma_m - \sigma_\infty)\,. \tag{65}$$

Each of the three tensile components of the principal stresses at the craze tip separately induce a contribution to the negative pressure inside the domain. These are all linearly additive. Thus, if we term any one of the principal components as σ_i, the induced negative pressure σ_{ST} can be obtained readily from the analysis of Goodier [58] or Eshelby [59] (see Mura [60])

$$\frac{\sigma_{ST}}{\sigma_i} = \frac{1}{3} \left[1 + \frac{2(1 - 4v_c + v_c^2)\left(\frac{K_{PB}}{K_c} - 1\right)}{(1 + v_c)\left[2(1 - 2v_c) + (1 + v_c)\frac{K_{PB}}{K_c}\right]} \right], \tag{66}$$

where K_{PB} and K_c are the bulk moduli of the PB and the overall composite, and v_c is the Poisson's ratio of the composite. These quantities themselves are readily calculable from composite theory [61] and are

$$\frac{K_c}{K_{PS}} = \left[1 + \frac{c\left(\frac{K_{PB}}{K_{PS}} - 1\right)}{1 + \left(\frac{K_{PB}}{K_{PS}} - 1\right)(1 - c)\frac{(1 + v)}{3(1 - v)}} \right] \tag{67}$$

$$v_c = \left(1 - \frac{E_c}{3K_c}\right)\frac{1}{2} \tag{68}$$

$$\frac{E_c}{E_{PS}} = \left[1 + \frac{c\left(\frac{K_{PB}}{K_{PS}} - 1\right)}{3 + (1 - c)\left(\frac{1 + v}{1 - v}\right)\left(\frac{K_{PB}}{K_{PS}} - 1\right)} - \frac{2c}{3 - (1 - c)\frac{2(4 - 5v)}{5(1 - v)}} \right] \tag{69}$$

where K_{PS}, E_{PS}, v are the bulk modulus, Young's modulus, and Poisson's ratio respectively of PS, c is the volume fraction of PB, and E_c is the Young's modulus of the composite. In the calculation of the elastic constants of the composite, explicit use was made of the fact that the shear modulus of PB is negligible when compared with that of PS.

The negative pressure σ_{TH} induced in the PB spheres due to the thermal mismatch between them and the composite surroundings is

$$\sigma_{TH} = \frac{2(\gamma_c - \gamma_{PB}) K_{PB} \Delta T}{2 + \frac{K_{PB}}{K_c}\left(\frac{1 + v_c}{1 - 2v_c}\right)}, \tag{70}$$

where γ_c and γ_{PB} are the volumetric coefficients of expansion of the composite and the PB respectively, and the temperature difference is to be reckoned from the temperature where the two phases coexist without stress. This is taken as the T_g for PS, i.e. 368 K. Once again the volumetric coefficient of the composite, obtained by resorting to composite theory [62], is

$$
\gamma_c = \left[\gamma_{PS} + \frac{c \left(\dfrac{K_{PB}}{K_{PS}} \right) (\gamma_{PB} - \gamma_{PS})}{1 + \left(\dfrac{K_{PB}}{K_{PS}} - 1 \right) \left[(1 - c) \dfrac{1 + v}{(3(1 - v))} + c \right]} \right]
\tag{71}
$$

where γ_{PS} is the volumetric coefficient of expansion of PS.

The peak traction σ_m can now be obtained from the condition that at this site the PB domains will cavitate under the action of the negative pressures induced by the three principal stresses $\sigma_{rr}, \sigma_{\theta\theta}, \sigma_{zz}$, and the added negative pressure σ_{TH}, due to the thermal mismatch by reaching the cavitational strength of PB that we have estimated earlier to be 60 MPa at room temperature. Thus,

$$
\sigma_{PBcav} = \sigma_{TH} + [\sigma_\infty + 2(1 + v_c)(\sigma_m - \sigma_\infty)](\sigma_{ST}/\sigma_i),
\tag{72}
$$

where σ_{TH} and σ_{ST}/σ_i are given by Eqs. (70) and (66), respectively. The various parameters: particle diameter δ, mean particle spacing in the volume Λ, the composite elastic constants K_c, E_c, v_c, the coefficient of expansion of the composite γ_c, the composite athermal deformation resistance $\hat{Y}_c$, the thermal stress σ_{TH}, the induced negative pressure σ_{ST} due to a principal stress σ_i, the maximum craze tip traction σ_m, and the pre-exponential factor D_2 of the craze velocity are tabulated in Table 3 for the twelve different blends in which the craze velocities of Fig. 23 a–c were measured.

From the stress dependence of the craze velocities shown in Fig. 23 a–c and the level of the velocity for a specific stress (chosen to be 20 MPa), the two less well known parameters of the craze velocities, i.e. B/kT and λ'_n (Eqs. (60), (61)), were determined to give the best fit to the data. These values, also given in Table 3, show that the constant B/kT for all cases is closely clustered around the average value of 43.58 with a very small coefficient of variation of $C_v = 0.034$, while the effective extention ratio λ'_n (as defined by Eq. (51)) varies somewhat more from a low value of 1.254 to a high value of 1.906, giving for the whole set an average of 1.575 with a coefficient of variation of 0.152. In the corresponding case of craze growth in homo-polystyrene by the interface convolution process, Argon and Salama [13] found for these same two parameters the values of B/kT = 44.72 and $\lambda'_n = 1.853$ that gave the best fit to the observed stress dependence of craze velocity at room temperature. Argon [20] and Argon and Bessonov [7] have provided a theoretical framework for the scale factor B of the activation free energy in a detailed yield theory for glassy polymers, where the make up of this parameter B is discussed in terms of dimensions and angles of rotation of molecular segments. On the basis of this theory, with regard to yield experiments conducted in compression, a value of 62.16 would be expected for B/kT. Thus, the average value of 43.58 found in our experiments is only 0.70 of the theoretical value based on compression experiments. As discussed earlier by Argon and Salama [13] for the corresponding case of homo-PS, where almost the same ratio was found between the observed and expected values, the apparent discrepancy is almost entirely due to the strength differential effect in PS between com-

Table 3. Critical Morphological and Craze Growth Parameters for Di-Block Blends with Spherical Morphology

Designation	c	$\hat{Y}_c$ MPa	γ_c $\times 10^{-4}\ K^{-1}$	σ_{TH} MPa	$\frac{\sigma_{ST}}{\sigma_i}$	σ_m MPa	D_2 $\times 10^6$ m/s	$\frac{B}{kT}$ [a]	λ_n' [a]
SB7/S4a	0.058	216	2.24	33.4	0.356	40.9	12.31	43.5	1.32
SB7/S4b	0.058	216	2.24	33.4	0.356	40.9	9.85	44.9	1.47
SB8/S3a	0.08	209	2.34	32.2	0.356	42.1	8.80	41.5	1.43
SB8/S3b	0.08	209	2.34	32.2	0.356	42.1	13.64	43.1	1.48
SB7a	0.11	199	2.47	30.7	0.358	43.4	8.50	44.5	1.85
SB7b	0.11	199	2.47	30.7	0.358	43.4	6.95	41.8	1.25
SB9/S4	0.11	199	2.47	30.7	0.358	43.4	14.10	46.0	1.91
SBd3	0.13	193	2.55	29.7	0.358	44.4	6.75		
SB8/S3c	0.15	187	2.64	28.6	0.360	45.3	10.64	44.2	1.76
SB9/S4a	0.18	178	2.78	27.1	0.361	46.7	10.17	45.1	1.85
SB8	0.26	157	3.15	23.3	0.363	50.2	7.00	42.8	1.67
SB9/S4b	0.26	157	3.15	23.3	0.363	50.2	8.09	41.8	1.32

[a] Calculated from experimental data from Figs. 24a–b.

pression and tension experiments, which results from the dependence of the shear modulus on pressure. Thus, e.g. Argon et al. [63] have found the ratio of the plastic resistance of PS in tension compared to that in compression at room temperature to be 0.71 — or almost exactly what corresponds to the "best fit" value given above. The explanation of the relatively larger variation in the effective extension ratio λ_n' is more obscure. Normally, in macroscopic samples, the effective extension ratio is prescribed by the two positions of impending instability and the position of restored stabilities on the true tensile stress strain curve where $d\sigma/d\varepsilon = \sigma$, which in turn is governed by the entanglement density in the polymer and its terminal network strain. Neither of these should be very sensitive to the molecular weight of PS, provided that molecular weight is large. One possibility is the very small physical dimensions of the craze tufts and the total number of entanglements that can be entrapped in a given tuft. Thus, the smaller the interparticle spacing, and thus the smaller the tuft diameter, the larger the actual extension ratio λ_n should be [2]. Unfortunately, no correlation is found between λ_n' and δ or Λ.

4.3 Time to Fracture

As we have discussed in Section 2.2.2 above with regard to cases where craze initiation is governed by surface sources, as is the case in the films with spherical morphology discussed here, the overall craze strain rate and the toughness of the sheets is not influenced by the craze velocity. Hence the changes in the strain to fracture that are shown in Fig. 22a–c must be entirely a result of the processes of fracture in craze matter that were discussed in Section 2.2.4. Since the solutions from which the films were cast were filtered to exclude coarse foreign particles of larger than micron size, it is most likely that fractures (of the craze matter) in most cases were initiated from surface imperfections. This points out again the need to study the sources of fracture initiation and the processes that govern the time dependence of the fracture of craze matter under stress.

5 Toughness of Glassy Polymers with Composite Particles

5.1 General Principles of Morphology Control

As we have discussed in Section 3, large degrees of freedom are available for constructing composite particles with a wide range of morphologies and volume fraction
by emulsification of block copolymers in high molecular weight homo-polystyrene
without and with blending of additional PB into the block copolymer phase of the
particles to further control their morphology. In this manner, e.g. by starting with
di-block copolymers having spherical, rod, and lamellar morphologies it is possible
to construct composite particles with the same inner morphology as the di-block
phase when the former is emulsified in high molecular weight PS. The introduction
of additional PB of lower molecular weight than that in the block components of the
di-block can then accomplish a coarsening of the inner morphological wave length
of the particles or, in some instances as we will discuss below, alter the morphology
altogether. The composite particle size can be controlled separately by a procedure
of over-coarsening the particle size first by coagulation in solutions of critical concentration, followed by reduction of the large particle sizes by a step of stirring as has
been demonstrated by Turley and Keskula [64] until the desired particle size is reached.
All along, the control of particle size in the solution of critical concentration can be
achieved accurately by active light scattering measurements [65]. Thus, the particle
morphology, volume fraction, and size can be brought fully under control to investigate
their effect on toughness. We will discuss here a selection of examples in which the
toughness ranges from very low strains to fracture to levels that far exceed those of
any commercial product.

5.2 Emulsification of K-Resins in Polystyrene

Gebizlioglu et al. [66, 6] have used the commercial K-Resins (KRO-1 and KRO-3)
to form micron size composite particles in high molecular weight ($M_w = 268$ kg/mole)
PS. Figure 25a and b show the obtained result where the familiar rod and lamellar
morphologies of these two resins are clearly evident in the particles. Volume fractions
of particles ranging from 1 % to 21.7 % and particle size ranging from 0.3 µm to 10 µm
for KRO-1 particles and corresponding range of volume fractions for KRO-3 Resin
but with a narrower size range were investigated.

The strains to fracture in both of these materials proved to be disappointingly
small as Fig. 26, summarizing some of the experiments, shows. The plastic strains to
fracture in no case exceeded 5 %. Nevertheless, the craze yield stresses were found to
be around 30 MPa, or comparable to those in HIPS. Examination of the specimens
just prior to fracture showed that crazing was initiated from the cut edges of the cast
films and was distributed very inhomogeneously in packets. Examination of the
morphology of the partially crazed samples disclosed, as shown in Fig. 27a and b,
that the particles were not very effective in nucleating crazes. For the case of the KRO-1
particles which were of the right size range, i.e. 1–10 µm that is usually considered
necessary for craze initiation, the conclusion was clear. In the case of the KRO-3
particles, the conclusion was less clear as few of these particles had the appropriate
size (i.e. greater than 1.0 µm). Naturally, a list of other factors might also have been

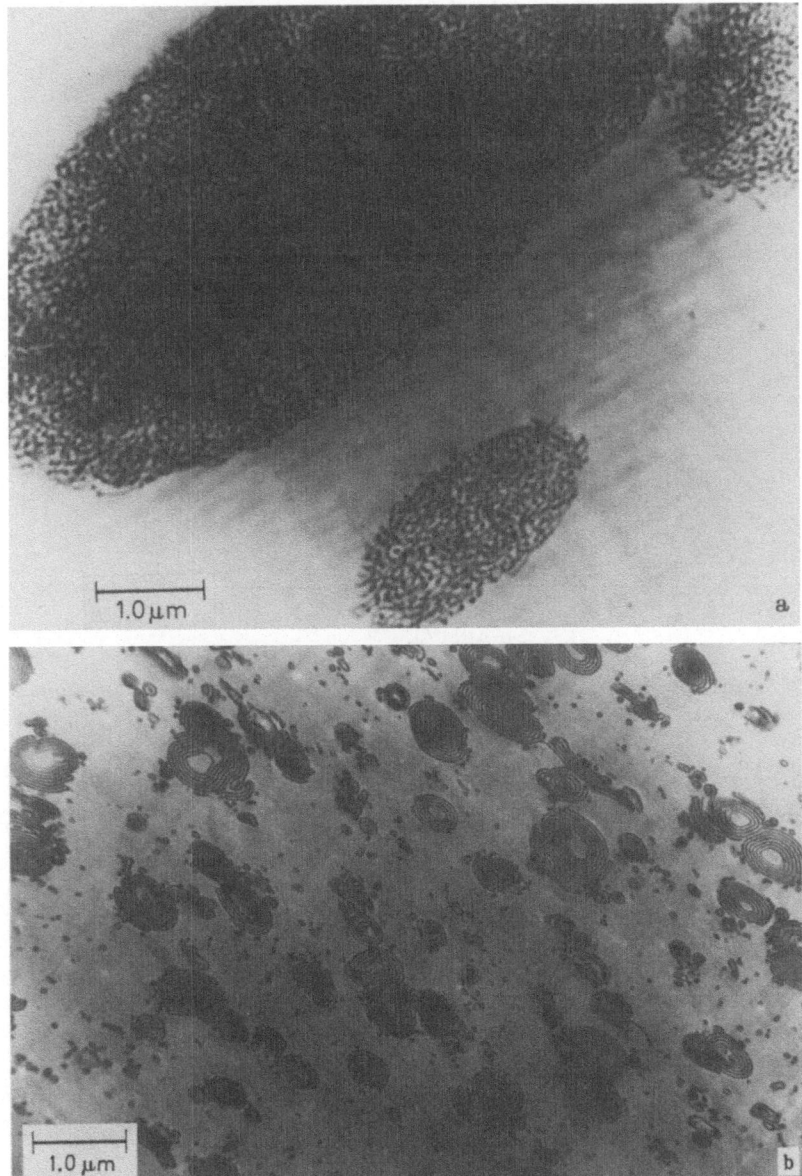

Fig. 25a and b. K-Resin particles in high molecular weight PS: **a** KRO-1 particles; **b** KRO-3 particles

responsible for the failure such as: an inadequate level of mismatch between the elastic and thermal properties of the particles and the matrix for producing the required concentrations of deviatoric stress and negative pressure outside the particle, and finally inadequate strength of the particles to support the prevailing stress necessary for craze nucleation and growth. Thus, without further modification to develop better

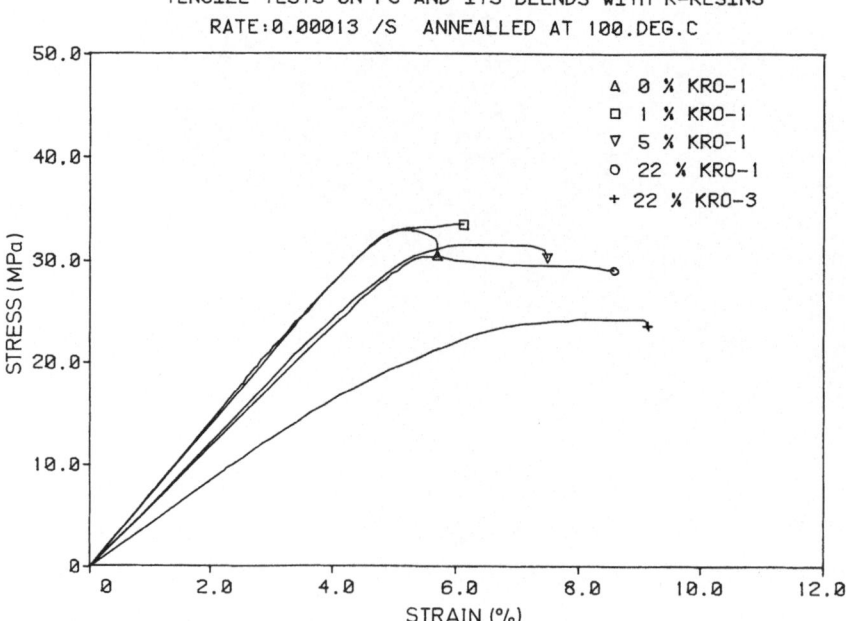

Fig. 26. Stress strain curves to fracture of PS with K-Resin particles (from Argon et al. [6]; courtesy of the Plastics and Rubber Institute)

mechanistic understanding, the mere incorporation of block copolymers with interesting morphologies into PS has not been promising.

5.3 Blending Additional PB and PS into K-Resins Particles

To probe some of the possibilities for inadequate performance listed above additional low molecular weight, homo PB can be blended into the particles to coarsen the inner wave length at constant composition to preserve the average mechanical properties of the particles. Alternatively, additional PB or additional PS could be blended into the particles to increase the thickness or scale of only one of the components and simultaneously also either increase or decrease the stiffness and thermal expansion mismatch between the particles and the surrounding matrix.

Table 1 indicates that the molecular weight M_w of KRO-1 is 179 kg/mole, indicating that for a volume fraction of 0.23 and 0.77 of PB and PS the individual molecular weights of these block components should be 41 and 138 respectively. Figure 28 and 29 show the result of blending additional narrow molecular weight range PB and PS of molecular weights 22 kg/mole and 64 kg/mole respectively into the composite. In both cases it is clear that the majority of the additional low molecular weight PB and PS is indeed associated with the K-Resin particles, but it is also clear that it has not dissolved into the separate component blocks at all but has instead formed separate internal domains. The reason for this is clearly in the stars shaped nature of the K-Resin molecules, where the individual block components are further divided up into segments

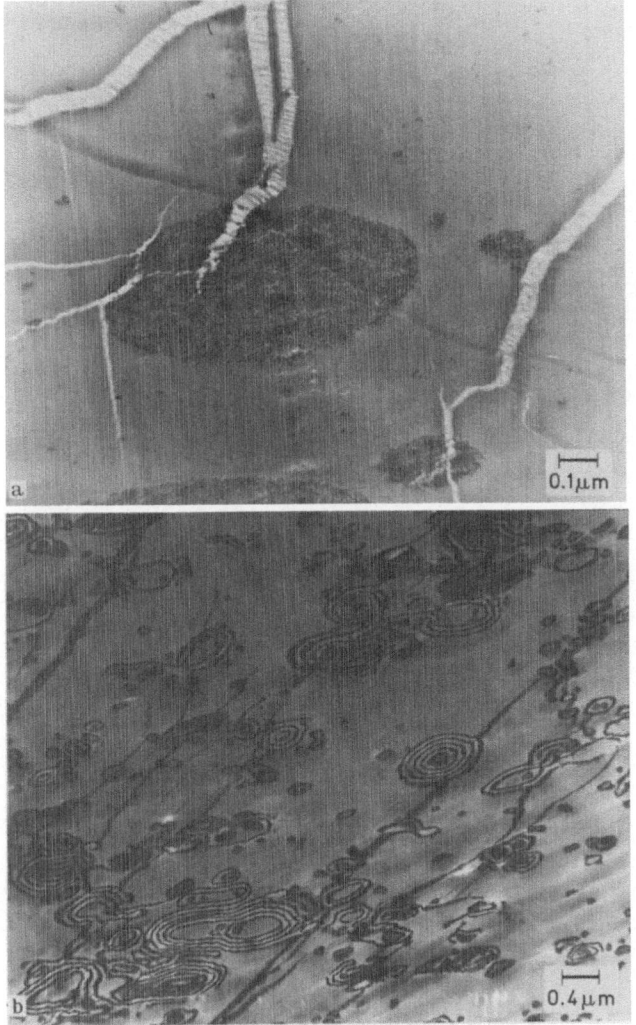

Fig. 27a and b. Micrographs of crazes around K-Resin particles; **a** KRO-1 Resin particles in PS; **b** KRO-3 Resin particles in PS (from Argon et al. [6]; courtesy of Plastics and Rubber Institute)

that are then tied together. The number of such connected segments has not been disclosed, but from the results of the blending experiment reported above, are clearly larger than 2. This requires a further reduction in the molecular weight of the solubilizing species of the homopolymers. In any event, the toughness of these modified polymers (Figs. 28 and 29) was even poorer than those with the unmodified K-Resin particles.

Table 4 summarizes the different conditions and compositions that were achieved in a blending study of narrow molecular weight PB with inly $M_w = 3$ kg/mole incorporated into the KRO-1 particles in PS. As the table indicates in the blending, the total volume fraction of KRO-1 Resin and PB_{3000} was kept constant to maintain a constant volume fraction of composite particles at 21.7%. In addition, pain was taken to keep the average composite particle size also roughly constant by the use

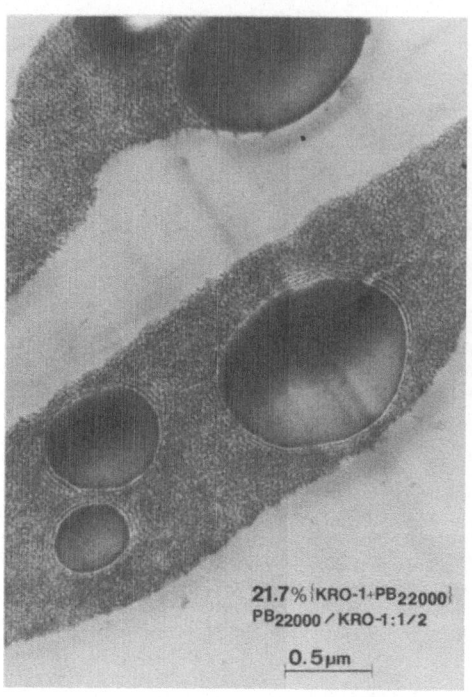

21.7% |KRO-1+PB22000|
PB22000 / KRO-1 : 1/2

0.5 µm

Fig. 28. Attempt to blend additional 22 K PB into the KRO-1 particles in PS. Note the lack of success with the additional PB being precipitated out inside the particles

Table 4. Blends of KRO-1/PS Emulsions with Additional Low Molecular Weight (3 kg/mole) PB, for a Constant Volume Fraction of Particulate Phase at 0.217

$PB_{3000}/KRO-1$	% PB_{total}	% PB_{3000}	% PB_{KRO-1}
0	4.99	—	4.99
1/8	6.85	2.41	4.43
1/4	8.35	4.34	4.01
1/3	9.17	5.43	3.74
1/2	10.54	7.22	3.32
1	13.35	10.85	2.50

of the technique described in Section 5.1 above. Thus the incorporation of PB_{3000} into the PB particle changes volume fraction of PB *inside* the particles, coarsens the PB phase dimension, and simultaneously increases the stiffness mismatch and the thermal expansion mismatch of the particles with their surroundings. A further and somewhat unplanned result of the blending was a change of morphology that is shown in Fig. 30 for the case $BP_{3000}/KRO-1$ of 1/3. Clearly, the low molecular weight PB has gone in solution into the PB block component, but has also resulted in a change of morphology from wavy rods to concentric spherical shells. The result of the blending over the range of solubility of the PB_{3000} in the PB phase of the KRO-1 Resin on the toughness of the composite is shown in the sequence of three stress strain curves of Fig. 31a–c for $PB_{3000}/KRO-1$ ratios of 0.125, 0.250, and 0.500. The craze flow stress for these three cases decreases monotonically from 10 to 8.5 to 8.0 MPa while the

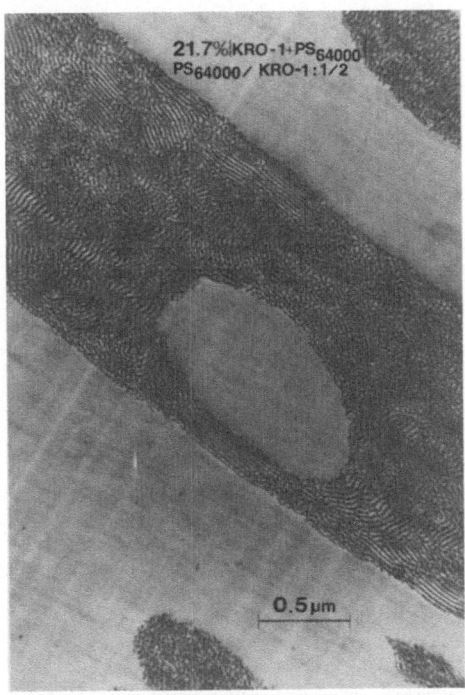

Fig. 29. Attempt to blend additional 64K PS into KRO-1 particles in PS. Note the lack of success with the additional PS being precipitated out inside the particles

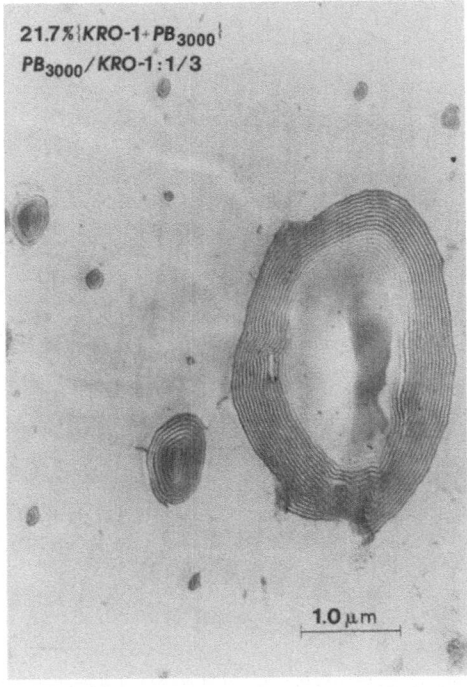

Fig. 30. Change of morphology inside KRO-1 particles from wavy rods of PB to spherical shells of PB upon the successful incorporation of additional 3K PB into the particles

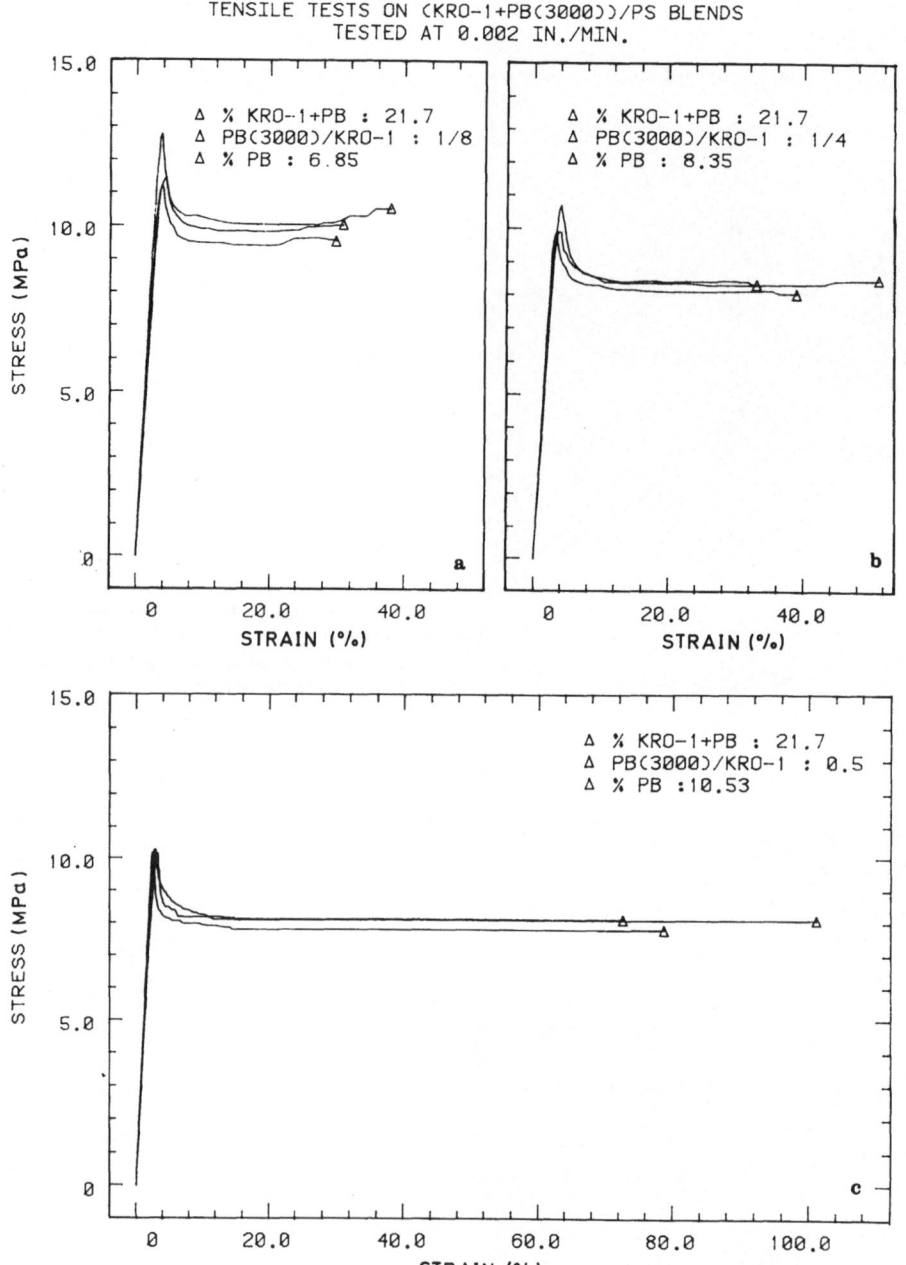

Fig. 31a–c. Stress-strain curves to fracture of heterogeneous polymer incorporating increasing levels of 3K PB into the KRO-1 Resin particles in high molecular weight PS, all at 21.7% by volume of particles: **a** PB (3000)/KRO-1, 1/8; **b** PB (3000)/KRO-1, 1/4, **c** PB (3000)/KRO-1, 1/2

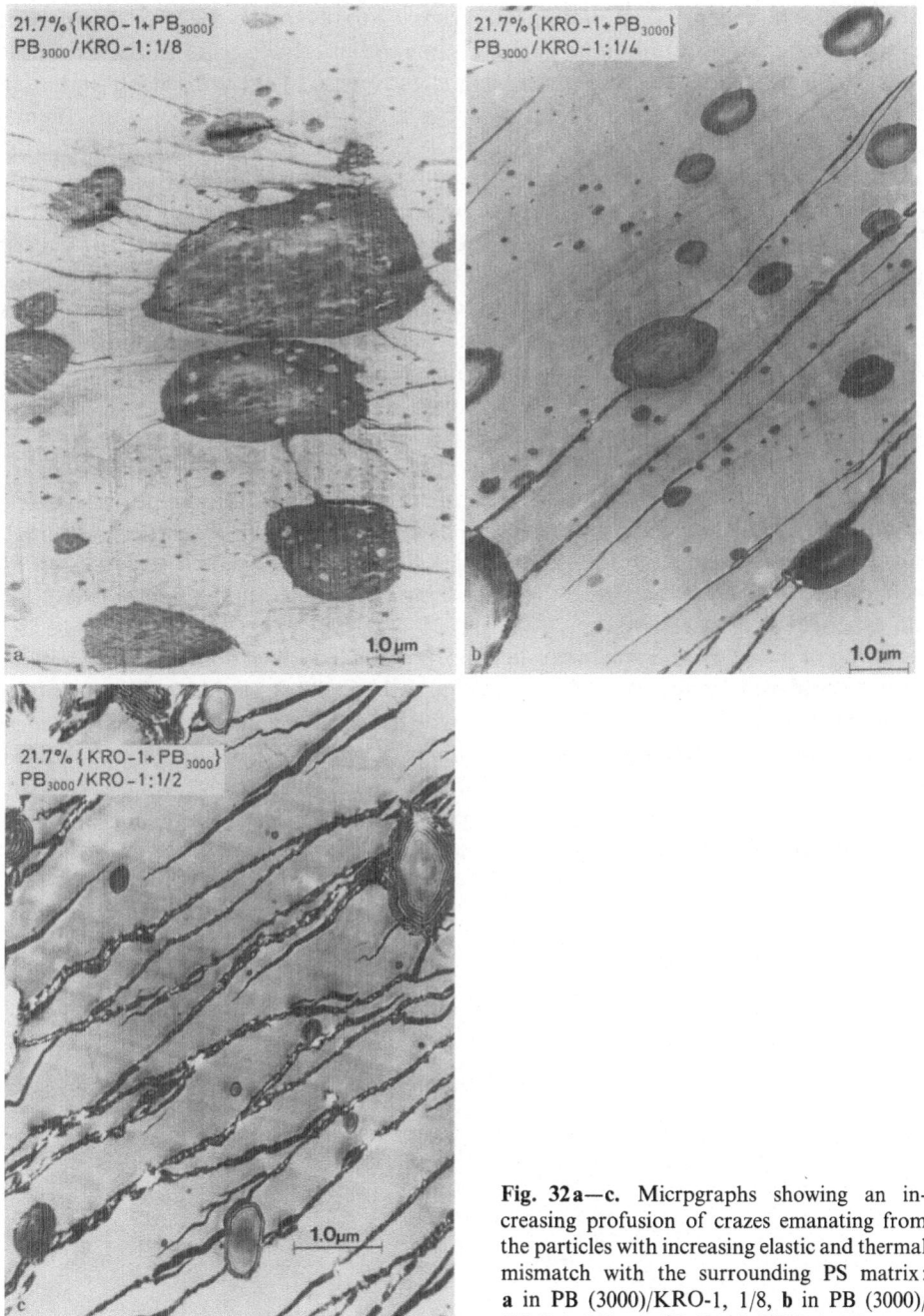

Fig. 32a—c. Micrpgraphs showing an increasing profusion of crazes emanating from the particles with increasing elastic and thermal mismatch with the surrounding PS matrix: **a** in PB (3000)/KRO-1, 1/8, **b** in PB (3000)/KRO-1, 1/4, **c** in PB (3000)/KRO-1, 1/2

strain to fracture increases monotonically from 0.3 to 0.4 to about 0.9 — all, roughly, at the same composite modulus and the same volume fraction of particles. These results suggest a monotonically increasing effectiveness of the particles in initiating crazes and maintaining a high level of active craze front density per unit volume, permitting the growth of these crazes at lower levels of flow stress, in response to the imposed strain rate. That this is indeed so could be seen in the rather uniform mode of whitening of the samples upon extension, and is more convincingly demonstrated in the micrographs of Fig. 32a–c that show the craze distribution in the fractured samples corresponding to the stress strain curves of Figs. 31a–c, respectively.

Ascertaining the causes of the increased effectiveness of the composite particles in craze initiation requires analysis of the mechanical properties of the particles. A preliminary study of this type is presented below.

5.4 Mechanical Properties of Composite Particles

Consider as a first approximation the microstructure of the composite particle smoothed out and replaced with averaged elastic and thermal properties. Although this approach is acceptable for the typical KRO-1 particles shown in Fig. 25a, as well as the usual HIPS particles, the smoothing should clearly be inappropriate for the spherical shell particles shown in Fig. 30 and Fig. 32a–c. Such more specific modeling of particles with spherical shell morphology has been done by Cunningham et al. [67] and will be reported elsewhere. Paranthetically it should suffice to point out that the elastic moduli of HIPS particles which consist of PS occlusions surrounded by a topologically interconnected, albeit minority phase of PB, cannot be modeled accurately by the technique of Chow [61] already mentioned above, while the elastic properties of the randomly intermixed morphology of the KRO-1 particles is obtained best by the standard approaches of the so-called self-consistent method (see Mura [60]).

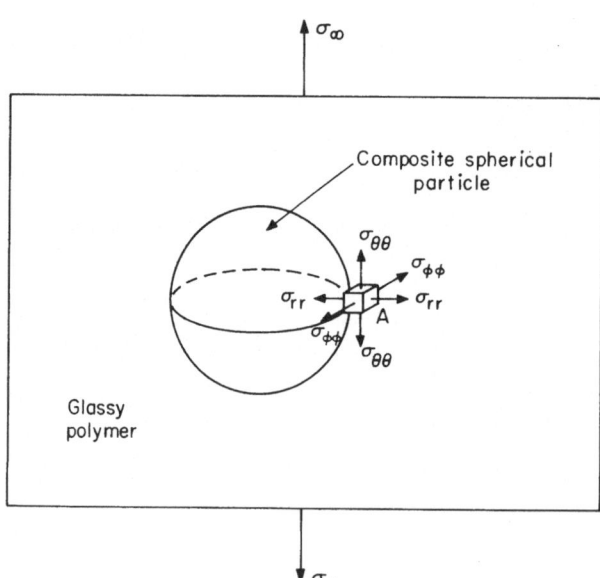

Fig. 33. Equatorial point A around a spherical particle under a tensile stress, where craze initiation is most likely

The average thermal expansion coefficient of the particles are best approximated by the method of Chow [62] even though they are strictly speaking appropriate only to the HIPS morphology. We will consider here two limiting cases of the unmodified KRO-1 particle and a particle made of pure PB, which comfortably bracket the behavior of the blends of spherical shell morphology. We are aware that in practice, particles of pure PB result in rather poor performance. This, however, is due to totally different reasons than their stiffness and thermal expansion properties, which we will demonstrate should be most attractive.

Consider two cases of spherical particles, one of KRO-1 morphology of a volume fraction of 0.23 of randomly wavy PB rods, and the other pure PB — both occupying a volume fraction of 0.22 in PS. We are interested in the craze initiation condition for these two particles at room temperature under a uniaxial tensile stress σ_∞. We consider the state of stress at a typical equatorial point A along the particle interface, on the PS side of the particle, as shown in Fig. 33. We determine first by standard methods the elastic properties of the particles and their thermal expansion coefficients, together with the elastic properties and thermal expansion coefficients of the composite matrix as a whole consisting of particles and the majority phase of PS.

KRO-1 Particles in PS

We denote the properties of the particles by a subscript P and those of the matrix by M, and obtain the required values from the developments of Chow [61, 62] as generalized forms of Eqs. (67)–(70) and (71). We find for the above description of particles with a volume fraction of 0.22, or a total rubber volume fraction of 0.05 the following values:

$$K_P = 0.861\ K_{PS} = 2.87\ \text{GPa}$$

$$E_P = 0.712\ E_{PS} = 2.31\ \text{GPa}$$

$$\nu_P = 0.335$$

$$K_M = 0.967\ K_{PS} = 3.22\ \text{GPa}$$

$$E_M = 0.928\ E_{PS} = 3.02\ \text{GPa}$$

$$\nu_M = 0.308$$
$$\gamma_P = 3.00 \times 10^{-4}/°C$$

$$\gamma_M = 2.21 \times 10^{-4}/°C$$

At room temperature $(\Delta T = -75\ °C)$ this results in an average thermally induced negative pressure $\sigma_{TH} = 7.22\ \text{MPa}$ inside the particles. At the equatorial point A (as well as at any other point on the periphery) this negative pressure sets up stresses in the surrounding matrix that are

$$\sigma_{rr} = \sigma_{TH}, \qquad \sigma_{\theta\theta} = -\sigma_{TH}/2, \qquad \sigma_{\varphi\varphi} = -\sigma_{TH}/2 \qquad \text{(73a, b, c)}$$

When a tensile stress σ_∞ is applied to the solid containing the particle direct stresses $\sigma_{\theta\theta}$ and $\sigma_{\varphi\varphi}$ appear at point A. They are [58]

$$\sigma_{\theta\theta} = \frac{3(9 - 5v_M)}{2(7 - 5v_M)}\sigma_\infty , \qquad \sigma_{\varphi\varphi} = \frac{3(5v_M - 1)}{(27 - 5v_M)}\sigma_\infty . \qquad (74\,a, b)$$

In addition, σ_∞ also sets up a negative pressure σ_P inside the particle that is Eq. (66)

$$\sigma_P = \frac{\sigma_\infty}{3}\left[1 + \frac{2(1 - 4v_M + v_M^2)\left(\dfrac{K_P}{K_M} - 1\right)}{(1 + v_M)\left[2(1 - 2v_M) + (1 + v_M)\dfrac{K_P}{K_M}\right]}\right]. \qquad (75)$$

This negative pressure induces additional principal stresses at the equatorial point A that are

$$\sigma_{rr} = \sigma_P , \qquad \sigma_{\theta\theta} = -\sigma_P/2 , \qquad \sigma_{\varphi\varphi} = -\sigma_P/2 . \qquad (76\,a, b, c)$$

Normalizing stresses with the yield stress of PS ($= 100$ MPa at RT) for the purpose of evaluating the craze initiation criterion of Eq. (12) we obtain for the total normalized deviatoric shear stress s/Y and normalized total negative pressure σ/Y the following quantities, based on results of Eqs. (73a, b, c), (74a, b), and (76a, b, c) and the calculated level of $\sigma_{TH} = 7.22$ MPa,

$$\frac{s}{Y} = \left\{\frac{1}{6}\left[\left(-1.545\,\frac{\sigma_\infty}{Y} + 0.1345\right)^2 + 3.625\left(\frac{\sigma_\infty}{Y}\right)^2\right.\right.$$
$$\left.\left. + \left(0.359\,\frac{\sigma_\infty}{Y} + 0.1345\right)^2\right]\right\}^{1/2} \qquad (77)$$

$$\frac{\sigma}{Y} = 0.732\,\frac{\sigma_\infty}{Y}. \qquad (78)$$

Utilizing the values of $AQ = 0.127$, and $Q \ln (t/\tau) = 0.282$ (for a craze initiation time of 10^2 sec) that were given in connection with the craze initiation condition of Argon and Hannoosh [12], in connection with Eq. (12), we find from

$$\left(\frac{s}{Y}\right)\left(Q \ln (t/\tau) + \frac{3}{2}\frac{\sigma}{Y}\right) - 1 = 0 , \qquad (79)$$

the value of

$$\sigma_\infty = 0.243\, Y = 24.3 \text{ MPa} . \qquad (80)$$

This falls somewhat short of the measured value of about 30 MPa. We attribute this discrepancy to the difficulty of obtaining accurate elastic constants for the particle by the method of Chow, where we used a random spherical block component morphology that differs significantly from the randomly wavy rod morphology of the KRO-1 particles. We conclude nevertheless that the craze yield stress is high due to the minimal elastic and thermal expansion mismatch of the particle with its surroundings. With such small mismatch requiring such high stresses to initiate crazes, many surface irregularities should prove to be better sites, for craze initiation and the role of the particles are thus minimized. In addition, the produced craze matter will have to sustain the very large tensile stresses that will hasten its breakdown and lead to early fracture. Thus the overall strain to fracture is small.

PB Particles in PS

The corresponding elastic and thermal properties of the particle (P) and the matrix (M) in this case are found to be for a volume fraction of 0.22 as before,

$$K_P = K_{PB} = 1.77 \text{ GPa}$$

$$E_P = E_{PB} = \text{negligible}$$

$$K_M = 0.878 \, K_{PS} = 2.92 \text{ GPa}$$

$$E_M = 0.744 \, E_{PS} = 2.42 \text{ GPa}$$

$$\nu_M = 0.330$$

$$\gamma_P = 7.5 \times 10^{-4}/°C$$

$$\gamma_M = 2.867 \times 10^{-4}/°C$$

At room temperature ($\Delta T = -75 °C$) the average negative pressure resulting from the coefficient of expansion mismatch is 28.11 MPa. We follow here the same development as for the KRO-1 particle with one significant exception, which requires us to take note of the increased particle interactions that amplify the local stresses at the equatorial point A. As has been modeled by Broutman and Panizza [68], the particle interactions for a small content of rubber are quite small but become substantial when the rubber content exceeds 20 % for a regular lattice of particles. Thus, while no corrections were necessary for the KRO-1 particle with an overall rubber content of only 0.05, a correction becomes necessary in this case where the rubber content has increased to 0.22. As pointed out by Bucknall [3], the critical interactions in a blend with randomly positioned particles, where the near distances of approach between particles count more heavily in initiating crazes than the mean distances, additional concentrations should be present. Thus we estimate the additional concentration of stress at the equatorial point due to particle interactions to be roughly that corresponding to a 30 % volume fraction of regularly spaced particles. The results of Broutman and Panizza then lead to a factor of 1.26 that must be applied to all components of stress equally. When this is done, we find for the normalized deviatoric shear stress s/Y and the normalized negative pressure at the equatorial point the following values:

$$\frac{s}{Y} = \left\{ \frac{1}{6} \left[\left(-1.923 \, \frac{\sigma_\infty}{Y} + 0.534 \right)^2 + 4.468 \left(\frac{\sigma_\infty}{Y} \right) \right. \right.$$
$$\left. \left. + \left(0.455 \, \frac{\sigma_\infty}{Y} + 0.534 \right)^2 \right] \right\}^{1/2} \tag{81}$$

$$\frac{\sigma}{Y} = 0.942 \, \frac{\sigma_\infty}{Y} \tag{82}$$

The use of Eq. (79) gives us in this case

$$\sigma_\infty = 0.111 \, Y = 11.1 \text{ MPa} \tag{83}$$

This compares well with the upper craze yield stress of 10.5 MPa shown in Fig. 31 c for the modified spherical shell particle that approximates to the solid PB particle.

From the above comparison we note that the elastic and thermal expansion mismatch are two very vital factors in producing particles with a high effectiveness for craze initiation.

The role of the elastic mismatch has long been recognized. The vital feature of the thermal mismatch, however, seems to have escaped general attention although some diffuse awareness of the role of thermal stresses exists among some manufacturers of high impact polymers. The reason for this lack of awareness would appear to be that thermally induced negative pressures do not set up around the particles any negative pressure at all (see Eqs. (73 a, b, c)) in the surrounding matrix. They contribute however, in a very important way to the deviatoric stress which plays the key role in initiating microporosity. Thus, the thermally induced negative pressure sets up a significant level of deviatoric stress, which, together with what results from the applied stresses, initiates micropores. The negative pressure then that results from the applied stresses alone expands these pores to initiate crazes. The generally good agreement that has been demonstrated above for the KRO-1 particle and the spherical shell particle albeit approximated by a pure PB particle, through the use of the craze initiation theory of Argon and Hannoosh [12], is further direct confirmation that this theory is quantitatively predictive even in very complex stress fields as that which exists around the composite particles under combined thermal stresses and applied stresses.

Acknowledgement: The research on craze mechanics in homopolymers discussed here has been supported by the NSF under Grant No. DMR-8012920. The research on heterogeneous polymers has been supported by the NSF/MRL under Grant No. DMR 78-24185, and administered through the Center for Materials Science and Engineering at M.I.T. We are grateful, moreover: to the Monsanto Plastics and Resins Co. of Springfield, Massachusetts, for a post-doctoral fellowship to O.S.G. that has made the research on emulsified particles possible; to the Michelin Foundation for a pre-doctoral fellowship to C.E.S.; to Dr. F. Bates for synthesizing many of the di-blocks discussed in Sect. D, and to Prof. D. M. Parks for help in some computational problems.

6 References

1. Argon, A. S. in: "Glass: Science and Technology", (eds.) Uhlmann, D. R., Kreidl, N. J., 79, New York, Academic Press, 1980
2. "Advances in Polymer Science", (ed.) Kausch, H. H., a) E. J. Kramer, Chapter 1; b) M. Dettenmaier, Chapter 2 (this volume)
3. Bucknall, C. B.: "Toughened Polymers", London, Applied Science Publ. 1977
4. Kawai, H., et al., *J. Macromol. Sci. — Phys. B17(3)*, 427 (1980)

5. Argon, A. S., et al.: in "Toughening of Plastics", 16-1, London, Plastics and Rubber Institute, 1978
6. Argon, A. S. et al., in: "Deformation, Yield and Fracture of Polymers", 28-1, London, Plastics and Rubber Institute, 1982
7. Argon, A. S., Bessonov, M. I.: *Phil. Mag.*, *35*, 917 (1977)
8. Wellinghoff, S. T., Baer, E.: *J. Appl. Polymer Sci.*, *22*, 2025 (1978)
9. Argon, A. S.: *Acta Met.*, *27*, 47 (1979)
10. Argon, A. S.: *Phys. Chem. Solids*, *43*, 945 (1982)
11. Megusar, J., et al., in: Rapidly Solidified Amorphous and Crystalline Solids, (eds.) Kear, B. H., Giessen, B. C., and Cohen, M., New York, North Holland, 282, 1982
12. Argon, A. S., Hannoosh, J. G.: *Phil. Mag.*, *36*, 1195 (1977)
13. Argon, A. S., Salama, M. M.: *Phil. Mag.*, *36*, 1217 (1977)
14. Donald, A. M., Kramer, E. J.: *Phil. Mag.*, *43*, 857 (1981)
15. Argon, A. S., et al.: *J. Polymer Sci.*, *19*, 253 (1981)
16. Murray, J., Hull, D.: *J. Polymer Sci.*, (Part A-2) 8, 1521 (1970)
17. Murray, J., Hull, D.: *Polymer Letters*, *8*, 159 (1970)
18. Brown, N.: *Phil. Mag.*, *32*, 1041 (1975)
19. Haward, R. N., Thackray, G.: *Proc. Roy. Soc. A302*, 453 (1968)
20. Argon, A. S.: *Phil. Mag.*, *28*, 839 (1973)
21. Lauterwasser, B. D., Kramer, E. J.: *Phil. Mag. 39*, 469 (1979)
22. Argon, A. S., Salama, M. M.: *Mater. Sci. Engng.*, *23*, 219 (1976)
23. Kambour, R. P.: *Polymer Sci. Macromol. Rev.*, *7*, 1 (1973)
24. Sternstein, S. S., Ongchin, L.: *Polymer Prepr.*, *10*, 117 (1969)
25. Argon, A. S.: *J. Macromol. Sci.-Phys.*, *B8(3–4)*, 573 (1973)
26. Argon, A. S.: *Pure Appl. Chem.*, *43*, 247 (1975)
27. Bucknall, C. B.: *J. Mater.*, *4*, 214 (1969)
28. Bucknall, C. B. et al.: *J. Mater. Sci.*, *7*, 1443 (1972)
29. Oxborough, R. J., Bowden, P. B.: *Phil. Mag.*, *28*, 547 (1973)
30. Oxborough, R. J., Bowden, P. B.: *Phil. Mag.*, *30*, 171 (1974)
31. Bates, F. S. et al.: *Macromolecules*, in the press
32. Green, A. E., Zerna, W.: "Theoretical Elasticity", 346, Oxford, Clarendon Press, 1954
33. Andrews, E. H., Bevan, S.: *Polymer*, *13*, 337 (1972)
34. Bevan, S.: *J. Appl. Polymer Sci.*, *27*, 4263 (1982)
35. Verheulpen-Heymans, N., Bauwens, J. C.: *J. Mater. Sci.*, *11*, 1 (1976); *Ibid.*, *11*, 7 (1976)
36. Beahan, P. et al.: *Phil. Mag.*, *24*, 1267 (1971)
37. Brown, H. R., Kramer, E. J.: *J. Macromol. Sci. Phys.*, *B19*, 487 (1981)
38. Schwier, C. E.: "Crazing in Block Copolymers", Ph. D. Thesis, M.I.T., Cambridge, Mass. 1983
39. Andersson, H., Bergkvist, H.: *J. Mech. Phys. Solids*, *18*, 1 (1970)
40. Bird, R. J. et al.: *Polymer*, *12*, 742 (1971)
41. Doyle, M. J. et al.: *Proc. Roy. Soc.*, *A329*, 137 (1972)
42. Doyle, M. J.: *J. Maters. Sci.*, *17*, 760 (1982)
43. Murray, J., Hull, D.: *J. Polymer Sci.*, (Part A-2) *8*, 583 (1970)
44. Doyle, M. J.: *J. Mater. Sci.*, *17*, 204 (1982)
45. Helfand, E., Wasserman, Z. R. in: "Developments in Block Copolymers", (ed.) Goodman, I., London, Appl. Sci. Publ., 99, 1982; *Macromolecules*, *11*, 960 (1978)
46. Meier, D. J.: *J. Polymer Sci.*, *C26*, 81 (1969)
47. Krause, S.: *Macromolecules*, *3*, 84 (1970)
48. Cohen, R. E.: *ACS Symposium Series*, *193*, 489 (1982)
49. Argon, A. S. et al.: *J. Polymer Sci.*, *19*, 253 (1981)
50. Kawai, E. et al., in: "Progress in Polymer Science, Japan", (eds.) Imahori, K., and Iwakura, Y., vol. 4, New York, Halstead Press, 1972
51. Leibler, L.: *Macromol. Chem., Rapid Communication*, *2*, 393 (1981)
52. Reiss, G. et al.: *J. Macromol. Sci.*, B17(2), 355 (1980)
53. Gebizlioglu, O. S. et al.: *SPE-ANTEC*, *40*, 126 (1982)
54. Echte, A.: *Angew. Makromol. Chem.*, *58/59*, 175 (1977)
55. Argon, A. S. in: "Polymeric Materials, Relationship Between Structure and Mechanical Behavior", 411, Metals Park, Ohio, ASM, 1975

56. Bohn, L.: *Angew. Makromol. Chem.*, *20*, 129 (1971)
57. Ferry, J. D.: "Viscoelastic Properties of Polymers", (2nd. Ed.), 586, New York, J. Wiley and Sons, 1970
58. Goodier, J. N.: *ASME Trans.*, *55*, 39 (1933)
59. Eshelby, J. D.: *Proc. Roy. Soc.*, *A 241*, 376 (1957)
60. Mura, T.: "Micromechanics of Defects in Solids", The Hague, Martinus Nijhoff Publ., 1982
61. Chow, T. S.: *J. Polymer Sci.-Phys.*, *16*, 959 (1978)
62. Chow, T. S.: *J. Polymer Sci.-Phys.*, *16*, 967 (1978)
63. Argon, A. S. et al.: *J. Appl. Phys.*, *39*, 1899 (1968)
64. Turley, S. G., Keskkula, H.: *Polymer*, *21*, 466 (1980)
65. Gebizlioglu, O. S., Tanaka, T.: to be published
66. Gebizlioglu, O. S. et al.: *Polymer Preprints*, *22*, (No. 2), 257 (1981)
67. Cunningham, M. et al.: to be published
68. Broutman, J. J., Panizza, G.: *Int. J. Poly. Mater.*, *1*, 95 (1971)

Received March 1, 1983
H. Kausch (editor)

List of Symbols

Polymer symbols

ABS	Acrylonitrile-butadiene-styrene copolymer
HDPE	High-density Polyethylene
LDPE	Low-density Polyethylene
PA	Polyamide
PA 6	Polyamide 6 (Polycaprolactam)
PA 66	Polyamide 66 (Polyhexamethylene adipamide)
PB-1	Polybutylene-1
PC	Polycarbonate
PCTFE	Polychlorotrifluoroethylene
PE	Polyethylene
PES	Polyethersulfone
PETP	Polyethylene Terephtalate
PMMA	Polymethyl Methacrylate
POM	Polyoxymethylene (Polyacetal)
PP	Polypropylene
PP 1120	Polypropylene with 5 % atatic content
PP 1320	Polypropylene with 20 % atatic content
PPO	Poly(phenylene oxide)
PS	Polystyrene
PSF	Polysulfone
PTFE	Polytetrafluoroethylene
PVC	Poly(vinyl chloride)
PVF_2	Poly(vinylidene fluoride)

Miscellaneous

CT	Compact tension-specimen
MW	Molecular Weight
SEM	Scanning Electron Microscopy
SEN	Single Edge Notch-Specimen
TEM	Transmission Electron Microscopy

Author Index Volumes 1–52/53

Subject Index

Key words set in *italics* refer to the headings.